Advanced Sciences and Technologies for Security Applications

Advanced Sciences and Technologies for Security Applications

The series Advanced Sciences and Technologies for Security Applications focuses on research monographs in the areas of
—Recognition and identification (including optical imaging, biometrics, authentication, verification, and smart surveillance systems)
—Biological and chemical threat detection (including biosensors, aerosols, materials detection and forensics),
and
—Secure information systems (including encryption, and optical and photonic systems).

The series is intended to give an overview at the highest research level at the frontier of research in the physical sciences.

The editors encourage prospective authors to correspond with them in advance of submitting a manuscript. Submission of manuscripts should be made to the Editor-in-Chief or one of the Editors.

Firooz Sadjadi
Bahram Javidi
Editors

Physics of Automatic Target Recognition

 Springer

Firooz Sadjadi
Staff Scientist, LMC
3400 Highcrest Road
Saint Anthony, MN 55418, USA
sadja001@tc.umn.edu

Bahram Javidi
Distinguished Professor of Electrical and Computer Engineering
Electrical and Computer Engineering Department
University of Connecticut
371 Fairfax Road, Unit 1157
Storrs, CT 06269-1157, USA
bahram@engr.uconn.edu

Library of Congress Control Number: 2006928730

ISBN-10: 0-387-36742-X   e-ISBN-10: 0-387-36943-0
ISBN-13: 978-0-387-36742-2   e-ISBN-13: 978-0-387-36943-3

Printed on acid-free paper.

springer.com

Firooz Sadjadi:
For my mother Fakhri

Bahram Javidi:
For my wife Bethany

Contents

Preface

This book is an attempt to provide a view of automated target recognition from the physics of the signatures. Automated target recognition (ATR) can be viewed as an inverse problem in electromagnetic and acoustics. Targets of interest are sensed by reflection/scattering, refraction or emission; the sensed signatures are then transmitted through some transmitting medium before being intercepted by detectors. The main goal of ATR is to use these signatures to classify the original objects. Our experience has indicated that only through understanding of the direct problem, the transformations that a signal goes through before being detected, one can hope of solving the inverse problem. So beside attention paid to abstract pattern classification techniques, one needs to also concentrate on the physics of the signatures, be it from a polarimetric infrared, hyperspectral, or ultra-wide band radar. Viewed in this context, ATR can be seen as an exciting physics problem whose general solution remains elusive but no too far off.

With the rapid advances in sensor designs, high speed computers and low cost platforms such as unmanned vehicles, the problem of ATR is becoming more pertinent to a much wider group of scientist and engineers than before. This book is intended to address this need by providing a cross sectional view of various issues related to the physics of the ATR.

We are grateful to the contributors, many of who are our friends and colleagues for many years for their outstanding contributions. All are imminent in their fields and most have been active in the ATR field for decades. We would like to thank the Springer Editor and staff Ms.Virginia Lipscy and Ms. Vasudha Gandi, for their support, and assistance in the preparation of this manuscript.

Contributors

N. Bertaux
Institut Fresnel - Equipe Φ-T
Unité Mixte de Recherche
  6133
Domaine Universitaire de Saint
  Jérôme
13397 Marseille, France

C.S.L. Chun
Physics Innovations Inc.
P.O. Box 2171
Inver Grove Heights
MN 55076-81-71

L. Cohen
City University of New York
695 Park Avenue
New York, NY 10021

Dr. W. de Jong
TNO Defence, Security and
  Safety
Oude Waalsdorperweg 63
2597 AK The Hague
The Netherlands

Y. Frauel
Instituto de Investigaciones en
  Matemáticas Aplicadas y en
  Sistemas
Universidad Nacional Autónoma de
  México
México, DF, Mexico

B. Gaunaurd
Army Research Laboratory
2800 Powder Mill Road
Adelphi, MD 20783-1197

G.A. Geri
Link Simulation & Training
6030 S. Kent
St. Mesa, AZ 85212

I. Gertner
Computer Science Department
City College of the City University
  of New York
138 Street, Convent Avenue
New York, NY 10031

R.I. Hammoud
Delphi Electronics and Safety
One Corporate Center
P.O. Box 9005
Kokomo, IN 46904-9005

D.W. Hansen
IT University, Copenhagen
Rued Langaardsvej 7
2300 Copenhagen S, Denmark

B. Javidi
Department of Electrical and
  Computer Engineering, U-2157
University of Connecticut
Storrs, CT 06269-2157

D. Kim
Electrical and Computer
   Engineering Department
University of Connecticut
Storrs, CT 06269-2157

H. Kwon
U.S. Army Research Laboratory
ATTN: AMSRL-SE-SE
2800 Powder Mill Road
Adelphi, MD 20783

P. Loughlin
University of Pittsburgh
348 Benedum Hall
Pittsburgh, PA 15261

I. Maslov
Department of Computer Science
CUNY/Graduate Center
365 Fifth Ave.
New York, NY 10016

M.S. Millan
Department of Optics and
   Optometry
Universitat Politècnica de
   Catalunya
Violinista Vellsolà 37
08222 Terrassa, Spain

I. Moon
Department of Electrical and
   Computer Engineering, U-2157
University of Connecticut
Storrs, CT 06269-2157

N.M. Nasrabadi
U.S. Army Research Laboratory
ATTN: AMSRL-SE-SE
2800 Powder Mill Road
Adelphi, MD 20783

E. Perez-Cabre
Department of Optics and
   Optometry
Universitat Politècnica de
   Catalunya
Violinista Vellsolà 37
08222 Terrassa, Spain

F. Sadjadi
Lockheed Martin Corporation
3400 High Crest Road
Saint Anthony, MN 55418

Dr. J.G.M. Schavemaker
TNO Defence, Security and Safety
Oude Waalsdorperweg 63
2597 AK The Hague
The Netherlands

H. Strifors
Swedish Defense Research Agency
   (FOI)
P.O. Box 1165
SE-58111 Linköping, Sweden

S. Yeom
Department of Electrical and
   Computer Engineering,
   U-2157
University of Connecticut
Storrs, CT 06269-2157

1

Kernel-Based Nonlinear Subspace Target Detection for Hyperspectral Imagery

Heesung Kwon and Nasser M. Nasrabadi

U.S. Army Research Laboratory ATTN: AMSRL-SE-SE 2800
Powder Mill Road Adelphi, MD 20783

Summary. In this book chapter, we compare several detection algorithms that are based on spectral matched (subspace) filters. Nonlinear (*kernel*) versions of these spectral matched detectors are also given and their performance is compared with the linear versions. The kernel-based matched filters and kernel matched subspace detectors exploit the nonlinear correlations between the spectral bands that is ignored by the conventional detectors. Nonlinear realization is mainly pursued to reduce data complexity in a high-dimensional feature space and consequently providing simpler decision rules for data discrimination. Several well-known matched detectors, such as matched subspace detector, orthogonal subspace detector, spectral matched filter, and adaptive subspace detector (adaptive cosine estimator), are extended to their corresponding kernel versions by using the idea of kernel-based learning theory. In kernel-based detection algorithms, the data are implicitly mapped into a high-dimensional kernel feature space by a nonlinear mapping which is associated with a kernel function. The detection algorithm is then derived in the feature space which is *kernelized* in terms of the kernel functions in order to avoid explicit computation in the high-dimensional feature space. Experimental results based on simulated toy-examples and real hyperspectral imagery show that the kernel versions of these detectors outperform the conventional linear detectors.

1.1 Introduction

Detecting signals of interest, particularly with wide signal variability, in noisy environments has long been a challenging issue in various fields of signal processing. Among a number of previously developed detectors, the well-known matched subspace detector (MSD) [1], orthogonal subspace detector (OSD) [1, 2], spectral matched filter (SMF) [3, 4], and adaptive subspace detectors (ASD) also known as adaptive cosine estimator (ACE) [5, 6], have been widely used to detect a desired signal (target).

Matched signal detectors, such as spectral matched filter and matched subspace detectors (whether adaptive or nonadaptive), only exploit second-order correlations, thus completely ignoring nonlinear (higher order) spectral interband correlations that could be crucial to discriminate between target and background. In this chapter, our aim is to introduce nonlinear versions of MSD, OSD, SMF and ASD detectors which effectively exploit the higher order spectral interband correlations in a high-(possibly infinite) dimensional-feature space associated with a certain nonlinear mapping via kernel-based learning methods [7]. A nonlinear mapping of the input data into a high-dimensional feature space is often expected to increase the data separability and reduce the complexity of the corresponding data structure [8].

The nonlinear versions of a number of signal processing techniques such as principal component analysis (PCA) [9], Fisher discriminant analysis [10], clustering in feature space [8], linear classifiers [11], nonlinear feature extraction based on kernel orthogonal centroid method [12], kernel-matched signal detectors for target detection [13, 14, 15], kernel-based anomaly detection [16], classification in the kernel-based nonlinear subspace [17], and classifiers based on kernel Bayes rule [18] have already been defined in kernel space. In [19] kernels were used as generalized dissimilarity measures for classification and in [20] kernel methods were applied to face recognition.

This chapter is organized as follows. Section 1.2 provides the background to the kernel-based learning methods and kernel trick. Section 1.3 introduces linear matched subspace and its kernel version. Orthogonal subspace detector is defined in section 1.4 as well as its kernel version. In Section 1.5, we describe the conventional spectral matched filter and its kernel version in the feature space and reformulate the the expression in terms of the kernel function using the kernel trick. Finally, in Section 1.6 the adaptive subspace detector and its kernel version are introduced. Performance comparison between the conventional and the kernel version of these algorithms is provided in Section 1.7. Conclusions and comparisons of these algorithms are given in Section 1.8.

1.2 Kernel Methods and Kernel Trick

Suppose that the input hyperspectral data are represented by the data space $(\mathcal{X} \subseteq \mathcal{R}^l)$ and $\mathcal{F}$ is a feature space associated with $\mathcal{X}$ by a nonlinear mapping function Φ

$$\Phi : \mathcal{X} \to \mathcal{F}, \mathbf{x} \mapsto \Phi(\mathbf{x}), \tag{1.1}$$

where $\mathbf{x}$ is an input vector in $\mathcal{X}$ which is mapped into a potentially much higher—(could be infinite)—dimensional feature space. Because of the high dimensionality of the feature space $\mathcal{F}$, it is computationally not feasible to implement any algorithm directly in the feature space. However, kernel-based learning algorithms use an effective kernel trick given by Eq. (1.2) to implement dot products in the feature space by employing kernel functions [7]. The idea in kernel-based techniques is to obtain a nonlinear version of an algorithm defined in the input space by implicitly redefining it in the feature space and then converting it in terms of dot products. The kernel trick is then used to implicitly compute the dot products in $\mathcal{F}$ without mapping the input vectors into $\mathcal{F}$; therefore, in the kernel methods, the mapping Φ does not need to be identified.

The kernel representation for the dot products in $\mathcal{F}$ is expressed as

$$k(\mathbf{x}_i, \mathbf{x}_j) = \Phi(\mathbf{x}_i) \cdot \Phi(\mathbf{x}_j), \tag{1.2}$$

where k is a kernel function in terms of the original data. There are a large number of Mercer kernels that have the kernel trick property; see [7] for detailed information about the properties of different kernels and kernel-based learning. Our choice of kernel in this chapter is the Gaussian RBF kernel and the associated nonlinear function Φ with this kernel generates a feature space of infinite dimensionality.

1.3 Linear Matched Subspace Detector and Kernel Matched Subspace Detector

In this model the target pixel vectors are expressed as a linear combination of target spectral signature and background spectral signature, which are represented by subspace target spectra and subspace background spectra, respectively. The hyperspectral target detection problem in a p-dimensional input space is expressed as two competing hypotheses $\mathbf{H_0}$ and $\mathbf{H_1}$

$$\mathbf{H}_0 : \mathbf{y} = \mathbf{B}\zeta + \mathbf{n}, \qquad\qquad \text{Target absent} \tag{1.3}$$

$$\mathbf{H}_1 : \mathbf{y} = \mathbf{T}\theta + \mathbf{B}\zeta + \mathbf{n} = \begin{bmatrix} \mathbf{T}\ \mathbf{B} \end{bmatrix} \begin{bmatrix} \theta \\ \zeta \end{bmatrix} + \mathbf{n}, \qquad \text{Target present}$$

where $\mathbf{T}$ and $\mathbf{B}$ represent orthogonal matrices whose p-dimensional column vectors span the target and background subspaces, respectively; θ and ζ are unknown vectors whose entries are coefficients that account for the abundances of the corresponding column vectors of $\mathbf{T}$ and $\mathbf{B}$, respectively; $\mathbf{n}$ represents Gaussian random noise ($\mathbf{n} \in \mathcal{R}^p$) distributed as $\mathcal{N}(0, \sigma^2\mathbf{I})$; and $\begin{bmatrix} \mathbf{T}\ \mathbf{B} \end{bmatrix}$ is a concatenated matrix of $\mathbf{T}$ and $\mathbf{B}$. The numbers of the column vectors of $\mathbf{T}$ and $\mathbf{B}$, N_t and N_b, respectively, are usually smaller than p ($N_t, N_b < p$).

1.3.1 Generalized Likelihood Ratio Test (GLRT) for Target Detection

Given the linear subspace detection model and the two hypotheses about how the input vector is generated, as shown by (1.3), the likelihood ratio test (LRT) is used to predict whether the input vector $\mathbf{y}$ includes the target and is defined by

$$l(\mathbf{y}) = \frac{p_1(\mathbf{y}|\mathbf{H}_1)}{p_0(\mathbf{y}|\mathbf{H}_0)} \overset{H_1}{\underset{H_0}{\gtrless}} \eta, \tag{1.4}$$

where $p_0(\mathbf{y}|\mathbf{H}_0)$ and $p_1(\mathbf{y}|\mathbf{H}_1)$ represent the class conditional probability densities of $\mathbf{y}$ given the hypotheses $\mathbf{H}_0$ and $\mathbf{H}_1$, respectively, and η is a threshold of the test. $p_0(\mathbf{y}|\mathbf{H}_0)$ and $p_1(\mathbf{y}|\mathbf{H}_1)$ can be expressed as Gaussian probability densities $\mathcal{N}(\mathbf{B}\boldsymbol{\zeta}, \sigma^2\mathbf{I})$ and $\mathcal{N}(\mathbf{T}\boldsymbol{\theta} + \mathbf{B}\boldsymbol{\zeta}, \sigma^2\mathbf{I})$, respectively, since $\mathbf{n}_i$ are assumed to be Gaussian random noise. The LRT is derived from the Neyman–Pearson criterion where the probability detection P_D is maximized while the probability of false alarm P_F is kept a constant [21]. $l(\mathbf{y})$ is compared to η to make a final decision about which hypothesis best relates to $\mathbf{y}$.

The LRT includes the unknown parameters $\boldsymbol{\zeta}$ and $\boldsymbol{\theta}$ that need to be estimated using the maximum likelihood principle. The generalized likelihood ratio test (GLRT) is directly obtained from $l(\mathbf{y})$ by replacing the unknown parameters with their maximum likelihood estimates (MLEs) $\hat{\boldsymbol{\zeta}}$ and $\hat{\boldsymbol{\theta}}$ and by taking $(P/2)$-root [1]

$$\mathbf{L}_2(\mathbf{y}) = (\hat{l}(\mathbf{y}))^{2/P} = \frac{\mathbf{y}^T(\mathbf{I} - \mathbf{P_B})\mathbf{y}}{\mathbf{y}^T(\mathbf{I} - \mathbf{P_{TB}})\mathbf{y}} \overset{H_1}{\underset{H_0}{\gtrless}} \eta, \tag{1.5}$$

where $\mathbf{P_B} = \mathbf{B}(\mathbf{B}^T\mathbf{B})^{-1}\mathbf{B}^T = \mathbf{B}\mathbf{B}^T$ is a projection matrix associated with the N_b-dimensional background subspace $< \mathbf{B} >$; $\mathbf{P_{TB}}$ is a projection matrix associated with the $(N_{bt} = N_b + N_t)$-dimensional target-and-background subspace $< \mathbf{TB} >$

$$\mathbf{P_{TB}} = \begin{bmatrix} \mathbf{T} & \mathbf{B} \end{bmatrix} \begin{bmatrix} \begin{bmatrix} \mathbf{T} & \mathbf{B} \end{bmatrix}^T \begin{bmatrix} \mathbf{T} & \mathbf{B} \end{bmatrix} \end{bmatrix}^{-1} \begin{bmatrix} \mathbf{T} & \mathbf{B} \end{bmatrix}^T. \tag{1.6}$$

1.3.2 Linear Subspace Models Defined in the Feature Space $\mathcal{F}$

The hyperspectral detection problem based on the target and background subspaces can be described in the feature space $\mathcal{F}$ as

$$\mathbf{H}_{0_\varPhi} : \varPhi(\mathbf{y}) = \mathbf{B}_\varPhi\boldsymbol{\zeta}_\varPhi + \mathbf{n}_\varPhi, \qquad\qquad\qquad \text{Target absent}$$

$$\mathbf{H}_{1_\varPhi} : \varPhi(\mathbf{y}) = \mathbf{T}_\varPhi\boldsymbol{\theta}_\varPhi + \mathbf{B}_\varPhi\boldsymbol{\zeta}_\varPhi + \mathbf{n}_\varPhi = [\mathbf{T}_\varPhi \mathbf{B}_\varPhi] \begin{bmatrix} \boldsymbol{\theta}_\varPhi \\ \boldsymbol{\zeta}_\varPhi \end{bmatrix} + \mathbf{n}_\varPhi, \quad \text{Target present}$$

$$\tag{1.7}$$

where $\mathbf{T}_\varPhi$ and $\mathbf{B}_\varPhi$ represent full-rank matrices whose column vectors span target and background subspaces $< \mathbf{B}_\varPhi >$ and $< \mathbf{T}_\varPhi >$ in $\mathcal{F}$, respectively; $\boldsymbol{\theta}_\varPhi$ and $\boldsymbol{\zeta}_\varPhi$ are unknown vectors whose entries are coefficients that account for the abundances of the corresponding column vectors of $\mathbf{T}_\varPhi$ and $\mathbf{B}_\varPhi$, respectively; $\mathbf{n}_\varPhi$ represents Gaussian random noise; and $[\mathbf{T}_\varPhi \mathbf{B}_\varPhi]$ is a concatenated matrix of $\mathbf{T}_\varPhi$ and $\mathbf{B}_\varPhi$. In general, any sets of basis vectors that span the corresponding subspace can be used as the column vectors of $\mathbf{T}_\varPhi$ and $\mathbf{B}_\varPhi$. In the proposed method we use the significant eigenvectors of the target and background covariance matrices ($\mathbf{C_{T_\Phi}}$ and $\mathbf{C_{B_\Phi}}$) in $\mathcal{F}$ as the the column vectors of $\mathbf{T}_\Phi$ and $\mathbf{B}_\Phi$, respectively. $\mathbf{C_{T_\Phi}}$ and $\mathbf{C_{B_\Phi}}$ are based on the zero mean (centered) target and background sample sets ($\mathbf{Z_T}$ and $\mathbf{Z_B}$), respectively:

$$\mathbf{C_{T_\Phi}} = \frac{1}{N} \sum_{i=1}^{N} \Phi(\mathbf{y}_i)\Phi(\mathbf{y}_i)^T, \text{ for } \mathbf{y}_i \in \mathbf{Z_T},$$

$$\mathbf{C_{B_\Phi}} = \frac{1}{M} \sum_{i=1}^{M} \Phi(\mathbf{y}_i)\Phi(\mathbf{y}_i)^T, \text{ for } \mathbf{y}_i \in \mathbf{Z_B},$$

where N and M represent the number of centered training samples in $\mathbf{Z_T}$ and $\mathbf{Z_B}$, respectively.

Using a similar reasoning as described in the previous subsection, the GLRT of the hyperspectral detection problem depicted by the model in (1.7) is given by

$$\mathbf{L}_2(\Phi(\mathbf{y})) = \frac{\Phi(\mathbf{y})^T(\mathbf{P_{I_\Phi}} - \mathbf{P_{B_\Phi}})\Phi(\mathbf{y})}{\Phi(\mathbf{y})^T(\mathbf{P_{I_\Phi}} - \mathbf{P_{T_\Phi B_\Phi}})\Phi(\mathbf{y})}, \tag{1.8}$$

where $\mathbf{P_{I_\Phi}}$ represents an identity projection operator in $\mathcal{F}$; $\mathbf{P_{B_\Phi}} = \mathbf{B}_\Phi(\mathbf{B}_\Phi^T\mathbf{B}_\Phi)^{-1}\mathbf{B}_\Phi^T = \mathbf{B}_\Phi\mathbf{B}_\Phi^T$ is a background projection matrix; and $\mathbf{P_{T_\Phi B_\Phi}}$ is a joint target-and-background projection matrix in $\mathcal{F}$

$$\mathbf{P_{T_\Phi B_\Phi}} = \begin{bmatrix} \mathbf{T}_\Phi & \mathbf{B}_\Phi \end{bmatrix} \left[\begin{bmatrix} \mathbf{T}_\Phi & \mathbf{B}_\Phi \end{bmatrix}^T \begin{bmatrix} \mathbf{T}_\Phi & \mathbf{B}_\Phi \end{bmatrix} \right]^{-1} \begin{bmatrix} \mathbf{T}_\Phi & \mathbf{B}_\Phi \end{bmatrix}^T$$

$$= \begin{bmatrix} \mathbf{T}_\Phi & \mathbf{B}_\Phi \end{bmatrix} \begin{bmatrix} \mathbf{T}_\Phi^T\mathbf{T}_\Phi & \mathbf{T}_\Phi^T\mathbf{B}_\Phi \\ \mathbf{B}_\Phi^T\mathbf{T}_\Phi & \mathbf{B}_\Phi^T\mathbf{B}_\Phi \end{bmatrix}^{-1} \begin{bmatrix} \mathbf{T}_\Phi^T \\ \mathbf{B}_\Phi^T \end{bmatrix}. \tag{1.9}$$

1.3.3 Kernelizing MSD in Feature Space

To kernelize (1.8) we will separately kernelize the numerator and the denominator. First consider its numerator,

$$\Phi(\mathbf{y})^T(\mathbf{P_{I_\Phi}} - \mathbf{P_{B_\Phi}})\Phi(\mathbf{y})\Phi(\mathbf{y})^T\mathbf{P_{I_\Phi}}\Phi(\mathbf{y}) - \Phi(\mathbf{y})^T\mathbf{B}_\Phi\mathbf{B}_\Phi^T\Phi(\mathbf{y}). \tag{1.10}$$

Using (1.59), as shown in Appendix I, $\mathbf{B}_\Phi$ and $\mathbf{T}_\Phi$ can be written in terms of their corresponding data spaces as

$$\mathbf{B}_\Phi = \begin{bmatrix} \mathbf{e_b}^1 & \mathbf{e_b}^2 & \dots & \mathbf{e_b}^{N_b} \end{bmatrix} = \Phi_{\mathbf{Z_B}}\tilde{\mathcal{B}}, \tag{1.11}$$

$$\mathbf{T}_\Phi = \begin{bmatrix} \mathbf{e_t}^1 & \mathbf{e_t}^2 & \dots & \mathbf{e_t}^{N_t} \end{bmatrix} = \Phi_{\mathbf{Z_T}}\tilde{\mathcal{T}}, \tag{1.12}$$

where $\mathbf{e_b}^i$ and $\mathbf{e_t}^j$ are the significant eigenvectors of $\mathbf{C_{B_\Phi}}$ and $\mathbf{C_{T_\Phi}}$, respectively; $\Phi_{\mathbf{Z_B}} = [\Phi(\mathbf{y}_1)\Phi(\mathbf{y}_2)\dots\Phi(\mathbf{y}_M)]$, $\mathbf{y}_i \in \mathbf{Z_B}$ and $\Phi_{\mathbf{Z_T}} = [\Phi(\mathbf{y}_1)\Phi(\mathbf{y}_2)\dots\Phi(\mathbf{y}_N)]$, $\mathbf{y}_i \in \mathbf{Z_T}$; the column vectors of $\tilde{\mathcal{B}}$ and $\tilde{\mathcal{T}}$ represent only the significant normalized eigenvectors ($\tilde{\beta}_1, \tilde{\beta}_2, \cdots, \tilde{\beta}_{N_b}$) and ($\tilde{\alpha}_1, \tilde{\alpha}_2, \cdots, \tilde{\alpha}_{N_t}$) of the background centered kernel matrix $\mathbf{K}(\mathbf{Z_B}, \mathbf{Z_B}) = (\mathbf{K})_{ij} = k(\mathbf{y}_i, \mathbf{y}_j)$, $\mathbf{y}_i, \mathbf{y}_j \in \mathbf{Z_B}$ and target centered kernel matrix $\mathbf{K}(\mathbf{Z_T}, \mathbf{Z_T}) = (\mathbf{K})_{ij} = k(\mathbf{y}_i, \mathbf{y}_j)$, $\mathbf{y}_i, \mathbf{y}_j \in \mathbf{Z_T}$, respectively.

Using (1.11) the projection of $\varPhi(\mathbf{y})$ onto $\mathbf{B}_\phi$ becomes

$$
\mathbf{B}_\phi^T \varPhi(\mathbf{y}) = \left[\mathbf{e_b}^1 \ \mathbf{e_b}^2 \cdots \mathbf{e_b}^{N_b} \right]^T \varPhi(\mathbf{y}) = \begin{bmatrix} \tilde{\boldsymbol{\beta}}_1^T \varPhi_{\mathbf{Z_B}}^T \varPhi(\mathbf{y}) \\ \tilde{\boldsymbol{\beta}}_2^T \varPhi_{\mathbf{Z_B}}^T \varPhi(\mathbf{y}) \\ \cdot \\ \tilde{\boldsymbol{\beta}}_{N_b}^T \varPhi_{\mathbf{Z_B}}^T \varPhi(\mathbf{y}) \end{bmatrix}
$$

$$
= \tilde{\boldsymbol{\mathcal{B}}}^T \mathbf{k}(\mathbf{Z_B}, \mathbf{y}), \tag{1.13}
$$

and, similarly, using (1.12) the projection onto $\mathbf{T}_\phi$ is

$$
\mathbf{T}_\phi^T \varPhi(\mathbf{y}) = \left[\mathbf{e_t}^1 \ \mathbf{e_t}^2 \cdots \mathbf{e_t}^{N_t} \right]^T \varPhi(\mathbf{y}) = \begin{bmatrix} \tilde{\boldsymbol{\alpha}}_1^T \varPhi_{\mathbf{Z_T}}^T \varPhi(\mathbf{y}) \\ \tilde{\boldsymbol{\alpha}}_2^T \varPhi_{\mathbf{Z_T}}^T \varPhi(\mathbf{y}) \\ \tilde{\boldsymbol{\alpha}}_{N_t}^T \varPhi_{\mathbf{Z_T}}^T \varPhi(\mathbf{y}) \end{bmatrix}
$$

$$
= \tilde{\boldsymbol{\mathcal{T}}}^T \mathbf{k}(\mathbf{Z_T}, \mathbf{y}), \tag{1.14}
$$

where $\mathbf{k}(\mathbf{Z_B}, \mathbf{y})$ and $\mathbf{k}(\mathbf{Z_T}, \mathbf{y})$, referred to as the empirical kernel maps in the machine learning literature [7], are column vectors whose entries are $k(\mathbf{x}_i, \mathbf{y})$ for $\mathbf{x}_i \in \mathbf{Z_B}$ and $\mathbf{x}_i \in \mathbf{Z_T}$, respectively. Now using (1.13) we can write

$$
\varPhi(\mathbf{y})^T \tilde{\mathbf{B}}_\phi \tilde{\mathbf{B}}_\phi^T \varPhi(\mathbf{y}) = \mathbf{k}(\mathbf{Z_B}, \mathbf{y})^T \tilde{\boldsymbol{\mathcal{B}}} \tilde{\boldsymbol{\mathcal{B}}}^T \mathbf{k}(\mathbf{Z_B}, \mathbf{y}). \tag{1.15}
$$

The projection onto the identity operator $\varPhi(\mathbf{y})^T \mathbf{P}_{\mathbf{I}_\phi} \varPhi(\mathbf{y})$ also needs to be kernelized. $\mathbf{P}_{\mathbf{I}_\phi}$ is defined as $\mathbf{P}_{\mathbf{I}_\phi} := \Omega_\phi \Omega_\phi^T$, where $\Omega_\phi = \left[\mathbf{e_q}^1 \ \mathbf{e_q}^2 \cdots \right]$ is a matrix whose columns are all the eigenvectors with $\lambda \neq 0$ that are in the span of $\varPhi(\mathbf{y}_i)$, $\mathbf{y}_i \in \mathbf{Z_T} \cup \mathbf{Z_B} := \mathbf{Z_{TB}}$. From (1.59) Ω_ϕ can be expressed as

$$
\Omega_\phi = [\mathbf{e_q}^1 \ \mathbf{e_q}^2 \ \cdots \ \mathbf{e_q}^{N_{bt}}] = \varPhi_{\mathbf{Z_{TB}}} \tilde{\Delta}, \tag{1.16}
$$

where $\varPhi_{\mathbf{Z_{TB}}} = \varPhi_{\mathbf{Z_T}} \cup \varPhi_{\mathbf{Z_B}}$ and $\tilde{\Delta}$ is a matrix whose columns are the eigenvectors (κ_1, $\kappa_2, \cdots, \kappa_{N_{bt}}$) of the centered kernel matrix $\mathbf{K}(\mathbf{Z_{TB}}, \mathbf{Z_{TB}}) = (\mathbf{K})_{ij} = k(\mathbf{y}_i, \mathbf{y}_j)$, $\mathbf{y}_i, \mathbf{y}_j \in \mathbf{Z_{TB}}$ with nonzero eigenvalues, normalized by the square root of their associated eigenvalues. Using $\mathbf{P}_{\mathbf{I}_\phi} = \Omega_\phi \Omega_\phi^T$ and (1.16)

$$
\varPhi(\mathbf{y})^T \mathbf{P}_{\mathbf{I}_\phi} \varPhi(\mathbf{y}) = \varPhi(\mathbf{y})^T \varPhi_{\mathbf{Z_{TB}}} \tilde{\Delta} \tilde{\Delta}^T \varPhi_{\mathbf{Z_{TB}}}^T \varPhi(\mathbf{y})
$$

$$
= \mathbf{k}(\mathbf{Z_{TB}}, \mathbf{y})^T \tilde{\Delta} \tilde{\Delta}^T \mathbf{k}(\mathbf{Z_{TB}}, \mathbf{y}), \tag{1.17}
$$

where $\mathbf{k}(\mathbf{Z_{TB}}, \mathbf{y})$ is a concatenated vector $\left[\mathbf{k}(\mathbf{Z_T}, \mathbf{y})^T \ \mathbf{k}(\mathbf{Z_B}, \mathbf{y})^T \right]^T$. The kernelized numerator of (1.8) is now given by

$$
\mathbf{k}(\mathbf{Z_{TB}}, \mathbf{y})^T \tilde{\Delta} \tilde{\Delta}^T \mathbf{k}(\mathbf{Z_{TB}}, \mathbf{y}) - \mathbf{k}(\mathbf{Z_B}, \mathbf{y})^T \tilde{\boldsymbol{\mathcal{B}}} \tilde{\boldsymbol{\mathcal{B}}}^T \mathbf{K}(\mathbf{Z_B}, \mathbf{y}). \tag{1.18}
$$

We now kernelize $\Phi(\mathbf{y})^T \mathbf{P}_{\mathbf{T}_\Phi \mathbf{B}_\Phi} \Phi(\mathbf{y})$ in the denominator of (1.8) to complete the kernelization process. Using (1.9), (1.11) and (1.12)

$$\Phi(\mathbf{y})^T \mathbf{P}_{\mathbf{T}_\Phi \mathbf{B}_\Phi} \Phi(\mathbf{y})$$

$$= \Phi(\mathbf{y})^T \begin{bmatrix} \mathbf{T}_\Phi & \mathbf{B}_\Phi \end{bmatrix} \begin{bmatrix} \mathbf{T}_\Phi^T \mathbf{T}_\Phi & \mathbf{T}_\Phi^T \mathbf{B}_\Phi \\ \mathbf{B}_\Phi^T \mathbf{T}_\Phi & \mathbf{B}_\Phi^T \mathbf{B}_\Phi \end{bmatrix}^{-1} \begin{bmatrix} \mathbf{T}_\Phi^T \\ \mathbf{B}_\Phi^T \end{bmatrix} \Phi(\mathbf{y})$$

$$= \begin{bmatrix} \mathbf{K}(\mathbf{Z}_T, \mathbf{y})^T \tilde{\mathcal{T}} & \mathbf{K}(\mathbf{Z}_B, \mathbf{y})^T \tilde{\mathcal{B}} \end{bmatrix} \begin{bmatrix} \tilde{\mathcal{T}}^T \mathbf{K}(\mathbf{Z}_T, \mathbf{Z}_T) \tilde{\mathcal{T}} & \tilde{\mathcal{T}}^T \mathbf{K}(\mathbf{Z}_T, \mathbf{Z}_B) \tilde{\mathcal{B}} \\ \tilde{\mathcal{B}}^T \mathbf{K}(\mathbf{Z}_B, \mathbf{Z}_T) \tilde{\mathcal{B}} & \tilde{\mathcal{B}}^T \mathbf{K}(\mathbf{Z}_B, \mathbf{Z}_B) \tilde{\mathcal{B}} \end{bmatrix}^{-1}$$

$$\times \begin{bmatrix} \tilde{\mathcal{T}}^T \mathbf{K}(\mathbf{Z}_T, \mathbf{y}) \\ \tilde{\mathcal{B}}^T \mathbf{K}(\mathbf{Z}_B, \mathbf{y}) \end{bmatrix}. \tag{1.19}$$

Finally, substituting (1.15), (1.17) and (1.19) into (1.8) the kernelized GLRT is given by

$$\mathbf{L}_{2\mathbf{K}} = (\mathbf{k}(\mathbf{Z}_{TB}, \mathbf{y})^T \tilde{\Delta}\tilde{\Delta}^T \mathbf{k}(\mathbf{Z}_{TB}, \mathbf{y}) - \mathbf{k}(\mathbf{Z}_B, \mathbf{y})^T \tilde{\mathcal{B}}\tilde{\mathcal{B}}^T \mathbf{k}(\mathbf{Z}_B, \mathbf{y})) /$$

$$\left(\mathbf{k}(\mathbf{Z}_{TB}, \mathbf{y})^T \tilde{\Delta}\tilde{\Delta}^T \mathbf{k}(\mathbf{Z}_{TB}, \mathbf{y}) - \begin{bmatrix} \mathbf{k}(\mathbf{Z}_T, \mathbf{y})^T \tilde{\mathcal{T}} & \mathbf{k}(\mathbf{Z}_B, \mathbf{y})^T \tilde{\mathcal{B}} \end{bmatrix} \Lambda_1^{-1} \right.$$

$$\left. \times \begin{bmatrix} \tilde{\mathcal{B}}^T \mathbf{k}(\mathbf{Z}_T, \mathbf{y}) \\ \tilde{\mathcal{B}}^T \mathbf{k}(\mathbf{Z}_B, \mathbf{y}) \end{bmatrix} \right), \tag{1.20}$$

where

$$\Lambda_1 = \begin{bmatrix} \tilde{\mathcal{T}}^T \mathbf{K}(\mathbf{Z}_T, \mathbf{Z}_T) \tilde{\mathcal{T}} & \tilde{\mathcal{T}}^T \mathbf{K}(\mathbf{Z}_T, \mathbf{Z}_B) \tilde{\mathcal{B}} \\ \tilde{\mathcal{B}}^T \mathbf{K}(\mathbf{Z}_B, \mathbf{Z}_T) \tilde{\mathcal{T}} & \tilde{\mathcal{B}}^T \mathbf{K}(\mathbf{Z}_B, \mathbf{Z}_B) \tilde{\mathcal{B}} \end{bmatrix}.$$

In the above derivation (1.20) we assumed that the mapped input data were centered in the feature space $\Phi_c(\mathbf{x}_i) = \Phi(\mathbf{x}_i) - \hat{\boldsymbol{\mu}}_{b_\Phi}$, where $\hat{\boldsymbol{\mu}}_{b_\Phi}$ represents the estimated mean in the feature space given by $\hat{\boldsymbol{\mu}}_\Phi = \frac{1}{N} \sum_{i=1}^N \Phi(\mathbf{x}_i)$. However, the original data are usually not centered and the estimated mean in the feature space cannot be explicitly computed, therefore, the kernel matrices have to be properly centered as shown in (1.70). The empirical kernel maps $\mathbf{k}(\mathbf{Z}_T, \mathbf{y})$, $\mathbf{k}(\mathbf{Z}_B, \mathbf{y})$, and $\mathbf{k}(\mathbf{Z}_{TB}, \mathbf{y})$ have to be centered by removing their corresponding empirical kernel map mean (e.g., $\tilde{\mathbf{k}}(\mathbf{Z}_T, \mathbf{y}) = \mathbf{k}(\mathbf{Z}_T, \mathbf{y}) - \frac{1}{N} \sum_{i=1}^N k(\mathbf{y}_i, \mathbf{y})$, $\mathbf{y}_i \in \mathbf{Z}_T$).

1.4 OSP and Kernel OSP Algorithms

1.4.1 Linear Spectral Mixture Model

The OSP algorithm [2] is based on maximizing the SNR (signal-to-noise ratio) in the subspace orthogonal to the background subspace and depends only on the noise second-order statistics. It also does not provide an estimate of the abundance measure for the desired end member in the mixed pixel. In [22] it was shown that the OSP classifier is related to the unconstrained least-squares estimate or the MLE (similarly derived by [1]) of the unknown signature abundance by a scaling factor.

A linear mixture model for pixel $\mathbf{y}$ consisting of p spectral bands is described by

$$\mathbf{y} = \mathbf{M}\boldsymbol{\alpha} + \mathbf{n}, \qquad (1.21)$$

where the $(p \times l)$ matrix $\mathbf{M}$ represents l endmembers spectra, $\boldsymbol{\alpha}$ is a $(l \times p)$ column vector whose elements are the coefficients that account for the proportions (abundances) of each endmember spectrum contributing to the mixed pixel, and $\mathbf{n}$ is an $(p \times p)$ vector representing an additive zero-mean Gaussian noise with covariance matrix $\sigma^2 \mathbf{I}$ and $\mathbf{I}$ is the $(l \times l)$ identity matrix. Assuming now we want to identify one particular signature (e.g., a military target) with a given spectral signature $\mathbf{d}$ and a corresponding abundance measure α_l, we can represent $\mathbf{M}$ and $\boldsymbol{\alpha}$ in partition form as $\mathbf{M} = (\mathbf{U} : \mathbf{d})$ and $\boldsymbol{\alpha} = \begin{bmatrix} \boldsymbol{\gamma} \\ \alpha_l \end{bmatrix}$ then model (1.21) can be rewritten as

$$\mathbf{r} = \mathbf{d}\alpha_l + \mathbf{B}\boldsymbol{\gamma} + \mathbf{n}, \qquad (1.22)$$

where the columns of $\mathbf{B}$ represent the undesired spectral signatures (background signatures or eigenvectors) and the column vector $\boldsymbol{\gamma}$ is the abundance measures for the undesired spectral signatures. The reason for rewriting model (1.21) as (1.22) is to separate $\mathbf{B}$ from $\mathbf{M}$ in order to show how to annihilate $\mathbf{B}$ from an observed input pixel prior to classification.

To remove the undesired signature, the background rejection operator is given by the $(l \times l)$ matrix

$$\mathbf{P}_B^\perp = \mathbf{I} - \mathbf{B}\mathbf{B}^\# \qquad (1.23)$$

where $\mathbf{B}^\# = (\mathbf{B}^T \mathbf{B})^{-1} \mathbf{B}^T$ is the pseudoinverse of $\mathbf{B}$. Applying $\mathbf{P}_B^\perp$ to the model (1.22) results in

$$\mathbf{P}_B^\perp \mathbf{r} = \mathbf{P}_B^\perp \mathbf{d}\alpha_l + \mathbf{P}_B^\perp \mathbf{n}. \qquad (1.24)$$

The operator $\mathbf{w}$ that maximizes the SNR of the filter output $\mathbf{w}\mathbf{P}_B^\perp \mathbf{y}$,

$$\mathrm{SNR}(\mathbf{w}) = \frac{[\mathbf{w}^T \mathbf{P}_B^\perp \mathbf{d}]\alpha_l^2[\mathbf{d}^T \mathbf{P}_B^\perp \mathbf{w}]}{\mathbf{w}^T \mathbf{P}_B^\perp E[\mathbf{n}\mathbf{n}^T]\mathbf{P}_B^\perp \mathbf{w}}, \qquad (1.25)$$

as shown in [2], is given by the matched filter $\mathbf{w} = \kappa \mathbf{d}$, where κ is a constant. The OSP operator is now given by

$$\mathbf{q}_{\mathrm{OSP}}^T = \mathbf{d}^T \mathbf{P}_B^\perp \qquad (1.26)$$

which consists of a background signature rejecter followed by a matched filter. The output of the OSP classifier is given by

$$D_{\text{OSP}} = \mathbf{q}_{\text{OSP}}^T \mathbf{r} = \mathbf{d}^T \mathbf{P}_{\mathbf{B}}^{\perp} \mathbf{y}. \tag{1.27}$$

1.4.2 OSP in Feature Space and Its Kernel Version

The mixture model in the high dimensional feature space $\mathcal{F}$ is given by

$$\Phi(\mathbf{r}) = \mathbf{M}_\phi \boldsymbol{\alpha}_\phi + \mathbf{n}_\phi, \tag{1.28}$$

where $\mathbf{M}_\phi$ is a matrix whose columns are the endmembers spectra in the feature space; $\boldsymbol{\alpha}_\phi$ is a coefficient vector that accounts for the abundances of each endmember spectrum in the feature space; $\mathbf{n}_\phi$ is an additive zero-mean noise. The model (1.28) can also be rewritten as

$$\Phi(\mathbf{r}) = \Phi(\mathbf{d})\alpha_{p_\phi} + \mathbf{B}_\phi \boldsymbol{\gamma}_\phi + \mathbf{n}_\phi, \tag{1.29}$$

where $\Phi(\mathbf{d})$ represent the spectral signature of the desired target in the feature space with the corresponding abundance α_{p_ϕ} and the columns of $\mathbf{B}_\phi$ represent the undesired background signatures in the feature space which are implemented by the eigenvectors of the background covariance matrix.

The output of the OSP classifier in the feature space is given by

$$D_{\text{OSP}_\phi} = \mathbf{q}_{\text{OSP}_\phi}^T \mathbf{r} = \Phi(\mathbf{d})^T (\mathbf{I}_\phi - \mathbf{B}_\phi \mathbf{B}_\phi^T) \Phi(\mathbf{r}) \tag{1.30}$$

where $\mathbf{I}_\phi$ is the identity matrix in the feature space. This output (1.30) is very similar to the numerator of (1.8). It can easily be shown that the kernelized version of (1.30) is now given by

$$D_{\text{KOSP}} = \mathbf{k}(\mathbf{Z}_{Bd}, \mathbf{d})^T \tilde{\boldsymbol{\Upsilon}} \tilde{\boldsymbol{\Upsilon}}^T \mathbf{k}(\mathbf{Z}_{Bd}, \mathbf{y}) - \mathbf{k}(\mathbf{Z}_B, \mathbf{d})^T \tilde{\boldsymbol{\mathcal{B}}} \tilde{\boldsymbol{\mathcal{B}}}^T \mathbf{k}(\mathbf{Z}_B, \mathbf{y}) \tag{1.31}$$

where $\mathbf{Z}_B = [\mathbf{x}_1 \ \mathbf{x}_2 \ \cdots \ \mathbf{x}_N]$ correspond to N input background spectral signatures and $\tilde{\boldsymbol{\mathcal{B}}} = (\tilde{\boldsymbol{\beta}}^1, \tilde{\boldsymbol{\beta}}^2, \cdots, \tilde{\boldsymbol{\beta}}_{N_b})^T$ are the N_b significant eigenvectors of the centered kernel matrix (Gram matrix) $\mathbf{K}(\mathbf{Z}_B, \mathbf{Z}_B)$ normalized by the square root of their corresponding eigenvalues. $\mathbf{k}(\mathbf{Z}_B, \mathbf{r})$ and $\mathbf{k}(\mathbf{Z}_B, \mathbf{d})$ are column vectors whose entries are $k(\mathbf{x}_i, \mathbf{y})$ and $k(\mathbf{x}_i, \mathbf{d})$ for $\mathbf{x}_i \in \mathbf{Z}_B$, respectively. $\mathbf{Z}_{Bd} = \mathbf{Z}_B \cup d$ and $\tilde{\boldsymbol{\Upsilon}}$ is a matrix whose columns are the N_{bd} eigenvectors ($\boldsymbol{v}_1, \boldsymbol{v}_2, \cdots, \boldsymbol{v}_{N_{bd}}$) of the centered kernel matrix $\mathbf{K}(\mathbf{Z}_{Bd}, \mathbf{Z}_{Bd}) = (\mathbf{K})_{ij} = k(\mathbf{x}_i, \mathbf{x}_j)$, $\mathbf{x}_i, \mathbf{x}_j \in \mathbf{Z}_B \cup \mathbf{d}$ with nonzero eigenvalues, normalized by the square root of their associated eigenvalues. Also $\mathbf{k}(\mathbf{Z}_{Bd}, \mathbf{y})$ is the concatenated vector $\left[\mathbf{k}(\mathbf{Z}_B, \mathbf{r})^T \ \mathbf{k}(d, \mathbf{y})^T \right]^T$ and $\mathbf{k}(\mathbf{Z}_{Bd}, \mathbf{d})$ is the concatenated vector $\left[\mathbf{k}(\mathbf{Z}_B, \mathbf{d})^T \ k(\mathbf{d}, \mathbf{d})^T \right]^T$. In the above derivation (1.31) we assumed that the mapped input data were centered in the feature space. For non-centered data the kernel matrices and the empirical kernel maps have to be properly centered as was shown in the previous Section 1.3.3

1.5 Linear Spectral Matched Filter and Kernel Spectral Matched Filter

1.5.1 Linear Spectral Matched Filter

In this section, we introduce the concept of linear SMF. The constrained least squares approach is used to derive the linear SMF. Let the input spectral signal $\mathbf{x}$ be $\mathbf{x} = [x(1), x(2), \cdots, x(p)]^T$ consisting of p spectral bands. We can model each spectral observation as a linear combination of the target spectral signature and noise

$$\mathbf{x} = a\mathbf{s} + \mathbf{n}, \tag{1.32}$$

where a is an attenuation constant (target abundance measure). When $a = 0$ no target is present and when $a > 0$ target is present, vector $\mathbf{s} = [s(1), s(2), \cdots, s(p)]^T$ contains the spectral signature of the target and vector $\mathbf{n}$ contains the added background clutter noise.

Let us define $\mathbf{X}$ to be a $p \times N$ matrix of the N mean-removed background reference pixels (centered) obtained from the input image. Let each observation spectral pixel to be represented as a column in the sample matrix $\mathbf{X}$

$$\mathbf{X} = [\mathbf{x}_1 \ \mathbf{x}_2 \ \cdots \ \mathbf{x}_N]. \tag{1.33}$$

We can design a linear matched filter $\mathbf{w} = [w(1), w(2), \dots, w(p)]^T$ such that the desired target signal $\mathbf{s}$ is passed through while the average filter output energy is minimized. The solution to this constrained least squares minimization problem is given by

$$\mathbf{w} = \frac{\hat{\mathbf{R}}^{-1}\mathbf{s}}{\mathbf{s}^T\hat{\mathbf{R}}^{-1}\mathbf{s}} \tag{1.34}$$

where $\hat{\mathbf{R}}$ represents the estimated correlation matrix for the reference data. The above expression is referred to as Minimum Variance Distortionless Response (MVDR) beamformer in the array processing literature [23, 24] and more recently the same expression was also obtained in [25] for hyperspectral target detection and was called Constrained Energy Minimization (CEM) filter. The output of the linear filter for the test input $\mathbf{r}$, given the estimated correlation matrix is given by

$$y_{\mathbf{r}} = \mathbf{w}^T\mathbf{r} = \frac{\mathbf{s}^T\hat{\mathbf{R}}^{-1}\mathbf{r}}{\mathbf{s}^T\hat{\mathbf{R}}^{-1}\mathbf{s}}. \tag{1.35}$$

If the observation data is centered a similar expression is obtained for the centered data which is given by

$$y_{\mathbf{r}} = \mathbf{w}^T\mathbf{r} = \frac{\mathbf{s}^T\hat{\mathbf{C}}^{-1}\mathbf{r}}{\mathbf{s}^T\hat{\mathbf{C}}^{-1}\mathbf{s}} \tag{1.36}$$

where $\hat{\mathbf{C}}$ represents the estimated covariance matrix for the reference centered data. Similarly, in [4] and [5] it was shown that using the GLRT a similar expression as in MVDR or CEM, (1.36), can be obtained if the $\mathbf{n}$ is assumed to be the background Gaussian random noise distributed as $\mathcal{N}(0, \mathbf{C})$ where C is the expected covariance matrix of only the background noise. This filter is referred to as matched filter in the signal processing literature or Capon method [26] in the array processing literature. In this book chapter, we implemented the matched filter given by the expression (1.36).

1.5.2 Spectral Matched Filter in Feacture Space and Its Kernel Version

We now consider a linear model in the kernel feature space which has an equivalent nonlinear model in the original input space

$$\Phi(\mathbf{x}) = a_\phi \Phi(\mathbf{s}) + \mathbf{n}_\phi, \tag{1.37}$$

where Φ is the non-linear mapping that maps the input data into a kernel feature space, a_ϕ is an attenuation constant (abundance measure), the high-dimensional vector $\Phi(\mathbf{s})$ contains the spectral signature of the target in the feature space, and vector $\mathbf{n}_\phi$ contains the added noise in the feature space.

Using the constrained least squares approach that was explained in the previous section it can easily be shown that the equivalent matched filter $\mathbf{w}_\phi$ in the feature space is given by

$$\mathbf{w}_\phi = \frac{\hat{\mathbf{R}}_\phi^{-1} \Phi(\mathbf{s})}{\Phi(\mathbf{s})^T \hat{\mathbf{R}}_\phi^{-1} \Phi(\mathbf{s})}, \tag{1.38}$$

where $\hat{\mathbf{R}}_\phi$ is the estimated correlation matrix in the feature space. The correlation matrix is given by

$$\hat{\mathbf{R}}_\phi = \frac{1}{N} \mathbf{X}_\phi \mathbf{X}_\phi{}^T \tag{1.39}$$

where $\mathbf{X}_\phi = [\Phi(\mathbf{x}_1)\,\Phi(\mathbf{x}_2)\,\cdots\,\Phi(\mathbf{x}_N)]$ is a matrix whose columns are the mapped input reference data in the feature space. The matched filter in the feature space (1.38) is equivalent to a non-linear matched filter in the input space and its output for the input $\Phi(\mathbf{r})$ is given by

$$y_{\Phi(\mathbf{r})} = \mathbf{w}_\phi^T \Phi(\mathbf{r}) = \frac{\Phi(\mathbf{s})^T \hat{\mathbf{R}}_\phi^{-1} \Phi(\mathbf{r})}{\Phi(\mathbf{s})^T \hat{\mathbf{R}}_\phi^{-1} \Phi(\mathbf{s})}. \tag{1.40}$$

If the data was centered the matched filter for the centered data in the feature space would be

$$y_{\Phi(\mathbf{r})} = \mathbf{w}_\phi^T \Phi(\mathbf{r}) = \frac{\Phi(\mathbf{s})^T \hat{\mathbf{C}}_\phi^{-1} \Phi(\mathbf{r})}{\Phi(\mathbf{s})^T \hat{\mathbf{C}}_\phi^{-1} \Phi(\mathbf{s})}. \tag{1.41}$$

We now show how to kernelize the matched filter expression (1.41) where the resulting non-linear matched filter is called the kernel matched filter. It was shown in Appendix I the pseudoinverse (inverse) of the estimated background covariance matrix can be written as

$$\hat{\mathbf{C}}_{\Phi}^{\#} = \mathbf{X}_{\Phi}\boldsymbol{\mathcal{B}}\Lambda^{-2}\boldsymbol{\mathcal{B}}^{T}\mathbf{X}_{\Phi}^{T} \tag{1.42}$$

Inserting Equation (1.42) into (1.41), it can be rewritten as

$$y_{\Phi(\mathbf{r})} = \frac{\Phi(\mathbf{s})^{T}\mathbf{X}_{\Phi}\boldsymbol{\mathcal{B}}\Lambda^{-1}\boldsymbol{\mathcal{B}}^{T}\mathbf{X}_{\Phi}^{T}\Phi(\mathbf{r})}{\Phi(\mathbf{s})^{T}\mathbf{X}_{\Phi}\boldsymbol{\mathcal{B}}\Lambda^{-1}\boldsymbol{\mathcal{B}}^{T}\mathbf{X}_{\Phi}^{T}\Phi(\mathbf{s})}. \tag{1.43}$$

Also using the properties of the Kernel PCA as shown in Appendix I, we have the relationship

$$\mathbf{K}^{-2} = \frac{1}{N^{2}}\boldsymbol{\mathcal{B}}\Lambda^{-2}\boldsymbol{\mathcal{B}}^{T}. \tag{1.44}$$

We denote $\mathbf{K} = \mathbf{K}(\mathbf{X}, \mathbf{X}) = (\mathbf{K})_{ij}$ an $N \times N$ Gram kernel matrix whose entries are the dot products $< \Phi(\mathbf{x}_i), \Phi(\mathbf{x}_j) >$. Substituting (1.44) into (1.43), the kernelized version of SMF is given by

$$y_{\mathbf{K}_{\mathbf{r}}} = \frac{\mathbf{k}(\mathbf{X}, \mathbf{s})^{T}\mathbf{K}^{-2}\mathbf{k}(\mathbf{X}, \mathbf{r})}{\mathbf{k}(\mathbf{X}, \mathbf{s})^{T}\mathbf{K}^{-2}\mathbf{k}(\mathbf{X}, \mathbf{s})} = \frac{\mathbf{k}_{\mathbf{s}}^{T}\mathbf{K}^{-2}\mathbf{k}_{\mathbf{r}}}{\mathbf{k}_{\mathbf{s}}^{T}\mathbf{K}^{-2}\mathbf{k}_{\mathbf{s}}} \tag{1.45}$$

where $\mathbf{k}_{\mathbf{s}} = \mathbf{k}(\mathbf{X}, \mathbf{s})$ and $\mathbf{k}_{\mathbf{r}} = \mathbf{k}(\mathbf{X}, \mathbf{r})$ are the empirical kernel maps for $\mathbf{s}$ and $\mathbf{r}$, respectively. As in the previous section, the kernel matrix $\mathbf{K}$ as well as the empirical kernel maps, $\mathbf{k}_{\mathbf{s}}$ and $\mathbf{k}_{\mathbf{r}}$ needs to be properly centered.

1.6 Adaptive Subspace Detector and Kernel Adaptive Subspace Detector

1.6.1 Linear Adaptive Subspace Detector

In this section, the GLRT under the two competing hypotheses ($\mathbf{H}_0$ and $\mathbf{H}_1$) for a certain mixture model is described. The subpixel detection model for a measurement $\mathbf{x}$ (a pixel vector) is expressed as

$$\mathbf{H}_0 : \mathbf{x} = \mathbf{n}, \qquad\qquad \text{Target absent} \tag{1.46}$$
$$\mathbf{H}_1 : \mathbf{x} = \mathbf{U}\boldsymbol{\theta} + \sigma\mathbf{n}, \qquad\qquad \text{Target present}$$

where $\mathbf{U}$ represents an orthogonal matrix whose column vectors are the eigenvectors that span the target subspace $<\mathbf{U}>$; $\boldsymbol{\theta}$ is an unknown vector whose entries are coefficients that account for the abundances of the corresponding column vectors of $\mathbf{U}$; $\mathbf{n}$ represents Gaussian random noise distributed as $\mathcal{N}(0, \mathbf{C})$.

In model, $\mathbf{x}$ is assumed to be a background noise under $\mathbf{H}_0$ and a linear combination of a target subspace signal and a scaled background noise, distributed as $\mathcal{N}(\mathbf{U}\boldsymbol{\theta}, \sigma^2\mathbf{C})$, under $\mathbf{H}_1$. The background noise under the two hypotheses is represented by the same covariance but different variances because of the existence of subpixel targets under $\mathbf{H}_1$. The GLRT for the subpixel problem described in [5], (so-called ASD), is given by

$$D_{\text{ASD}}(\mathbf{x}) = \frac{\mathbf{x}^T\hat{\mathbf{C}}^{-1}\mathbf{U}(\mathbf{U}^T\hat{\mathbf{C}}^{-1}\mathbf{U})^{-1}\mathbf{U}^T\hat{\mathbf{C}}^{-1}\mathbf{x}}{\mathbf{x}^T\hat{\mathbf{C}}^{-1}\mathbf{x}} \underset{H_0}{\overset{H_1}{\gtrless}} \eta_{\text{ASD}}, \qquad (1.47)$$

where $\hat{\mathbf{C}}$ is the MLE of the covariance $\mathbf{C}$ and η_{ASD} represents a threshold. Expression (1.47) has a constant false alarm rate (CFAR) property and is also referred to as the adaptive cosine estimator because (1.47) measures the angle between $\tilde{\mathbf{x}}$ and $< \tilde{\mathbf{U}} >$, where $\tilde{\mathbf{x}} = \hat{\mathbf{C}}^{-1/2}\mathbf{x}$ and $\tilde{\mathbf{U}} = \hat{\mathbf{C}}^{-1/2}\mathbf{U}$.

1.6.2 ASD in the Feature Space and Its Kernel Version

We define a new subpixel model by assuming that the input data have been implicitly mapped by a nonlinear function Φ into a high-dimensional feature space $\mathcal{F}$. The subpixel model in $\mathcal{F}$ is then given by

$$\mathbf{H}_{0_\phi} : \Phi(\mathbf{x}) = \mathbf{n}_\phi, \qquad \qquad \text{Target absent} \qquad (1.48)$$

$$\mathbf{H}_{1_\phi} : \Phi(\mathbf{x}) = \mathbf{U}_\phi\boldsymbol{\theta}_\phi + \sigma_\phi\mathbf{n}_\phi, \qquad \text{Target present}$$

where $\mathbf{U}_\phi$ represents a full-rank matrix whose M_1 column vectors are the eigenvectors that span target subspace $< \mathbf{U}_\Phi >$ in $\mathcal{F}$; $\boldsymbol{\theta}_\phi$ is unknown vectors whose entries are coefficients that account for the abundances of the corresponding column vectors of $\mathbf{U}_\phi$; $\mathbf{n}_\phi$ represents Gaussian random noise distributed by $\mathcal{N}(0, \mathbf{C}_\phi)$; and σ_ϕ is the noise variance under $\mathbf{H}_{1_\phi}$. The GLRT for the model (1.48) in $\mathcal{F}$ is now given by

$$D(\Phi(\mathbf{x})) = \frac{\Phi(\mathbf{x})^T\hat{\mathbf{C}}_\phi^{-1}\mathbf{U}_\phi(\mathbf{U}_\phi^T\hat{\mathbf{C}}_\phi^{-1}\mathbf{U}_\phi)^{-1}\mathbf{U}_\phi^T\hat{\mathbf{C}}_\phi^{-1}\Phi(\mathbf{x})}{\Phi(\mathbf{x})^T\hat{\mathbf{C}}_\phi^{-1}\Phi(\mathbf{x})}, \qquad (1.49)$$

where $\hat{\mathbf{C}}_\phi$ is the MLE of $\mathbf{C}_\phi$.

We now show how to kernelize the GLRT expression (1.49) in the feature space. The inverse (pseudoinverse) background covariance matrix in (1.49) can be represented by its eigenvector decomposition (see Appendix I) given by the expression

$$\hat{\mathbf{C}}_\phi^{\#} = \mathbf{X}_\phi\boldsymbol{\mathcal{B}}\Lambda^{-2}\boldsymbol{\mathcal{B}}^T\mathbf{X}_\phi^T, \qquad (1.50)$$

where $\mathbf{X}_\phi = [\Phi_c(\mathbf{x}_1) \ \Phi_c(\mathbf{x}_2) \cdots \Phi_c(\mathbf{x}_N)]$ represents the centered vectors in the feature space corresponding to N independent background spectral signatures $\mathbf{X} = [\mathbf{x}_1 \ \mathbf{x}_2 \ \cdots \ \mathbf{x}_N]$ and $\boldsymbol{\mathcal{B}} = [\boldsymbol{\beta}^1 \ \boldsymbol{\beta}^2 \ \cdots \ \boldsymbol{\beta}^{N_1}]$ are the nonzero eigenvectors of the centered kernel matrix (Gram matrix) $\mathbf{K}(\mathbf{X}, \mathbf{X})$ normalized by the square root of their cooresponding eigenvalues. Similarly, $\mathbf{U}_\phi$ is given by

$$\mathbf{U}_\phi = \mathbf{Y}_\phi\tilde{\boldsymbol{\mathcal{T}}}, \qquad (1.51)$$

where $\mathbf{Y}_\Phi = [\Phi_c(\mathbf{y}_1)\ \Phi_c(\mathbf{y}_2) \ldots \Phi_c(\mathbf{y}_M)]$ are the centered vectors in the feature space corresponding to the M independent target spectral signatures $\mathbf{Y} = [\mathbf{y}_1\ \mathbf{y}_2\ \cdots\ \mathbf{y}_M]$ and $\tilde{\mathcal{T}} = [\tilde{\alpha}^1\ \tilde{\alpha}^2\ \ldots\ \tilde{\alpha}^{M_1}]$, $M_1 < M$, is a matrix consisting of the M_1 eigenvectors of the kernel matrix $\mathbf{K}(\mathbf{Y}, \mathbf{Y})$ normalized by the square root of their corresponding eigenvalues. Now, the term $\Phi(\mathbf{x})^T \hat{\mathbf{C}}_\Phi^{-1} \mathbf{U}_\Phi$ in the numerator of (1.49) becomes

$$\Phi(\mathbf{x})^T \hat{\mathbf{C}}_\Phi^{-1} \mathbf{U}_\Phi = \Phi(\mathbf{x})^T \mathbf{X}_\Phi \mathcal{B} \Lambda^{-2} \mathcal{B}^T \mathbf{X}_\Phi{}^T \mathbf{Y}_\Phi \tilde{\mathcal{T}} \qquad (1.52)$$

$$= \mathbf{k}(\mathbf{x}, \mathbf{X})^T \mathbf{K}(\mathbf{X}, \mathbf{X})^{-2} \mathbf{K}(\mathbf{X}, \mathbf{Y}) \tilde{\mathcal{T}} \equiv \mathbf{K_x},$$

where $\mathcal{B}\Lambda^{-2}\mathcal{B}^T$ is replaced by $\mathbf{K}(\mathbf{X}, \mathbf{X})^{-2}$ using (1.69), as shown in Appendix I. Similarly,

$$\mathbf{U}_\Phi{}^T \hat{\mathbf{C}}_\Phi^{-1} \Phi(\mathbf{x}) = \tilde{\mathcal{T}}^T \mathbf{K}(\mathbf{X}, \mathbf{Y})^T \mathbf{K}(\mathbf{X}, \mathbf{X})^{-2} \mathbf{k}(\mathbf{x}, \mathbf{X}) = \mathbf{K_x}^T \qquad (1.53)$$

and

$$\mathbf{U}_\Phi{}^T \hat{\mathbf{C}}_\Phi^{-1} \mathbf{U}_\Phi = \tilde{\mathcal{T}}^T \mathbf{K}(\mathbf{X}, \mathbf{Y})^T \mathbf{K}(\mathbf{X}, \mathbf{X})^{-2} \mathbf{K}(\mathbf{X}, \mathbf{Y}) \tilde{\mathcal{T}}. \qquad (1.54)$$

The denominator of (1.49) is also expressed as

$$\Phi(\mathbf{x})^T \hat{\mathbf{C}}_\Phi^{-1} \Phi(\mathbf{x}) = \mathbf{k}(\mathbf{x}, \mathbf{X})^T \mathbf{K}(\mathbf{X}, \mathbf{X})^{-2} \mathbf{k}(\mathbf{x}, \mathbf{X}). \qquad (1.55)$$

Finally, the kernelized expression of (1.49) is given by

$$D_{\mathrm{KASD}}(\mathbf{x}) = \frac{\mathbf{K_x}[\tilde{\mathcal{T}}^T \mathbf{K}(\mathbf{X}, \mathbf{Y})^T \mathbf{K}(\mathbf{X}, \mathbf{X})^{-2} \mathbf{K}(\mathbf{X}, \mathbf{Y}) \tilde{\mathcal{T}}]^{-1} \mathbf{K_x}^T}{\mathbf{k}(\mathbf{x}, \mathbf{X})^T \mathbf{K}(\mathbf{X}, \mathbf{X})^{-2} \mathbf{k}(\mathbf{x}, \mathbf{X})}. \qquad (1.56)$$

As in the previous sections all the kernel matrices $\mathbf{K}(\mathbf{X}, \mathbf{Y})$ and $\mathbf{K}(\mathbf{X}, \mathbf{X})$ as well as the empirical kernel maps need to be properly centered.

1.7 Experimental Results

In this section, the kernel-based matched signal detectors, such as the kernel MSD (KMSD), kernel ASD (KASD), kernel OSP (KOSP), and kernel SMF (KSMF), as well as the corresponding conventional detectors are implemented on the basis of two different types of data sets—illustrative toy data sets and real hyperspectral images that contain military targets. The Gaussian RBF kernel, $k(\mathbf{x}, \mathbf{y}) = \exp(\frac{-\|\mathbf{x}-\mathbf{y}\|^2}{c})$, was used to implement the kernel-based detectors. c represents the width of the Gaussian distribution, and the value of c was chosen such that the overall data variations can be fully exploited by the Gaussian RBF function; the values of c were determined experimentally.

1.7.1 Illustrative Toy Examples

Figures 1.1 and 1.2 show contour and surface plots of the conventional detectors and the kernel-based detectors, on two different types of two-dimensional

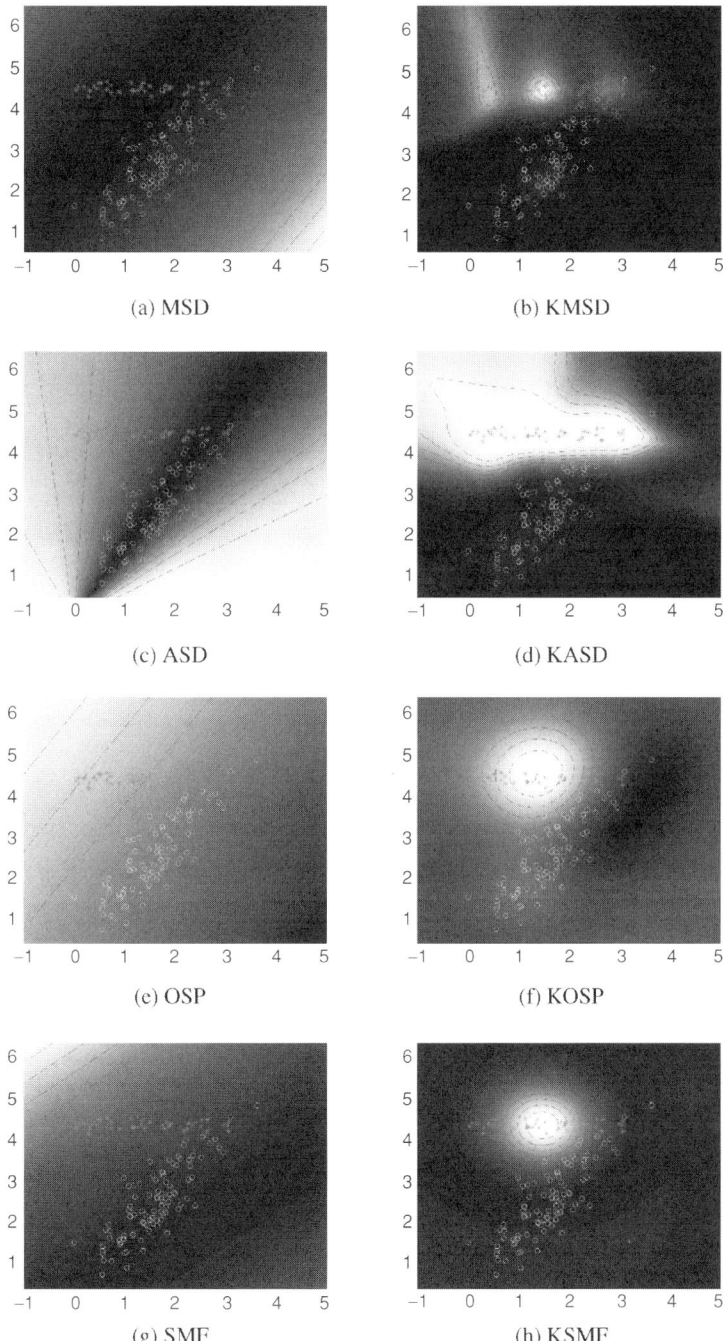

Figure 1.1. Contour and surface plots of the conventional matched signal detectors and their corresponding kernel versions on a toy data set (a mixture of Gaussian).

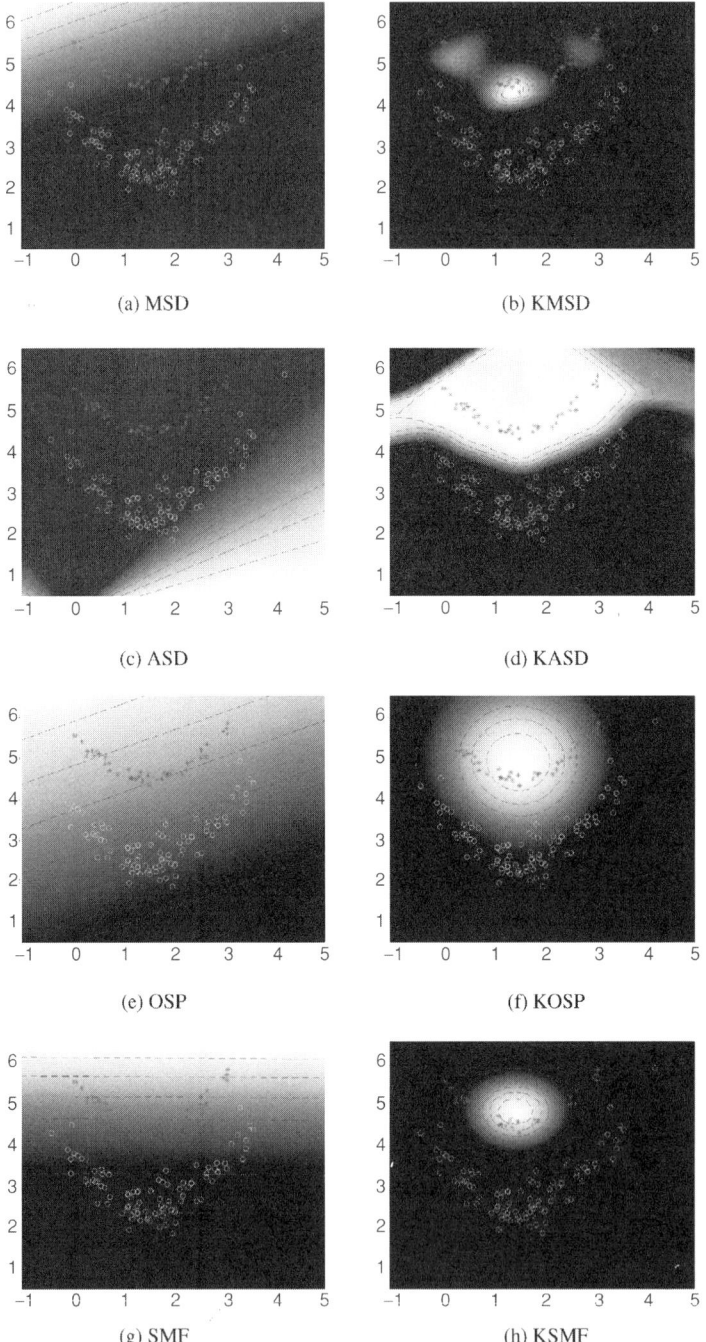

(a) MSD   (b) KMSD

(c) ASD   (d) KASD

(e) OSP   (f) KOSP

(g) SMF   (h) KSMF

Figure 1.2. Contour and surface plots of the conventional matched signal detectors and their corresponding kernel versions on a toy data set: in this toy example, the Gaussian mixture data shown in Fig. 1 was modified to generate nonlinearly mixed data.

toy data sets: a Gaussian mixture in Fig. 1.1 and nonlinearly mapped data in Fig. 1.2. In the contour and surface plots, data points for the desired target were represented by the star-shaped symbol and the background points were represented by the circles. In Fig. 1.2, the two-dimensional data points $\mathbf{x} = (x, y)$ for each class were obtained by nonlinearly mapping the original Gaussian mixture data points $\mathbf{x_0} = (x_0, y_0)$ in Fig. 1.1. All the data points in Fig. 1.2 were nonlinearly mapped by $\mathbf{x} = (x, y) = (x_0, x_0^2 + y_0)$. In the new data set the second component of each data point is nonlinearly related to its first component.

For both data sets, the contours generated by the kernel-based detectors are highly nonlinear and naturally following the dispersion of the data and thus successfully separating the two classes, as opposed to the linear contours obtained by the conventional detectors. Therefore, the kernel-based detectors clearly provided significantly improved discrimination over the conventional detectors for both the Gaussian mixture and nonlinearly mapped data. Among the kernel-based detectors, KMSD and KASD outperform KOSP and KSMF mainly because targets in KMSD and KASD are better represented by the associated target subspace than by a single spectral signature used in KOSP and KSMF. Note that the contour plots for MSD (Fig. 1.1(a) and Fig. 1.2 (a)) represent only the numerator of Eq. 1.5 because the denominator becomes unstable for the two-dimensional cases: i.e., the value inside the brackets $(\mathbf{I} - \mathbf{P_{TB}})$ becomes zero for the two-dimensional data.

1.7.2 Hyperspectral Images

In this section, HYDICE (HYperspectral Digital Imagery Collection Experiment) images from the Desert Radiance II data collection (DR-II) and Forest Radiance I data collection (FR-I) were used to compare detection performance between the kernel-based and conventional methods. The HYDICE imaging sensor generates 210 bands across the whole spectral range (0.4–2.5 μm) which includes the visible and short-wave infrared (SWIR) bands. But we use only 150 bands by discarding water absorption and low SNR bands; the spectral bands used are the 23rd–101st, 109th–136th, and 152nd–194th for the HYDICE images. The DR-II image includes 6 military targets along the road and the FR-I image includes total 14 targets along the tree line, as shown in the sample band images in Fig. 1.3. The detection performance of the DR-II and FR-I images was provided in both the qualitative and quantitative—the receiver operating characteristics (ROC) curves—forms. The spectral signatures of the desired target and undesired background signatures were directly

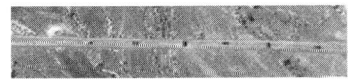 

Figure 1.3. Sample band images from (a) the DR-II image and (b) the FR-I image.

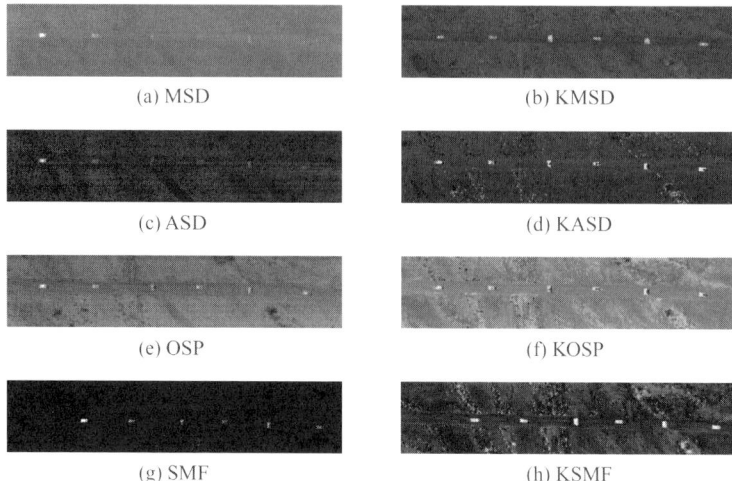

(a) MSD   (b) KMSD

(c) ASD   (d) KASD

(e) OSP   (f) KOSP

(g) SMF   (h) KSMF

Figure 1.4. Detection results for the DR-II image using the conventional detectors and the corresponding kernel versions.

collected from the given hyperspectral data to implement both the kernel-based and conventional detectors.

All the pixel vectors in a test image are first normalized by a constant, which is a maximum value obtained from all the spectral components of the spectral vectors in the corresponding test image, so that the entries of the normalized pixel vectors fit into the interval of spectral values between 0 and 1. The rescaling of pixel vectors was mainly performed to effectively utilize the dynamic range of Gaussian RBF kernel.

Figures 1.4–1.7 show the detection results including the ROC curves generated by applying the kernel-based and conventional detectors to the DR-II and FR-I images. In general, the detected targets by the kernel-based detectors are much more evident than the ones detected by the conventional detectors, as shown in Figs. 1.4 and 1.5. Figures 1.6 and 1.7 show the ROC curve plots for the kernel-based and conventional detectors for the DR-II and FR-I images; in general, the kernel-based detectors outperformed the conventional detectors. In particular, KMSD performed the best of all the kernel-based detectors detecting all the targets and significantly suppressing the background. The performance superiority of KMSD is mainly attributed to the utilization of both the target and background kernel subspaces representing the target and background signals in the feature space, respectively.

1.8 Conclusions

In this chapter, nonlinearversions of several matched signal detectors, such as KMSD, KOSP, KSMF, and KASD, have been implemented using the

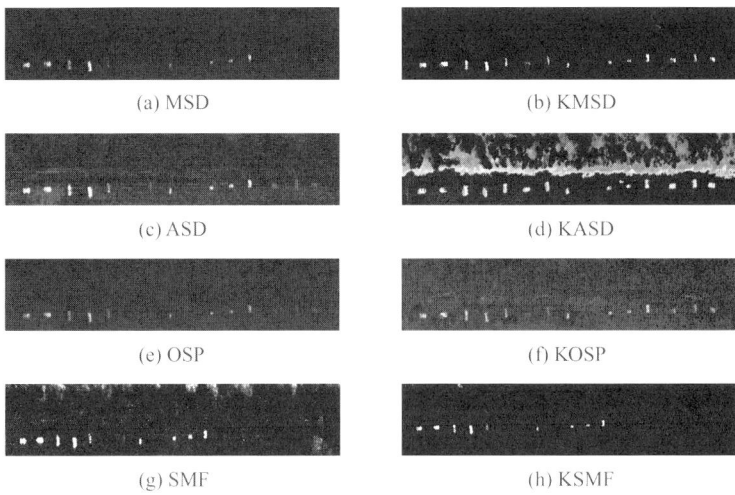

(a) MSD   (b) KMSD

(c) ASD   (d) KASD

(e) OSP   (f) KOSP

(g) SMF   (h) KSMF

Figure 1.5. Detection results for the FR-I image using the conventional detectors and the corresponding kernel versions.

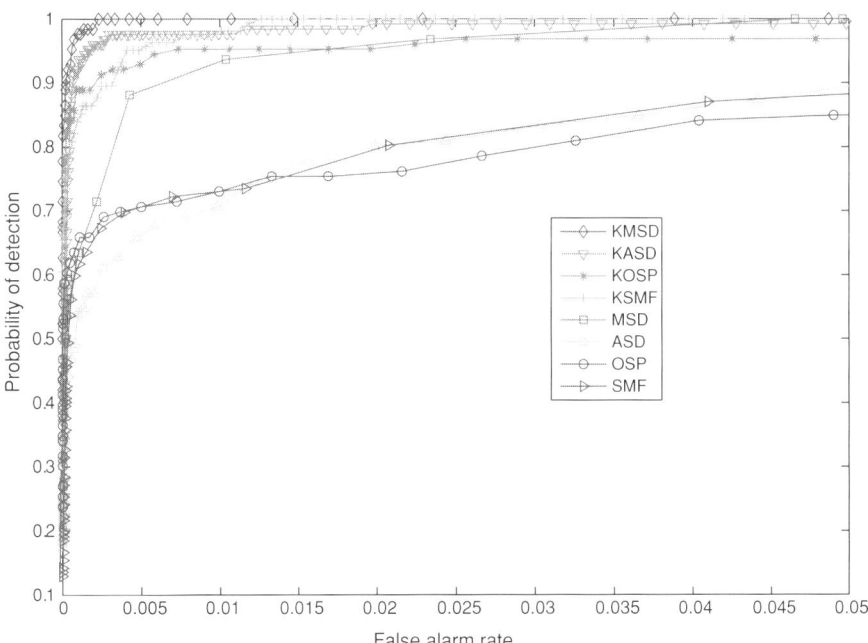

Figure 1.6. ROC curves obtained by conventional detectors and the corresponding kernel versions for the DR-II image (see color insert).

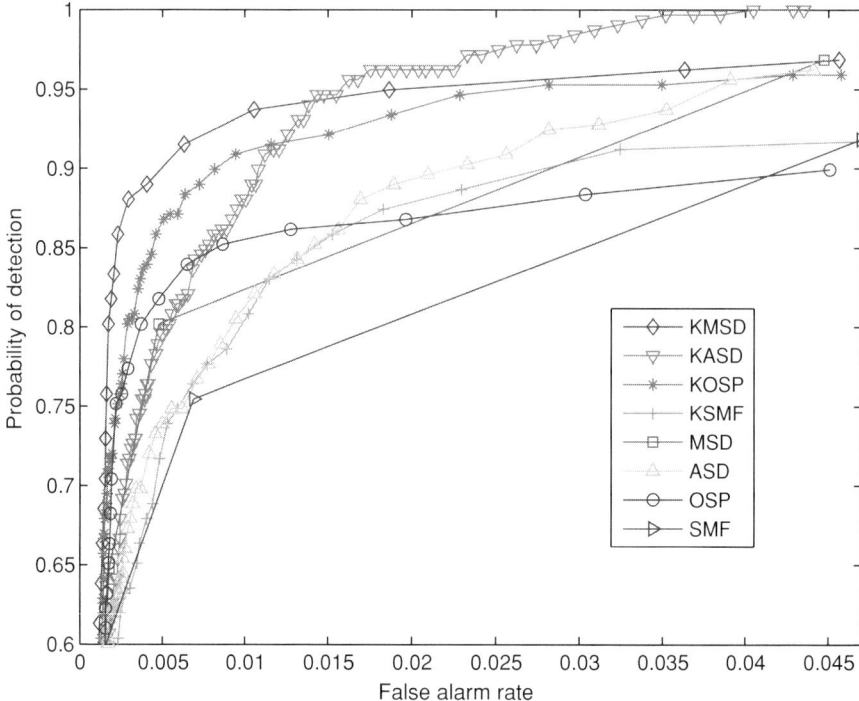

Figure 1.7. ROC curves obtained by conventional detectors and the corresponding kernel versions for the DR-II image (see color insert).

kernel-based learning theory. Performance comparison between the matched signal detectors and their corresponding nonlinear versions was conducted on the basis of two-dimensional toy-examples as well as real hyperspectral images. It is shown that the kernel-based nonlinear versions of these detectors outperform the linear versions.

1.9 Appendix I (Kernel PCA)

In this appendix we will represent derivation of Kernel PCA and its properties providing the relationship between the covariance matrix and the corresponding Gram matrix is presented. Our goal is to prove (1.50) and (1.69). To drive the Kernel PCA, consider the estimated background clutter covariance matrix in the feature space and assume that the input data has been normalized (centered) to have zero mean. The estimated covariance matrix in the feature space is given by

$$\hat{\mathbf{C}}_{\Phi} = \frac{1}{N}\mathbf{X}_{\Phi}\mathbf{X}_{\Phi}^{T}. \tag{1.57}$$

The PCA eigenvectors are computed by solving the eigenvalue problem

$$\lambda \mathbf{v}_\Phi = \hat{\mathbf{C}}_\Phi \mathbf{v}_\Phi \tag{1.58}$$

$$= \frac{1}{N} \sum_{i=1}^{N} \Phi(\mathbf{x}_i) \Phi(\mathbf{x}_i)^T \mathbf{v}_\Phi$$

$$= \frac{1}{N} \sum_{i=1}^{N} < \Phi(\mathbf{x}_i), \mathbf{v}_\Phi > \Phi(\mathbf{x}_i),$$

where $\mathbf{v}_\Phi$ is an eigenvector in $\mathcal{F}$ with a corresponding nonzero eigenvalue λ. Equation (1.58) indicates that each eigenvector $\mathbf{v}_\Phi$ with corresponding $\lambda \neq 0$ are spanned by $\Phi(\mathbf{x}_1), \cdots, \Phi(\mathbf{x}_N)$—i.e.

$$\mathbf{v}_\Phi = \sum_{i=1}^{N} \beta_i \Phi(\mathbf{x}_i) = \mathbf{X}_\Phi \boldsymbol{\beta} \tag{1.59}$$

where $\mathbf{X}_\Phi = \begin{bmatrix} \Phi(\mathbf{x}_1) \ \Phi(\mathbf{x}_2) \cdots \Phi(\mathbf{x}_N) \end{bmatrix}$ and $\boldsymbol{\beta} = (\beta_1, \beta_2, \cdots, \beta_N)^T$. Substituting (1.59) into (1.58) and multiplying with $\Phi(\mathbf{x}_n)^T$, yields

$$\lambda \sum_{i=1}^{N} \beta_i < \Phi(\mathbf{x}_n), \Phi(\mathbf{x}_i) > \tag{1.60}$$

$$= \frac{1}{N} \sum_{i=1}^{N} \beta_i \Phi(\mathbf{x}_n) \Phi(\mathbf{x}_i) \Phi(\mathbf{x}_i)^T \sum_{i=1}^{N} \Phi(\mathbf{x}_i)$$

$$= \frac{1}{N} \sum_{i=1}^{N} \beta_i < \Phi(\mathbf{x}_n), \sum_{j=1}^{N} \Phi(\mathbf{x}_j) < \Phi(\mathbf{x}_j), \Phi(\mathbf{x}_i) >>,$$

for all $n = 1, \cdots, N$.

We denote by $\mathbf{K} = \mathbf{K}(\mathbf{X}, \mathbf{X}) = (\mathbf{K})_{ij}$ the $N \times N$ kernel matrix whose entries are the dot products $< \Phi(\mathbf{x}_i), \Phi(\mathbf{x}_j) >$. Equation (1.60) can be rewritten as

$$N\lambda \boldsymbol{\beta} = \mathbf{K} \boldsymbol{\beta}, \tag{1.61}$$

where $\boldsymbol{\beta}$ turn out to be the eigenvectors with nonzero eigenvalues of the kernel matrix $\mathbf{K}$. Therefore, the Gram matrix can be written in terms of it eigenvector decomposition as

$$\mathbf{K} = \boldsymbol{\mathcal{B}} \Omega \boldsymbol{\mathcal{B}}^T, \tag{1.62}$$

where $\boldsymbol{\mathcal{B}} = [\boldsymbol{\beta}^1 \ \boldsymbol{\beta}^2 \ \dots \ \boldsymbol{\beta}^N]$ are the eigenvectors of the kernel matrix and Ω is a diagonal matrix with diagonal values equal to the eigenvalues of the kernel matrix $\mathbf{K}_b$. Similarly, from the definition of PCA in the feature space (1.58) the estimated background covariance matrix is decomposed as

$$\hat{\mathbf{C}}_\Phi = \mathbf{V}_\Phi \Lambda \mathbf{V}_\Phi^T, \tag{1.63}$$

where $\mathbf{V}_\Phi = [\mathbf{v}_\Phi^1 \; \mathbf{v}_\Phi^2 \; \ldots \; \mathbf{v}_\Phi^N]$ and Λ is a diagonal matrix with its diagonal elements being the eigenvalues of $\hat{\mathbf{C}}_\Phi$. From (1.58) and (1.61) the eigenvalues of the covariance matrix Λ in the feature space and the eigenvalues of the kernel matrix Ω are related by

$$\Lambda = \frac{1}{N}\Omega. \tag{1.64}$$

Substituting (1.64) into (1.62) we obtain the relationship

$$\mathbf{K} = N\boldsymbol{\mathcal{B}}\Lambda\boldsymbol{\mathcal{B}}^T, \tag{1.65}$$

where N is a constant representing the total number of background clutter samples, which can be ignored.

The sample covariance matrix in the feature space is rank deficient, therefore, its inverse cannot be obtained but its pseudoinverse can be written as [27]

$$\hat{\mathbf{C}}_\Phi^{\#} = \mathbf{V}_\Phi \Lambda^{-1} \mathbf{V}_\Phi{}^T \tag{1.66}$$

since the eigenvectors $\mathbf{V}_\Phi$ in the feature space can be represented as

$$\mathbf{V}_\Phi = \mathbf{X}_\Phi \boldsymbol{\mathcal{B}} \Lambda^{-1/2} = \mathbf{X}_\Phi \tilde{\boldsymbol{\mathcal{B}}}, \tag{1.67}$$

then the pseudoinverse background covariance matrix $\hat{\mathbf{C}}_\Phi^{\#}$ can be written as

$$\hat{\mathbf{C}}_\Phi^{\#} = \mathbf{V}_\Phi \Lambda^{-1} \mathbf{V}_\Phi{}^T = \mathbf{X}_\Phi \boldsymbol{\mathcal{B}} \Lambda^{-2} \boldsymbol{\mathcal{B}}^T \mathbf{X}_\Phi{}^T. \tag{1.68}$$

The maximum number of eigenvectors in the pseudoinverse is equal to the number of non-zero eigenvalues (or the number of independent data samples), which cannot be exactly determined due to round-off error in the calculations. Therefore, the effective rank [27] is determined by only including the eigenvalues that are above a small threshold. Similarly, the inverse Gram matrix $\mathbf{K}_b^{-1}$ can also be written as

$$\mathbf{K}^{-1} = \boldsymbol{\mathcal{B}}\Lambda^{-1}\boldsymbol{\mathcal{B}}^T. \tag{1.69}$$

If the data samples are not independent then the pseudoinverse of the Gram matrix has to be used, which is the same as (1.69) except only the eigenvectors with eigenvalues above a small threshold are included in order to obtain a numerically stable inverse.

In the derivation of the kernel PCA we assumed that the data have already been centered in the feature space by removing the sample mean. However, the sample mean cannot be directly removed in the feature space due to the high dimensionality of $\mathcal{F}$. That is the kernel PCA needs to be derived in terms of the original uncentered input data. Therefore, the kernel matrix $\hat{\mathbf{K}}$ needs to be properly centered [7]. The effect of centering on the kernel PCA can be seen by replacing the uncentered $\mathbf{X}_\Phi$ with the centered $\mathbf{X}_\Phi - \boldsymbol{\mu}_\Phi$ (where $\boldsymbol{\mu}_\Phi$ is the mean of the reference input data) in the estimation of the covariance matrix expression (1.57). The resulting centered $\hat{\mathbf{K}}$ is shown in [7] to be given by

$$\hat{\mathbf{K}} = (\mathbf{K} - \mathbf{1}_N \mathbf{K} - \mathbf{K} \mathbf{1}_N + \mathbf{1}_N \mathbf{K} \mathbf{1}_N), \qquad (1.70)$$

where the $N \times N$ matrix $(\mathbf{1}_N)_{ij} = 1/N$. In the above Eqs. (1.62) and (1.69) the kernel matrix $\mathbf{K}$ needs to be replaced by the centered kernel matrix $\hat{\mathbf{K}}$.

References

[1] L. L. Scharf and B. Friedlander. Matched subspace detectors. *IEEE Trans. Signal Process.*, 42(8):2146–2157, Aug. 1994.

[2] J. C. Harsanyi and C.-I. Chang. Hyperspectral image classification and dimensionality reduction: An orthogonal subspace projection approach. *IEEE Trans. Geosci. Remote Sensing*, 32(4):779–785, July 1994.

[3] D. Manolakis, G. Shaw, and N. Keshava. Comparative analysis of hyperspectral adaptive matched filter detector. In *Proc. SPIE*, volume 4049, pages 2–17, April 2000.

[4] F. C. Robey, D. R. Fuhrmann, and E. J. Kelly. A CFAR adaptive matched filter detector. *IEEE Trans. Aerospace Elect. Syst.*, 28(1):208–216, Jan. 1992.

[5] S. Kraut and L. L. Scharf. The CFAR adaptive subspace detector is a scale-invariant GLRT. *IEEE Trans. Signal Process.*, 47(9):2538–2541, Sept. 1999.

[6] S. Kraut, L. L. Scharf, and T. McWhorter. Adaptive subspace detectors. *IEEE Trans. Signal Process.*, 49(1):1–16, Jan. 2001.

[7] B. Schölkopf and A. J. Smola. *Learning with Kernels*. The MIT Press, 2002.

[8] Mark Girolami. Mercer kernel-based clustering in feature space. *IEEE Trans. Neural Networks*, 13(3):780–784, 2002.

[9] B. Schölkolpf, A. J. Smola, and K.-R. Müller. Kernel principal component analysis. *Neural Comput*, 10:1299–1319, 1998.

[10] G. Baudat and F Anouar. Generalized discriminant analysis using a kernel approach. *Neural Computation*, 12(12):2385–2404, 2000.

[11] A. Ruiz and E. Lopez-de Teruel. Nonlinear kernel-based statistical patten analysis. *IEEE Trans. Neural Networks*, 12:16–32, 2001.

[12] H. P. Park and H. Park. Nonlinear feature extraction based on centroids and kernel functions. *Pattern Recogn.*, 37:801–810, 2004.

[13] H. Kwon and N. M. Nasrabadi. Kernel-based subpixel target detection in hyperspectral images. In *Proc. of IEEE Joint Conference on Neural Networks*, pages 717–722, Budapest, Hungary, July 2004.

[14] H. Kwon and N. M. Nasrabadi. Kernel adaptive subspace detector for hyperspectral target detection. In *Proc. of IEEE International Conference on Acoustics, Speech and Signal Processing*, pages 681–684, Philadelphia, PA, March 2005.

[15] H. Kwon and N. M. Nasrabadi. Kernel spectral matched filter for hyperspectral target detection. In *Proc. of IEEE International Conference on Acoustics, Speech and Signal Processing*, pages 665–668, Philadelphia, PA, March 2005.

[16] H. Kwon and N. M. Nasrabadi. Kernel RX-algorithm: A nonlinear anomaly detector for hyperspectral imagery. *IEEE Trans. Geosci. Remote Sens.*, 43(2):388–397, Feb. 2005.

[17] E. Maeda and H. Murase. Multi-category classification by kernel-based nonlinear subspace. In *Proc. of IEEE International Conference on Acoustics, Speech and Signal Processing*, 1999.

[18] M. M. Dundar and D. A. Landgrebe. Toward an optimal supervised classification for the analysis of huperspectral data. *IEEE Trans. Geosci. Remote Sensing*, 42(1):271–277, Jan. 2004.

[19] E. Pekalska, P. Paclik, and R. P. W. Duin. A generalized kernel approach to dissimilarity-based classification. *J. Mach. Learn.*, 2:175–211, 2001.

[20] J. Lu, K.N. Plataniotis, and A.N. Venetsanopoulos. Face recognition using kernel direct discriminant analysis algorithm. *IEEE Trans. Neural Networks*, 14:117–126, 2003.

[21] H. L. Van Trees. *Detection, Estimation, and Modulation Theory*. John Wiley and Sons, Inc., 1968.

[22] J. J. Settle. On the relationship between spectral unmixing and subspace projection. *IEEE Trans. Geosci. Remote Sens.*, 34(4):1045–1046, July 1996.

[23] B. D. Van Veen and K. M. Buckley. Beamforming: A versatile approach to spatial filtering. *IEEE ASSP Magazine*, pages 4–24, Apr. 1988.

[24] D. H. Johnson and D. E. Dudgeon. *Array Signal Processing*. Prentice hall, 1993.

[25] J. C. Harsanyi. *Detection and Classification of Subpixel Spectral Signatures in Hyperspectral Image Sequences*. Ph D dissertation, Dept. Elect. Eng., Univ. of Maryland, Baltimore County, 1993.

[26] J. Capon. High-resolution frequency-wavenumber spectrum analysis. *Proc. of the IEEE*, 57:1408–1418, 1969.

[27] G. Strang. *Linear Algebra and Its Applications*. Harcourt Brace & Company, 1986.

Theory of Invariant Algebra and Its Use in Automatic Target Recognition

Firooz Sadjadi

Lockheed Martin Corporation, 3400 Highcrest Road, Saint Anthony, MN,
sadja001@umn.edu

2.1 Introduction

Automatic recognition of objects independent of size, orientation, position in the field of view, and color is a difficult and important problem in computer vision, image analysis, and automatic target recognition fields. A direct approach to this problem is by use of a large library of target signatures at all potential positions, viewing angles, spectral bands, and contrast conditions that can lead to a combinatorial explosion of models to be considered. Another approach is by development of composite template filters by means of which potential viewing instances of a target under differing size, orientation, spectral, and contrast variations are used to create a single composite template filter that is then used for detection and classification of that target.

In this chapter, we consider a third approach by exploring the theory of invariant algebra to develop solutions for this problem. Algebraic invariants of binary and ternary quantics are used to develop features that remain unchanged when the object undergoes linear geometrical and spectral transformations. Invariant algebra is a mathematical discipline that arises in relation with a number of problems in algebra and geometry. Extractions of algebraic expressions that remain unchanged under changes of coordinate systems are part of this discipline. Lagrange seems to be among the first mathematicians who first studied invariants.

In the following section, we provide a background on invariant algebra, develop the concept of object representation in terms of a probability density function (PDF) and its statistical moments, and present invariants of binary and ternary quantics. Section 2.3 provides an application of invariants of binary quantics in geometrical and spectral invariant object representation. Section 2.4 provides an application of ternary quantics invariants in 3D

Ladar target classification. An analysis of computational complexities of invariant-based classifications is made in Section 2.5, and compared with that of a typical noninvariant template-based classification method is made. Finally, in Section 2.6, we provide a summary of the main finding of this chapter.

2.2 Theory of Invariant Alagebra

Studying the intrinsic properties of polynomials, which remain undisturbed under changes of variables, forms the domain of this theory. The study and derivation of the algebraic invariants has a long history, which goes back to Lagrange and Boole. However, its development as an independent discipline is due to the work of Cayley and Sylvester in the nineteenth century [1–5].

Consider a homogenous nth-order polynomial of m variables. In the parlance of invariant algebra this polynomial is referred to as an m-ary quantic of order n (or m-ary n-ic). The goal pursued under this theory is the derivation of those algebraic expressions of the coefficients of this quantic that remain invariant when the m variables undergo a linear transformation. The coefficients of the transformation act as a multiplying factor. When this factor is eliminated the invariants are referred to as absolute invariants.

As an example consider a ternary quantic of order m

$$f(x_1, x_2, x_3) = \sum_{\substack{p,q,r=0 \\ p+q+r=m}}^{m} \frac{m!}{p!q!r!}\, a_{pqr} x_1^p x_2^q x_3^r.$$

A homogenous polynomial $I(a)$ of coefficients is called an invariant of an algebraic form f if, after transforming its set of variables from x to x' and constructing a corresponding polynomial $I(a')$ of the new coefficients the following holds true:

$$I(a) = \Lambda I(a'). \tag{2.1}$$

Λ is independent of the $f(x)$ and depends only on the transformation. For homogenous polynomials considered here $\Lambda = \Delta^\omega$, where Δ is the determinant of the transformation and ω is called the weight of the invariant. The invariant is called absolute when $\omega = 0$.

Any object in a multidimensional coordinate system $(x_1, x_2, x_3, \ldots)$ can be represented in terms of a probability density function $\rho(x_1, x_2, x_3, \ldots)$ by proper normalization. Moreover, it is well known that any PDF can be uniquely defined in terms of its infinite statistical moments [3].

Multidimensional moment of order $p + q + r + \cdots$ of a density $\rho(x_1, x_2, x_3, \cdots)$ is defined as the Riemann integral:

$$m_{pqr\ldots} = \int\limits_{-\infty}^{\infty} \int\limits_{-\infty}^{\infty} \int\limits_{-\infty}^{\infty} \cdots x_1^p x_2^q x_3^r \cdots \rho(x_1, x_2, x_3, \ldots)\mathrm{d}x_1 \mathrm{d}x_2 \mathrm{d}x_3 \cdots. \tag{2.2}$$

The sequence of $\{m_{pqr...}\}$ determines uniquely $r(x_1, x_2, x_3, \ldots)$.

Using the definition of moment generating function of multidimensional moments and expanding it into a power series one has the following:

$$M(u_1, u_2, u_3, \ldots) = \int\limits_{-\infty}^{\infty}\int\limits_{-\infty}^{\infty}\int\limits_{-\infty}^{\infty} \cdots \sum_{p=0}^{\infty} \frac{1}{p!}(u_1 x_1 + u_2 x_2 + u_3 x_3 \cdots)^p$$

$$\times \rho(x_1, x_2, x_3, \ldots) \mathrm{d}x_1 \mathrm{d}x_2 \mathrm{d}x_3 \cdots . \tag{2.3}$$

This by a few algebraic manipulations is reduced to an n-ary quantic of order m similar to (2.1).

2.2.1 Fundamental Theorem of Moment Invariants

If an m-ary p-ic (a homogeneous polynomial of order p in m variables) has an invariant

$$f(a_{p..0}, \ldots, a_{0..p}) = \Delta^\omega f(a_{p..0}, \ldots, a_{0..p}). \tag{2.4}$$

Then the moment of order p has an algebraic invariant

$$f(\mu_{p...0}, \ldots, \mu_{0...P}) = |J|\,\Delta^\omega f(\mu_{p...0}, \ldots, \mu_{0...p}), \tag{2.5}$$

where J is the Jacobian of the transformation.

For the case of binary quantic, the following invariants can be derived [6]:

$$
\begin{aligned}
\phi_{1+} &= \eta_{20} + \eta_{02} \\
\phi_2 &= (\eta_{20} - \eta_{02})^2 + 4\eta_{11}^2 \\
\phi_3 &= (\eta_{03} - 3\eta_{12})^2 + (3\eta_{21} + \eta_{03})^2 \\
\phi_4 &= (\eta_{30} + \eta_{12})^2 + (\eta_{21} + \eta_{03})^2 \\
\phi_5 &= (\eta_{30} - 3\eta_{12})(\eta_{03} + \eta_{30})[(\eta_{30} + \eta_{12})^2 - 3(\eta_{21} + \eta_{03})^2] \\
&\quad + (3\eta_{21} - \eta_{03})(\eta_{21} + \eta_{03})[3(\eta_{30} + \eta_{12})^2 - (\eta_{21} + \eta_{03})^2] \\
\phi_6 &= (\eta_{20} - \eta_{02})[(\eta_{30} + \eta_{12})^2 - (\eta_{21} + \eta_{03})^2] \\
&\quad + 4\eta_{11}(\eta_{30} + \eta_{12})(\eta_{21} + \eta_{03})
\end{aligned}
\tag{2.6}
$$

A seventh invariant can be added that will change sign under "improper" rotation.

$$
\begin{aligned}
\phi_7 &= (3\eta_{12} - \eta_{30})(\eta_{30} + \eta_{12})[(\eta_{30} + \eta_{12})^2 - 3(\eta_{21} + \eta_{03})^2] \\
&\quad + (3\eta_{21} - \eta_{03})(\eta_{21} + \eta_{03})[3(\eta_{30} + \eta_{12})^2 - (\eta_{21} + \eta_{03})^2]
\end{aligned}
\tag{2.7}
$$

ϕs are related to μ by the following normalization factor that will make the central moments invariant under size change:

$$\eta_{pq} = \frac{\mu_{pq}}{\mu_{00}^{\frac{(p+q)}{2}} + 1}. \tag{2.8}$$

For the case of ternary quadratics, the following invariants are derived [2]:

$$
\begin{aligned}
J_{1\mu} &= \mu_{200} + \mu_{020} + \mu_{002} \\
J_{2\mu} &= \mu_{020}\mu_{002} - \mu_{001}^2 + \mu_{200}\mu_{002} - \mu_{101}^2 \\
&\quad + \mu_{200}\mu_{020} - \mu_{110}^2 \\
\Delta_{2\mu} &= \det \begin{pmatrix} \mu_{200} & \mu_{110} & \mu_{101} \\ \mu_{110} & \mu_{020} & \mu_{011} \\ \mu_{101} & \mu_{011} & \mu_{002} \end{pmatrix}
\end{aligned}
\tag{2.9}
$$

where μ_{pqr} denote the centralized moments. The following absolute invariants are then obtained by simple algebraic manipulations:

$$
I_3 = \frac{J_1 J_2}{\Delta_2} \qquad I_1 = \frac{J_1^2}{J_2} \quad \text{or} \quad I_1 = \frac{J_1^2}{J_2} \qquad I_2 = \frac{\Delta_2}{J_1^3}
\tag{2.10}
$$

2.3 Applications of Invariants of Binary Quantics

2.3.1 Two-Dimensional Geometrical Invariancy

Figure 2.1 shows an airborne view of a military truck viewed in three different field of view geometries. Fig. 2.1(a) shows the first view. Fig. 2.1(b) shows the same scene when the field of view is rotated 90°. Fig. 2.1(c) shows the same scene again when the field of view is rotated 180°. The corresponding invariant expressions for these three scenes are computed using the relations (2.6) and (2.7). These values are tabulated in Table 2.1. As can be seen from this table the computed invariants remain mostly unchanged when the field of the view of the scene is changed by 90° and 180°.

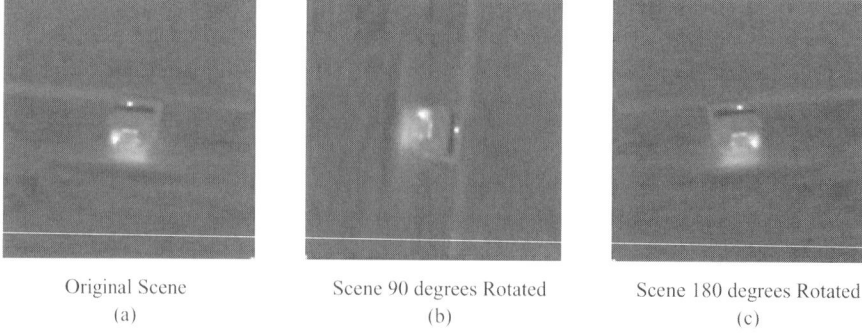

| Original Scene | Scene 90 degrees Rotated | Scene 180 degrees Rotated |
| (a) | (b) | (c) |

Figure 2.1. A scene under three different rotation states: (a) Original scene, (b) scene 90° rotated, (c) scene 180° rotated.

Table 2.1. Invariant values for a scene under three different geometries.

Invariants	Original Scene	Scene Rotated by 90°	Scene Rotated by 180°
Φ_1	5344.64	5344.64	5344.64
Φ_2	1112.63	1112.63	1112.63
Φ_3	7.63	7.63	7.69
Φ_4	0.72	0.72	0.72
Φ_5	−1.69	−1.69	−1.69
Φ_6	−16.90	−16.90	−16.90
Φ_7	−1.21	1.21	−1.21

2.3.2 Joint Geometrical and Material Invariancy

From the Planck's Law [7] (which in its common form does not have the spectral emissivity included in it), one has the following relationship between the emissivity, temperature, wavelength, and the spectral radiant emittance W_λ:

$$W_\lambda = \frac{2\pi hc^2}{\lambda^5} \frac{\varepsilon_\lambda}{e^{\frac{ch}{\lambda KT}} - 1}, \tag{2.11}$$

where W_λ is the spectral radiant emittance in watts per square centimeter per micrometer; T is the absolute temperature in degree Kelvin; ε_λ is the spectral emissivity; h is the Planck's constant $= (6.6256 \pm 0.0005) \times 10^{-34}$ W s^2; λ is the wavelength in centimeters; K is the Boltzman constant $= (1.38054 \pm 0.00018) \times 10^{-23}$ W s/K; and c is the speed of light in centimeter per second. In (2.1) only emissivity is material dependent. Emissivity, defined as the ratio of the radiant emittance of the illuminating source to the radiant emittance of the black body, is dimensionless and has a value between 0 and 1.

The scene radiation is obtained by integrating (2.12) over different wavelength bands. From a wide range of wavelength bands, different spectral images corresponding to the same scene are obtained.

When the effects of the radiation reflectance are negligible or ignored and we assume that the scene is in thermal equilibrium, the radiation varies only with ε_λ in aparticular scene. The variations of ε_λ with frequency for different materials and paints are well documented.

The output of a focal plane array (FPA), in general, is a linear function of the incidence photons emanating from the scene

$$N_{ij} = K_{ij} \int_{\lambda_1}^{\lambda_2} \zeta_{ij}(\lambda)\{\varepsilon_\lambda(i,j)W_\lambda(T_{ij}) + [1 - \varepsilon_\lambda(i,j)W_\lambda(T_b)]\}d\lambda + N_{ij}^d, \tag{2.12}$$

where N_{ij} is the total number of accumulated electrons at pixel ij, K_{ij} is a coefficient that is dependent on the active pixel area, optical transmission, frame time, and pixel angular displacement from the optical axis, and the

f-number of the optics. The quantum efficiency of the ij pixel is denoted by $\zeta_{ij}(\lambda)$, and the background radiant reflectance is shown by $W_\lambda(T_{\rm b})$. $T_{\rm b}$ is the background temperature in degree Kelvin, and λ_1 and λ_2 define the spectral band of the sensor. Finally, the dark charge for the pixel ij is denoted by N_{ij}^d. The $\varepsilon_\lambda(i,j)$ indicates the spectral emissivity at the pixel location ij.

At each pixel location ij, consider $\varepsilon(i,j)$ as an n-dimensional vector (n being the number of wavelengths used). Then, consider the probability of it being from a material π_k, k being the number of different materials in the scene, be denoted as $p(\varepsilon\,|\,\pi_k)$ and the probability of material occurrence π_k as $p(\pi_k)$. Then according to the Bayes decision rule, one has to select the following:

$$\max_k\{p(\pi_k\,|\varepsilon)\} = \max_k\left\{\frac{p(\varepsilon\,|\pi_k)p(\pi_k)}{p(\varepsilon)}\right\}, \tag{2.13}$$

where $p(\varepsilon|\pi_k)$ is assumed to be known for each frequency k at a range of temperatures of interest. This assumption is not restrictive since for different material (and paints) the emissivity as function of frequency and temperature has been documented. The $p(\varepsilon)$ is obtained from the following:

$$p(\varepsilon) = \sum_{\pi_k} p(\varepsilon\,|\pi_k)p(\pi_k). \tag{2.14}$$

Once for each pixel location, a material label has been chosen a new image is formed. This image is formed by replacing the value of each pixel with its most likely emissivity label (iron = 1, water = 2, etc.). Denoting each pixel as $\pi_k(i,j)$, the information content of the image varies by the frequency of occurrence of the emissivity label in the image.

In the above case, the invariant expressions (2.6) and (2.7) depend on the material that the targets and scene are made of. Moreover, they are k-ary quantics (homogeneous polynomials of k variables). Consequently, for any linear transformation in π_k (changes in material mixtures) there exist a set of algebraic expressions that will remain unchanged. These second-order invariants will be invariant under scene rotation, scale, translation, and material mixture transformations. The expressions $\phi_1(\pi_k)$ to $\phi_7(\pi_k)$ are polynomials of order 1 to 4 in terms of π_k for various k. Each of different k values indicates a different material. Any linear transformation of π_k indicates a change of material mixture in the scene such as changing the paint on a target, or having the objects on a dry land versus wetland, or for a target being on a grass verses being on a concrete background.

Example: Consider an object whose image is represented analytically as by the following function:

$$f(x,y,p,q) = (p+q)\mathrm{e}^{-x-y}, \tag{2.15}$$

where x and y represent the axes for the object's geometry and p and q represent the axis for each pixel's material mixtures. In the following, we will refer to the object's change of orientation, scale, and position as its geometrical transformation to distinguish this type of change from those associated with the object's surface material.

Changes in the object's orientation, scale, and positions in the field of view will lead to changes in the object image, and consequently in its representation, as has been expressed in Eq. (2.15). Similarly, any change in the surface material of the object (for example, a differing surface paint) also causes a change in the object's image representation as expressed in Eq. (2.15).

In the following, we will derive the geometrical invariants for this object and from them will extract the material invariants associated for each pixel on the object. We will then change the material mixture arbitrarily and show the invariancy of these expressions to the joint geometrical and material transformations.

Deriving the geometrical invariants for the object, whose image is represented by Eq. (2.13), one can obtain the following relationships by using Eqs. (2.6) and (2.7):

$$\text{GeoInvariant } 1 = \frac{(2(1.15(p+q) - 0.57(p+q)^2 + 0.09(p+q)^3))}{(1 + 0.33(p+q))} \tag{2.16}$$

$$\text{GeoInvariant } 2 = \frac{(4(p+q)^2(0.84 - 0.57(p+q) + (0.09(p+q)^2)^2)}{(1 + 0.33(p+q))^2} \tag{2.17}$$

$$\begin{aligned}
\text{GeoInvariant } 3 = (0.55(-1.10 + p+q))(-1.10 + p+q)(p+q)^2 \\
\times \frac{(17.60 - 7.75(p+q) + (p+q)^2))}{(5.07 + (p+q)^{1.5})^2}
\end{aligned} \tag{2.18}$$

$$\begin{aligned}
\text{GeoInvariant } 4 = (0.2788(p+q)^2(15.9731 - 7.9674(p+q) + (p+q)^2) \\
\times (18.1287 - 6.2169(p+q) + (p+q)^2) \\
\times \frac{(9.8189 - 3.5321(p+q) + (p+q)^2))}{(5.0739 + (p+q)^{1.5})^2}
\end{aligned} \tag{2.19}$$

$$\begin{aligned}
\text{GeoInvariant } 5 = (0.38(-3.8 + p+q)(-1.10 + p+q)(p+q)^4 \\
\times (17.60 - 7.75(p+q) + (p+q)^2)(15.42 - 5.60(p+q) \\
+ (p+q)^2)(13.27 - 5.02(p+q) + (p+q)^2) \\
\times \frac{(11.40 - 4.36(p+q) + (p+q)^2))}{(5.07 + (p+q)^{1.5})^4}
\end{aligned} \tag{2.20}$$

$$\begin{aligned}
\text{GeoInvariant } 6 = (-0.16(-3.83 + p+q)(-2.95 + p+q)(-2.95 + p+q) \\
\times (p+q)^4(14.72 - 7.67(p+q) + (p+q)^2) \\
\times (13.27 - 5.02(p+q) + (p+q)^2)(13.27 - 5.02(p+q) \\
+ (p+q)^2)(13.27 - 5.02(p+q) + (p+q)^2)) \\
\frac{}{((2.95 + (p+q))(5.07 + (p+q)^{1.5})^3)}
\end{aligned} \tag{2.21}$$

$$\begin{aligned}
\text{GeoInvariant } 7 = (0.69(-4.11 + p+q)(-3.95 + p+q)(-3.83 + p+q) \\
\times (-1.10 + p+q)(p+q)^4(17.60 - 7.75(p+q) + (p+q)^2) \\
\times (13.277 - 5.02(p+q) + (p+q)^2)(14.21 - 4.74(p+q) \\
+ (p+q)^2)) \\
\frac{}{(5.07 + (p+q)^{1.5})^4}
\end{aligned} \tag{2.22}$$

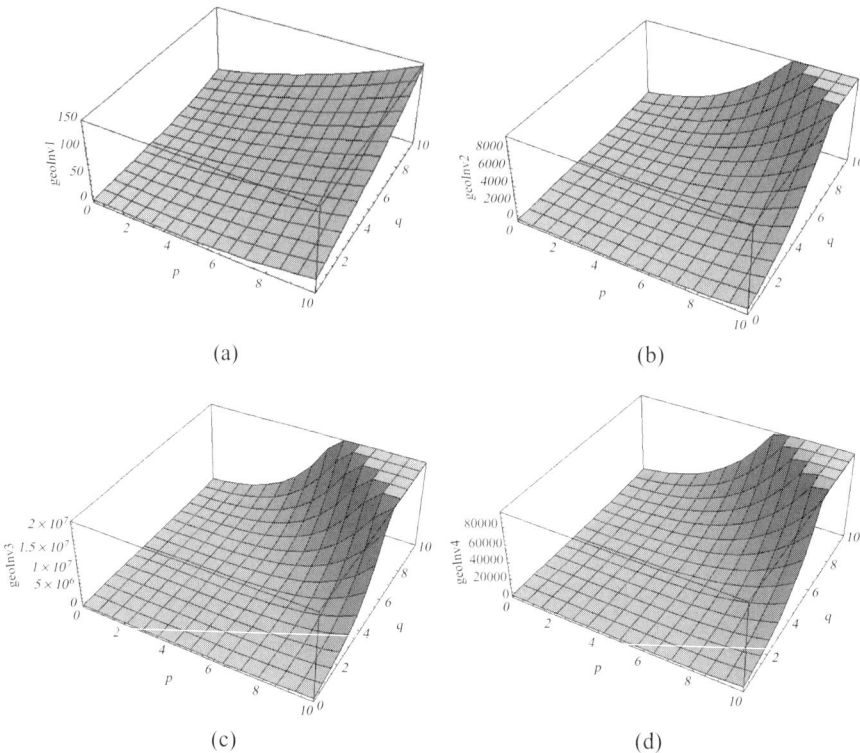

Figure 2.2. Plots of geometrical invariants associated with Eqs. (2.16) to (2.23).

Figure 2.2(a) to (g) shows the 3-D plots of the functions indicated by Eqs. (2.16) to (2.23), respectively. We next change the object's materials via a rotation of its basis vectors p and q by 180°. The resulting material invariant values, tabulated in Table 2.2, illustrate clearly that the material invariants remain unchanged to a high degree of precision. It should be noted that the material invariants, by virtue of being derived from the geometrical invariants, are also unchanged under object's rotation, translation, and scale change in the sensor's field of view.

2.4 Invariant of Ternary Quantics

A set of Ladar targets composed of a total of 35 tactical military targets encompassing tanks, trucks, APCs, self-propelled guns from the United States and other countries were used in the experiment. In this study mostly one resolution representing sensor-to-scene distance of 100 m was considered. Fig. 2.3 shows sample Ladar images of some of these targets. The ground

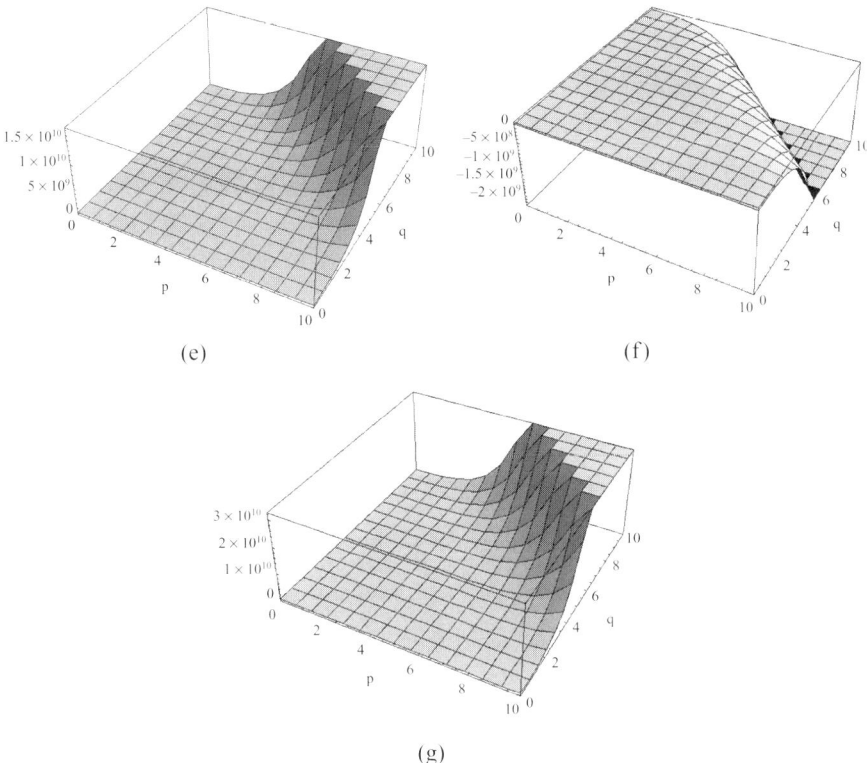

(e)   (f)

(g)

Figure 2.2. (*Continued*)

Table 2.2. Joint geometrical and material invariants.

	Inv1	Inv2	Inv3	Inv4	Inv5	Inv6	Inv7
GeoInv 1	11571.435	20314148	188349.36	27209.686	−1.946E+09	−122551302	89242710
	11571.435	20314148	188349.36	27209.686	−1.946E+09	−122551302	89242710
GeoInv 2	33861.86	88715529	250759.07	2678.8649	−68932078	−25108715	8707525.9
	33861.86	88715529	250759.07	2678.8649	−68932078	−25108715	8707525.9
GeoInv 3	11341.667	19448072	184326.73	27161.357	−1.921E+09	−119747842	54648854
	11341.667	19448072	184326.73	27161.357	−1.921E+09	−119747842	54648854
GeoInv 4	17021.829	17355576	615048.93	172176.8	−5.59E+10	−715782571	3.311E+09
	17021.829	17355576	615048.93	172176.8	−5.59E+10	−715782571	3.311E+09
GeoInv 5	16648.762	17988388	580883.58	156267.49	−4.699E+10	−661440009	2.845E+09
	16648.762	17988388	580883.58	156267.49	−4.699E+10	−661440009	2.845E+09
GeoInv 6	19571.193	16468142	783278.67	233664.72	−9.99E+10	−947133124	−2.735E+09
	19571.193	16468142	783278.67	233664.72	−9.99E+10	−947133124	−2.735E+09
GeoInv 7	28642.364	16119054	824001.26	247466.68	−1.115E+11	−989943688	5.617E+09
	28642.364	16119054	824001.26	247466.68	−1.115E+11	−989943688	5.617E+09

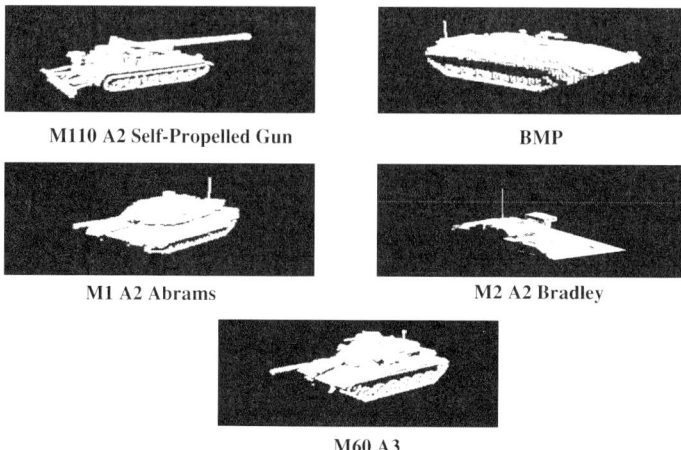

Figure 2.3. Sample of targets used in the experiments.

plane template for each of the image scenes is a 20 m × 20 m flat surface placed on the XY plane. The maximum height was 10 m. To explore the effects of noise, a Gaussian probability density function of zero mean with varying variances was added to the coordinates of each point in the point clouds. A standard deviation value of 1 corresponds to a distance of 1 m.

Figure 2.4 shows how the targets are distributed in terms of their absolute invariants. The ranges of variations for different invariants seem to be different from each others.

To test the effects of coordinate transformations on the 3D invariants, for five typical targets, M60, T72, M1A1 Abrams, BMP1, and M2A2 Bradley, the Ladar cloud points were rotated around z-axis, from $0°$ to $360°$, by

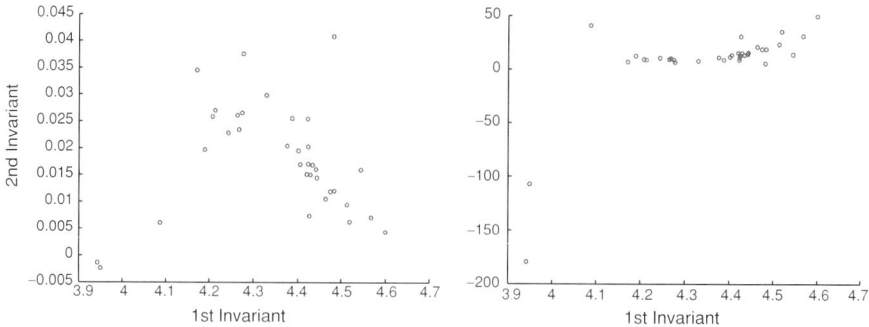

Figure 2.4. Distribution of the second and first and the third and first absolute invariants.

increments of $0.1°$ and at each state the set of 3-D invariants were computed. The results are shown in Fig. 2.5. The right column plots show the results for the first 3-D invariants. As can be seen for all the five targets the first invariant remains very much unchanged. The second column in Fig. 2.5, showing the plots for the second 3-D invariants, indicates that there some small disturbances for all of the targets. The third 3-D invariants, shown in the third column of the Fig. 2.5, display variations that even though small seem to be larger than those associated with the second invariants. We attribute these disturbances to the quantization errors that are produced in the data due to the rotation of the coordinate system. The variations in second and third 3-D invariants each show unique patterns that require further investigation.

By representing each target in terms of their first set of absolute invariant representation and using Frobenius distance as a similarity measure, we computed the probabilities of correct detection and false alarms for each of the 35 targets at various noise variances. Fig. 2.6 shows the receiver operating characteristic curves (ROC) for all 35 targets superimposed on top of each other, at various noise variances. For noise variances below or equal to 0.09, good results are obtained. However, as noise variance is increased to 0.25 and 1 (equivalent to 1 m error), for a good number of targets, the classification performances degrade. Similarly, results can be obtained for the second set of absolute invariants.

2.5 Complexity Analysis

To answer the issues of computational complexities and storage requirements needed in general 2-D model/template techniques, an analysis of the complexities of the algorithms are made.

2.5.1 Assumptions

1. The template size is m by n pixels.
2. Number of material types is M. The scene is composed of a mixture of two distinct materials types m_1 and m_2. The range of m_1 being M_1 and that of m_2 is M_2 is. Then M is equal to $M_1 M_2$.
3. The number of geometrical templates (not including their surface materials) per target needed for classification is equal to N.
4. The number of target classes is equal to T.
5. The number of templates per target class is equal to NM.
6. The number of templates for classification is equal to NMT.
7. Only one perspective view for each target is considered. This can be easily generalized to the case of P perspective views and the number of operations for the both cases scaled accordingly.

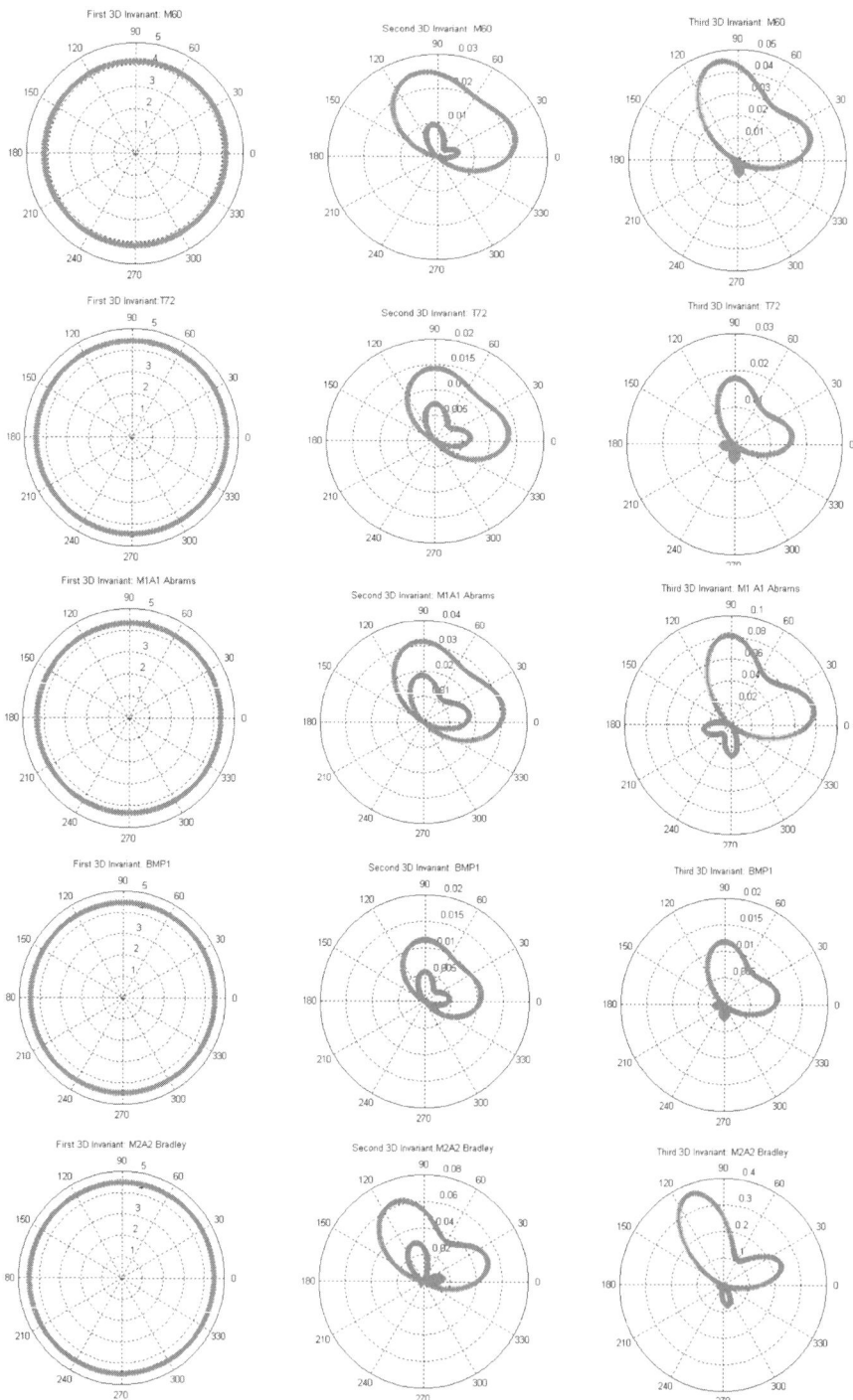

Figure 2.5. Plot of invariant values for five different targets as the targets rotate around z-axis from $0°$ to $360°$ (see color insert).

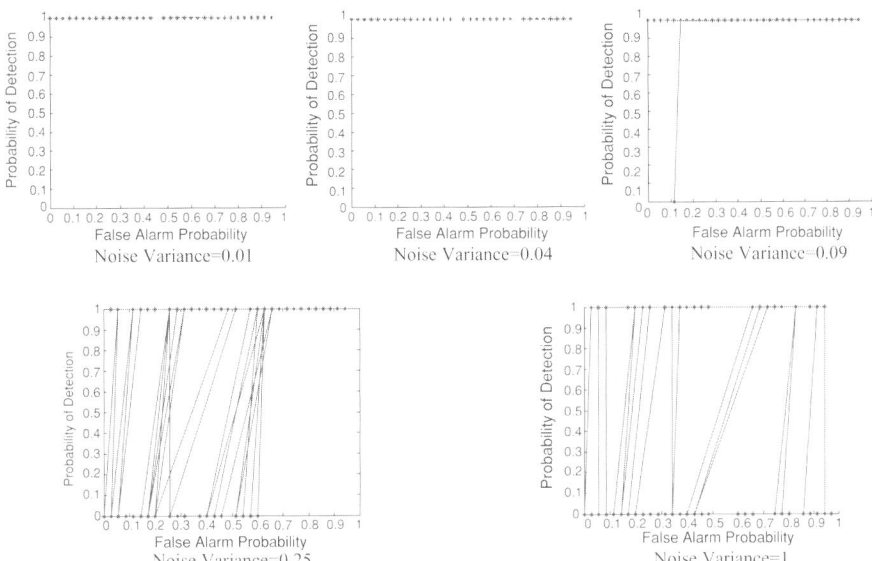

Figure 2.6. ROC curves for different noise variances for the first set of invariants.

2.5.2 Noninvariant Approach

For a noninvariant target classification, we will consider a typical template matching or correlation-based technique. In this case, one can show by straightforward calculation that the following holds:

$$\text{Number of operations for noninvariant (NIOP) technique}$$
$$= (2m^2n^2 + nm)NTM^{nm}. \tag{2.23}$$

The operations in the above relationship are defined as add, subtract, multiply, divide, read, and write.

2.5.3 Invariant Approach

For the case of the previously presented algorithm, one can similarly show that the following relation holds

$$\text{Number of operations for the invariant technique (IOP)}$$
$$= (790, 272)nmMT. \tag{2.24}$$

The relationships (2.24) and (2.25) show the definite advantages of the presented invariant technique over a typical noninvariant template matching technique. This observation is also demonstrated visually by means of Figs. 2.7 and 2.8.

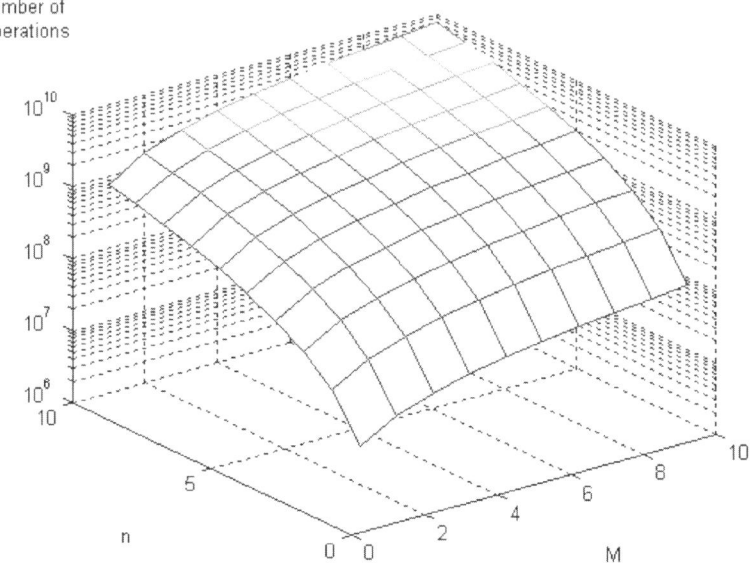

Figure 2.7. The number of operations (on a log scale) as a function of the template size and material classes for the presented approach.

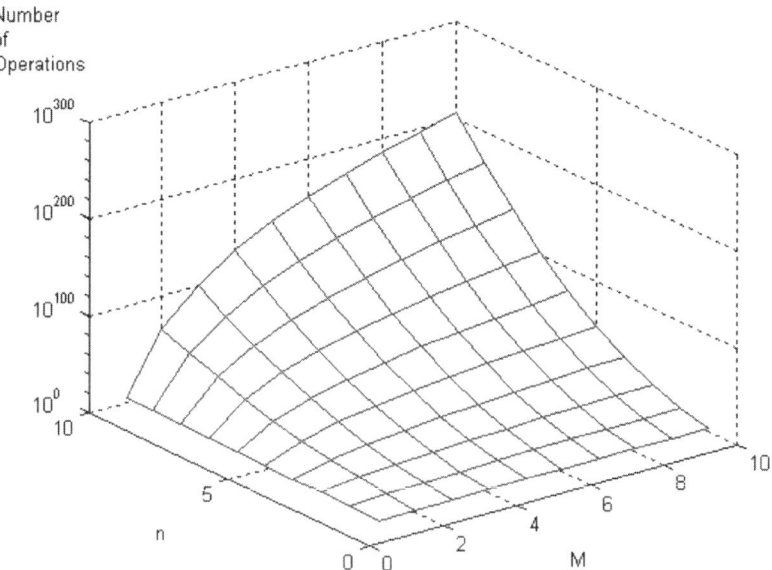

Figure 2.8. The number of operations (on a log scale) as a function of the template size and material classes for a typical noninvariant approach.

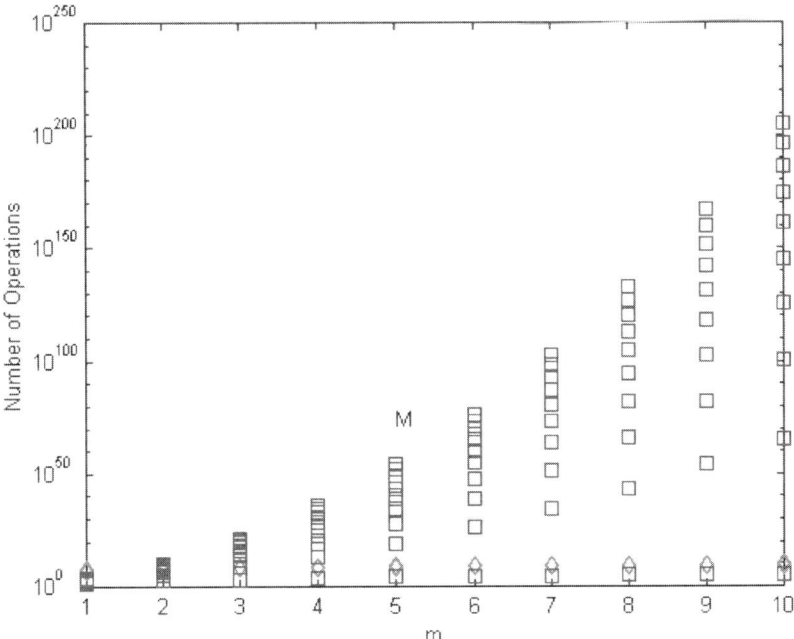

Figure 2.9. The 2-D plots of the number of operations (on a log scale) as functions of the template size and material classes for the presented approach and a typical noninvariant approach (see color insert).

Figures 2.7 and 2.8 show the 3-D plots of the number of operations (in logarithmic scale) as functions of the template size and material classes for the presented invariant and a typical noninvariant approaches. In these figures, the templates are assumed to be squares of size, $m = n$. The number of target classes, T, is assumed to be 10.

Figure 2.9 shows the 2-D plots of the number of operations (in logarithmic scale) as functions of the template size (horizontal axes) and material classes, for the presented approach, shown in more solid straight lines with diamond markers, and a typical noninvariant approach, displayed in a more spread curved lines, with square markers. These figure indicate that for the number of material classes, $M > 2$, and also for template size $m = n > 3$, the presented approach has clear advantage over a typical noninvariant technique. This advantage becomes more prominent as the number of material types and the template size increases.

2.6 Summary

In this chapter we explored the theory of invariant algebra and its applications for automatic classification of objects through their multidimensional signatures. The applications of binary quantics include derivation *2-D invariant shape attributes* and joint *geometrical and material invariant features*. The use of invariants of ternary quantics led to the derivation of a set of 3-D invariant shape descriptors that are then used for classification of Ladar signatures of a large class of land vehicles.

References

[1] G. B. Gurevich. *Fundamentals of the Theory of Algebraic Invariants*. Noordhoff Ltd, Groningen, The Netherlands, 1964.

[2] F. A. Sadjadi and E. Hall. Three-dimensional moment invariants. *IEEE Trans. Pattern Anal. Mach. Intell.*, 2:127–136, 1980.

[3] A. Papoulis. *Probability, Random Variables, and Stochastic Processes*. McGraw-Hill, New York, 1965.

[4] G. Salmon. *Lessons Introductory to the Modern Higher Algebra*. Dodges, Foster and Co., Dublin, Ireland, 1876.

[5] E. B. Elliott. *An Introduction to the Algebra of Quantics*. Oxford University Press, Oxford, 1913.

[6] F. A. Sadjadi and E. L. Hall. Numerical computations of moment invariants for scene analysis. In *Proceedings of the IEEE Conference on Pattern Recognition and Image Processing*, 1978.

[7] F. A. Sadjadi. Invariant algebra and the fusion of multi-spectral information. *Infor. Fusion J.*, 3(1):39–50, 2002.

3

Automatic Recognition of Underground Targets Using Time–Frequency Analysis and Optimization Techniques

Guillermo C. Gaunaurd[1] and Hans C. Strifors[2]

[1]Army Research Laboratory, Adelphi, MD 20783-1197
[2]Swedish Defense Research Agency (FOI), SE-58111 Linköping, Sweden

3.1 Introduction

Landmines have been laid in conflicts around the world, and they cause enormous humanitarian problems in more than 60 countries, killing, mutilating, or maiming the innocent every day. They do not differentiate between elderly, men, women, children, or animals, and they are triggered off by the victims themselves. Detection and clearance of landmines, however, have turned out to be an immensely challenging problem. A traditional means for detecting mines is the metal detector. The detector head is slowly moved over the suspicious terrain, and it gives out audible alerts when metal in the ground disturbs the magnetic field the detector generates. In general, the number of false alarms caused by various (man-made) metal objects in the ground is a great deal larger than the number of mines. This makes mine clearance a tiring task that demands the highest level of concentration. Moreover, a hazard involved is that the metal detector could be too insensitive to mines of low-metal content.

Ground-penetrating radar (GPR) can be used to detect mines of low-metal content or of no metal at all, and thus can provide a feasible technology for landmine detection. In particular, ultra-wideband (UWB) GPR systems have been of interest to scientists for many years because they are capable of extracting information from returned echoes that is useful for target recognition purposes [1–6]. Of the currently used GPR systems one is down-looking with the antenna elements (transmitting and receiving) placed near the surface of the ground. In general, adequate performance of the radar system requires that the coupling of the antenna elements and the background clutter be reduced. This can be accomplished by subtracting from the measurements separately recorded signals with the antennas positioned above an area of the ground with no target present, also called ground bounce removal. Since the ground bounce removal cannot entirely compensate for various clutter contributions, detection and classification algorithms must be sufficiently robust.

It was demonstrated that the burial depth of a target could be challenging enough if target signatures in the frequency domain were attempted for classification purposes [4]. The reason is due to multiple scatter between a target and the ground surface. The frequency spectrum of returned waveforms was affected by the burial depth to the extent that prominent peaks occurring at small depths split into two or more pointed peaks with increasing depth. Using, instead, a pseudo-Wigner distribution (PWD) [7] with a suitably selected smooth time-window increased target depth resulted in a later occurrence in time of the time–frequency signatures.

There are many a time–frequency distribution (TFD) to choose between, not all of them equivalent. Six candidate TFDs were surveyed with the objective to investigate their relative merits by Strifors et al. [6]. Of these six TFDs, the PWD demonstrated a distinct ability to extract the target's time–frequency signature also for targets at the largest burial depth. Recently, the PWD and the Baraniuk–Jones optimal kernel distribution (BJOKD) [8] were compared by Gaunaurd et al. [9], and it was found that these two TFDs generate quite similar features when extracted from returned radar echoes. The faster running computational results it produces is an obvious advantage of the PWD.

UWB radar emits an extremely short pulse, impulse, or a frequency modulated signal (e.g., sweep or stepped-frequency). The frequency content of the emitted signals is designed to match the size and kind of typical targets and environments. We studied the relative merits for an impulse radar and a stepped-frequency continuous-wave (SFCW) radar operating in the same frequency range [10]. This radar system transmits a sequential series of individual frequencies. The digital processing of the received signals gives in-phase (I) sampled signals and quadrature-phase (Q) sampled signals. The I and Q signals contain the amplitude and phase information about the received waveform that are necessary for synthesizing a pulse response, using the inverse Fourier transform. It followed from the analysis that synthesized waveforms generated in the SFCW radar can be compared to the UWB waveforms transmitted by impulse radar. However, in this chapter the calculations and predictions shown in the figures for target identification purposes are based on the SFCW type of radar.

The time–frequency signature of radar illuminated targets buried underground is affected by their burial depth and the composition and moisture content of the soil. This is particularly true for targets of low-metal content. In a series of papers [11–13] we have developed a methodology for transforming to standard burial conditions the time–frequency signatures generated by returned waveforms from targets buried at unknown depth in soil of unknown moisture content. However, it is assumed that the electromagnetic properties of the soil as a function of moisture content are known. The methodology was based on an approximate technique for transforming time–frequency signatures when a given target is buried in a soil of specified electromagnetic properties at different depths and moisture contents, which was called the target translation method (TTM). The electromagnetic backscattering from the targets was analyzed in the frequency domain using a method of moments (MoM) algorithm, and suitable waveforms in the time domain were synthesized. The single signature template needed for each target was computed using a PWD of the returned echo when the target was buried at standard conditions, i.e., a selected depth in the soil of a selected moisture content. Then, the PWD of any echo returned from a target was computed and the difference between this PWD and the template PWD computed using a suitable distance function was evaluated. Applying the TTM with the depth and soil moisture content taken to be unknown parameters, the distance function, used as the objective function or fitness function, was minimized using the differential evolution method (DEM) [14,15]. The template that gave the smallest value of the minimized objective function for a given target was taken to represent the correct classification, and the corresponding values of the depth and moisture parameters were taken to be estimates of the actual target's burial conditions. To make the computations more efficient, only two or three iterations in the DEM need to be used. A fast direct search method is then used for zooming in on the smallest value of the objective function [16].

This classification method has now been implemented for use with an SFCW GPR system. We demonstrate the ability to classify mines buried underground from measured GPR signals. The targets were buried in a test field at the Swedish Explosive Ordnance Disposal and Demining Center (SWEDEC) at Eksjö, Sweden.

3.2 Theoretical Models

3.2.1 Spline Models of Soil Properties

The electromagnetic properties of the soil in which targets are buried are generally only vaguely known. At best, the complex-valued electromagnetic (dielectric) properties of samples of the soil with measured moisture content have been experimentally determined as a function of frequency. Thus, the soil properties are known for discrete values of moisture content and frequency. The optimization method (the DEM) and the method for approximating target translation (the TTM) used in the classification procedure require both continuous functions to work. This can most conveniently be accomplished using spline functions [17].

B-Splines are the basic elements by which polynomial splines are constructed. Univariate B-spline functions can be generalized to bivariate functions, using the tensor product of two B-splines. Specifically, the spline model for a particular function value $f(u,\nu)$ is given by

$$f(u,\nu) = \sum_{i=0}^{m} \sum_{j=0}^{n} a_{ij} N_{ik}(u) N_{jl}(\nu), \qquad (3.1)$$

where a_{ij} denotes the coefficients that gives a good approximation to the function f.

The nondecreasing sequences of scalars $S = \{s_i\}, i = 0, \ldots, m+k$, and $T = \{t_j\}, j = 0, \ldots, n+l$, are called knots. The B-spline basis functions, $N_{ik}(u)$, are smooth piecewise polynomials with joints at the knots of the order k (or degree $k-1$) with $\sum_{i=0}^{m} N_{ik} = 1$ with the dependence on S implied. In this application cubic splines (i.e., order 4) are used.

A simple closed-form representation of a cubic B-spline can be written as

$$\beta^3(x) = \begin{cases} \dfrac{2}{3} - |x|^2 + \dfrac{|x|^3}{2}, & 0 \le |x| < 1 \\[2mm] \dfrac{(2-|x|)^3}{6}, & 1 \le |x| < 2 \\[2mm] 0, & 2 \le |x|. \end{cases} \qquad (3.2)$$

Original data are the given points g_{ij} with pairs of assigned parameter values (u_i, ν_j) for $i = 1, \ldots, N_u, j = 1, \ldots, N_\nu$. The coefficients are here determined by minimizing the Euclidean distance

$$Q_{\text{dist}} = \sum_{i=1}^{N_u} \sum_{j=1}^{N_\nu} \|g_{ij} - f(u_i, \nu_j)\|^2 \qquad (3.3)$$

in the mean-square sense. The above bivariate splines method is applied with the function f denoting the dielectric permittivity of the soil ε and the target's burial depth and soil moisture content (d, m) substituted for the variables (u, ν).

3.2.2 The Target Translation Method in Practice

In the TTM it is tacitly assumed that the returned echo from an underground target depends on the depth and soil moisture content only through a simple translation of the waveform [12]. In addition, the surface reflection at the air–soil interface is taken into account as the reflection at the soil-target assumed to be a flat surface whatever the shape of the target. To account for the latter kind of reflection is essential when the target is dielectric with permittivity close to that of the soil. The scattering interaction with the target of the interrogating waveform transmitted by the antenna is assumed to be otherwise independent on the depth and moisture parameters. In the frequency and time domains the general trend of the target signatures were shown to be preserved at target translation. Best agreement between translated target response and actually computed response was found in the time–frequency domain.

The (normalized) radar cross section (RCS) in terms of form-function can be written in the alternative forms

$$\frac{\sigma}{\pi a^2} = \left| f(m,d,\omega) e^{-2\alpha d} \right|^2 \equiv \left| \tilde{f}(m,d,\omega) \right|^2, \tag{3.4}$$

where ω is the angular frequency. The form-function $f(m,d,\omega)$ is assumed to completely incorporate the scattering interaction of the waveform and target, and $\tilde{f}(m,d,\omega)$ is the actually computed form-function at the depth d and moisture content m.

The back-scattered E-field, disregarding the air–soil reflections, in the time domain can then be written in the alternative forms

$$E_{\mathrm{sc}} = E_0 \frac{a}{2r} \mathbf{e}_x \frac{1}{2\pi} \int_{-\infty}^{+\infty} G(\omega) f(m,d,\omega) e^{i2\omega d/c - 2\alpha d} e^{-i(\omega t - kr)} \, d\omega$$

$$\equiv E_0 \frac{a}{2r} \mathbf{e}_x \frac{1}{2\pi} \int_{-\infty}^{+\infty} G(\omega) \tilde{f}(m,d,\omega) e^{-i(\omega t - kr)} \, d\omega, \tag{3.5}$$

where $G(\omega)$ is the Fourier transform of the target-illuminating pulse $g(t)$ as defined immediately below the upper surface of the soil. Moreover, real-valued quantities c and α are the phase velocity and attenuation in the soil medium and they can be written as

$$c = c_0 \left(\frac{|\mu_r \varepsilon_r| + \mathrm{Re}\,(\mu_r \varepsilon_r)}{2} \right)^{-1/2},$$
$$\alpha = \frac{\omega}{c_0} \left(\frac{|\mu_r \varepsilon_r| - \mathrm{Re}\,(\mu_r \varepsilon_r)}{2} \right)^{1/2}, \tag{3.6}$$

where $c_0 = 1/\sqrt{\mu_0 \varepsilon_0}$ is the wave speed in free space. These convenient relations can be obtained by introducing definitions $\mu = \mu_r \mu_0 \equiv (\mu_r' + i\mu_r'')\mu_0$ and $\varepsilon = \varepsilon_r \varepsilon_0 \equiv (\varepsilon_r' + i\varepsilon_r'')\varepsilon_r$, where a prime or double prime indicates the real or imaginary part of a complex number, into the combined relations:

$$k = k' + ik'',$$
$$k^2 = \omega^2 \mu \varepsilon. \tag{3.7}$$

Identifying the real and imaginary parts of the resulting expression then gives

$$k' = \omega \left(\frac{|\mu \varepsilon| + \mathrm{Re}(\mu \varepsilon)}{2} \right)^{1/2},$$
$$k'' = \omega \left(\frac{|\mu \varepsilon| - \mathrm{Re}(\mu \varepsilon)}{2} \right)^{1/2}. \tag{3.8}$$

Finally, on defining $k \equiv \omega/c + i\alpha$, Eqs. (3.6) follow.

The quantity $\mu_r \varepsilon_r$ is a complex-valued function of the angular frequency ω that is determined by the properties of the soil constituents and moisture content. In general, the soil medium is both dispersive (i.e., c is a function of ω) and dissipative or lossy (i.e., $\alpha > 0$).

Returned waveforms from illuminated targets are analyzed in the time–frequency domain using a PWD. Introducing a smooth time window, $w_f(\tau)$, the (real-valued) PWD assumes the form

$$\mathrm{PWD}_f(\omega, t) = 2 \int_{-\infty}^{+\infty} f(t + \tau) f^*(t - \tau) w_f(\tau) w_f^*(-\tau) \mathrm{e}^{-i2\omega\tau} \, \mathrm{d}\tau, \qquad (3.9)$$

which allows us to control the amount of cross-terms interference. Here, the window function is a Gaussian of the form: $w_f(t) = \exp(-\gamma t^2)$, where the parameter γ controls the width of the time window.

As a measure of the distance between the PWD signatures of two different echoes $\mathrm{PWD}_{f_1}(\omega, t)$ and $\mathrm{PWD}_{f_2}(\omega, t)$, we use the Euclidean distance defined by

$$d_{L^2}(\mathrm{PWD}_{f_1}, \mathrm{PWD}_{f_2}) = \left[\int \int (\mathrm{PWD}_{f_1}(\omega, t) - \mathrm{PWD}_{f_2}(\omega, t))^2 \mathrm{d}\omega \, \mathrm{d}t \right]^{1/2}, \qquad (3.10)$$

and for the norm of a PWD signature the L^2-norm

$$\|\mathrm{PWD}_f\|_2 = \left[\int\!\!\int (PWD_f(\omega, t))^2 \, \mathrm{d}\omega \, \mathrm{d}t \right]^{1/2}. \qquad (3.11)$$

The template of each target is computed when the target is buried at standard conditions (depth d_s and soil moisture m_s), which generates the PWD signature $\mathrm{PWD}_f^T(\omega, t; d_s, m_s)$, where the superscript T designates a template. Assuming the depth d and soil moisture content m to be unknown parameters, the PWD of any echo returned from a target can be translated from the conditions (d, m) to the standard conditions (d_s, m_s). Then, the signature difference expressed by

$$D_{\mathrm{sign}} = d_{L^2}(\mathrm{PWD}_f^T(\omega, t; d_s, m_s), \mathrm{PWD}_f(\omega, t; d, m)) \big/ \|\mathrm{PWD}_f^T\|_2 \qquad (3.12)$$

is a measure of the fitness of the translated target signature.

3.2.3 The Differential Evolution Method

The recently proposed DEM is a population-based evolutionary algorithm based on the use of a special recombination operator, which, to create one child, performs a linear combination of three different individuals and one parent, which is also to be replaced. The best of parent and child remains in the next population. The main differential algorithm [15] can be stated schematically as follows.

Given an objective function $f(\mathbf{x})$, defined on a space D of real-valued object variables $x_j, j = 1, 2, \ldots, D$, for which value $\mathbf{x}^*$ of $\mathbf{x}$ does $f(\mathbf{x})$ reach a minimum value? The DEM generates a randomly distributed initial population $P_0 = \{\mathbf{x}_{1,0}, \mathbf{x}_{2,0}, \ldots, \mathbf{x}_{i,0}, \ldots, \mathbf{x}_{NP,0}\}$, where $NP \geq 4$ is the number of members in the population. Evaluate $f(\mathbf{x}_{i,0}), i = 1, \ldots, NP$. With $G < G_{\max}$ denoting a generation, $CR \in [0, 1]$ and $F \in [0, 1+]$ being user-defined parameters, the algorithm can be written symbolically as

> For $G = 1$ to $G_{\max}$ Do
>
> For $i = 1$ to NP Do
>
> Select randomly : $r_1 \neq r_2 \neq r_3 \neq i \in \{1, 2, \ldots, NP\}$
>
> Select randomly once each i: $j_{rand} \in \{1, 2, \ldots, D\}$
>
> For $j = 1$ to D Do
>
> If $(rand_j[0, 1) < CR$ or $j = j_{rand})$:
>
> $\quad$ Then $u_{j,i,G+1} = x_{j,r_3,G} + F(x_{j,r_1,G} - x_{j,r_2,G})$ $\qquad$ (3.13)
>
> Else $u_{j,i,G+1} = x_{j,i,G}$
>
> End If
>
> End For
>
> If $f(\mathbf{u}_{i,G+1}) \leq f(\mathbf{x}_{i,G})$ Then $\mathbf{x}_{i,G+1} = \mathbf{u}_{i,G+1}$
>
> Else $\mathbf{x}_{i,G+1} = \mathbf{x}_{i,G}$
>
> End If
>
> End For
>
> $G = G + 1$
>
> End For

Here, $rand[0,1)$ is a function that returns a real number between 0 and 1. When feasible, the algorithm is made to stop when a specified stopping condition is met, e.g., $f_{\max} - f_{\min} \leq \varepsilon$, where ε is a small number, and $f_{\max}$ and $f_{\min}$ are, respectively, the current maximum and the minimum function value to reach.

The DEM is then used with the signature difference D_{sign} according to Eq. (3.12) as the objective function to be minimized by a suitable combination of the parameters (d, m), which are thought of as a member vector of a population. The number of pairs of (d, m) in the population is chosen to be $NP = 20$. To make the computations more efficient, the maximum number of generations is set to $G_{\max} = 3$. This number is far from sufficient for convergence of the DEM but, in general, good enough for reaching a proper classification of the target and providing start values for the fast Nelder–Mead simplex (direct search) method [16]. The resulting parameter values are those that most closely model the observed data to the specified target template.

Furthermore, the template that corresponds to the smallest value of the minimized objective function D_{sign} is a measure of the actual classification.

3.3 Classification of Test Field Targets

3.3.1 Radar Measurements

The SFCW radar transmits 55 different frequencies in the band 300 to 3000 MHz (in steps of 50 MHz). Having a duration of about 100 μs, each transmitted waveform has an extremely narrow band. The I and Q signals of 14 bits resolution give information of both the amplitude and the phase of the signal returned from a target. As a result, a complex-valued line spectrum of the target response is obtained. The present SFCW radar uses frequency sequences containing $N = 256$ samples. The recorded spectrum sequence is located from position $n = 6$ to $= 60$ (position $n = 0$, corresponding to the frequency $f = 0$). Folding the conjugate sequence ($n = 0$ to 127) about $n = N/2$ locates the mirrored conjugate sequence in positions $n = 196$ to 250. The remaining positions of the frequency sequence are set equal to zero. This conjugate symmetric frequency series constitutes an analytic signal, which is then used to synthesize the (real-valued) target response in the time-domain, using the inverse discrete-time Fourier transform [18]. The Nyquist frequency is 6.4 GHz and the Nyquist sampling rate 12.8 GS/s. The duration of the synthesized time-series is 20 ns, which has a length of about 6 m in air. Half of this length is the unambiguous range in air (i.e., 3 m). The possible clutter contributed from targets outside the unambiguous range by aliasing does not interfere noticeably with the correctly positioned signal components.

The radar measurements at the SWEDEC test Field at Eksjö, Sweden were performed during a few days in October 2003 by using a man held SFCW GPR system developed at FOI. This experimental GPR system HUMUS (HUmanitarian MUltiSensor) has an antenna unit with two parallel elements, transmitting and receiving, and is prepared to include also a metal detector. Fig. 3.1 shows this experimental GPR system HUMUS. The operator lets the antenna in front of him sweep across the width of the lane to be searched, approximately 1 m. At each scan, 75 waveforms are recorded. Directly upon each scan various pictures such as a depth view of the waveforms, detection alerts, and possible mine classification results are presented in easily comprehensible form to the operator on a computer screen. Classification using the present DEM-based algorithm with depth and moisture estimation was implemented later, and the capability of this method is demonstrated here using the previously recorded waveforms.

The theoretical studies referred to in Section 3.1 used analytical approximations of the complex-valued dielectric permittivity previously measured from soil samples of known moisture content taken from the Yuma Proving

Figure 3.1. The HUMUS GPR system and operator. The computer screen displays depth view of the waveforms, detection alerts, and possible mine classification results.

ground, AZ. A similar approach was used to model the properties of moist sand. Only preliminary results are available, and they are not derived for the present kind of soil. Instead, this preliminary algorithm is used for the soil of both the earth-layer and the road. The actual values of the moisture content in the earth-layer and road are not known, but a moisture content of 5% is assumed when generating the templates. The targets used to generate templates are all buried at a depth of 2 cm.

The mine targets used in these experiments were three types of dummy antitank mines: TMM1 (metallic), TMA1, and TMA5 (both of low-metal content) (see Figs. 3.2 to 3.4, right image). They were all buried some time in advance in a portion of an earth-layer (about 40 m^2) and a portion of a dirt road covered with gravel (about 120 m^2). The positions of the buried mines were unknown to us until after the test, but a small portion of both the earth-layer and the road had known targets at known depths that were used to record backscattered signals used to generate PWD templates of the targets. The targets used to generate the templates were all buried at the depth of 2 cm. The moisture content was assumed to be 5%, but the actual value was not known. Moreover, the value of the dielectric permittivity of the low-metallic mines was assumed to be 4 in both cases.

Besides the dummy mines the targets also included a few Coke cans. Various fragments such as pebbles, roots, and twigs were part of the soil background. The modulus of the normalized PWD templates of the three mine targets used in classification are displayed in the surface plots to the left in

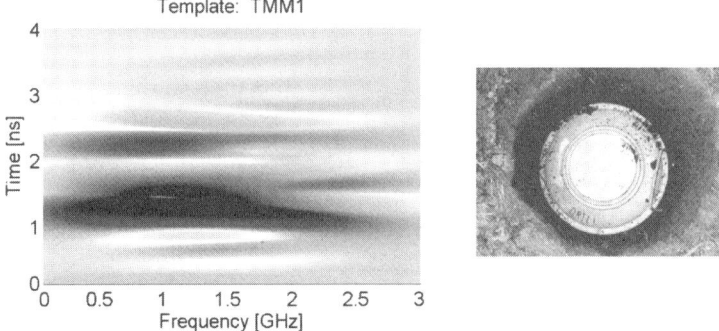

Figure 3.2. PWD template of the dummy antitank mine TMM1 (metal) (left) and picture of a TMM1 mine (right).

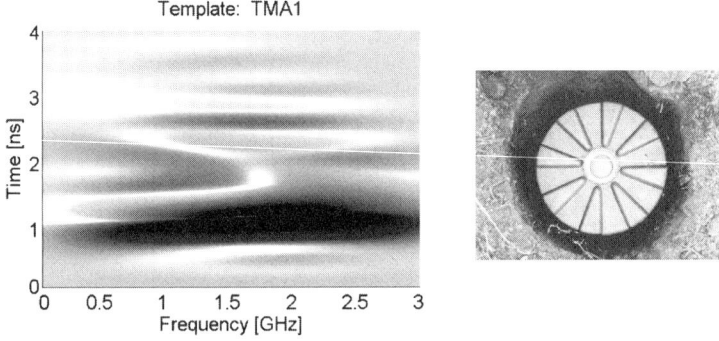

Figure 3.3. PWD template of the dummy antitank mine TMA1 (low-metal) (left) and picture of a TMA1 mine (right).

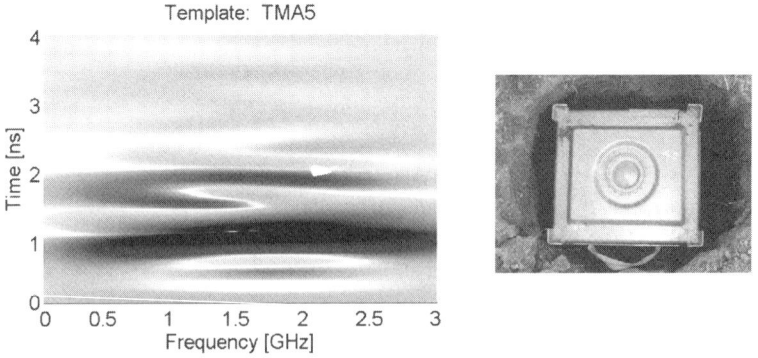

Figure 3.4. PWD template of the dummy antitank mine TMA5 (low-metal) (left) and picture of a TMA5 mine (right).

Figs. 3.2 to 3.4. These surface plots use progressively darker shades of gray (pseudocolors), and the value of the PWD is further emphasized by three-dimensional effects of light and shadow. Each PWD is normalized by setting its maximum value to unity, which corresponds to black in the images.

3.3.2 Classification of Underground Targets

As examples, three scans, each comprising 75 recorded and synthesized wave-forms, are used here for target classification. The first scan is taken close to a TMM1 metal mine buried in the earth-layer at a depth of about 20 cm. The other two scans are taken close to a TMA1 and a TMA5 low-metal mines buried below the road surface at unknown depths. The classification results along a 35-cm segment of the scan above the TMM1 mine are displayed in Fig. 3.5, left plot. The distance function for the mine templates is displayed using a heavy solid line for the TMM1 and a heavy dashed or a dash-dotted line for the TMA1 or TMA5. The distance function for the Coke can template or background template is displayed by a light solid or a dashed line. It can be seen in the left plot that the TMM1 target is correctly classified (smallest values of the distance function) with the Coke can (a metal object) as the classifier's second choice. The low-metal targets and background give even larger values of the distance function at this segment of the scan. The right plot displays the estimates of the burial depth and soil moisture content for the classified target, viz., the TMM1. The depth can be seen to be under-estimated as it is shown to be about 13 to –14 cm to be compared with the actual depth of 20 cm (see Fig. 3.5, right plot). The discrepancy is believed to be caused by a too unrefined fit of the dielectric and moisture properties of the present soil. For the portion of the earth-layer where the classified target is located a moisture content of about 2% can be read off from the right plot of Fig. 3.5.

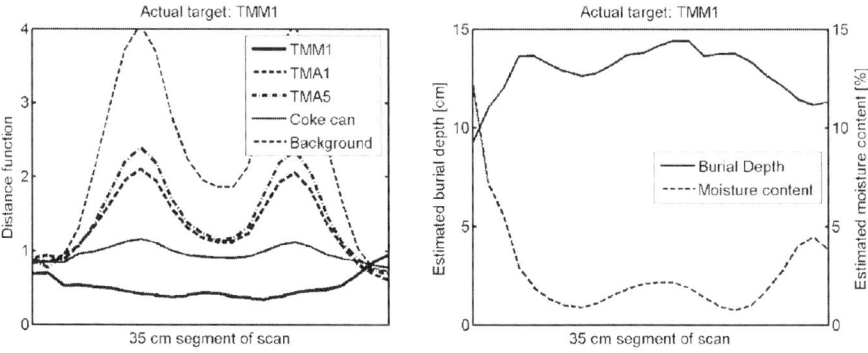

Figure 3.5. Distance function for target templates (left plot) and estimated burial depth and soil moister content for the TMM1 mine (right plot).

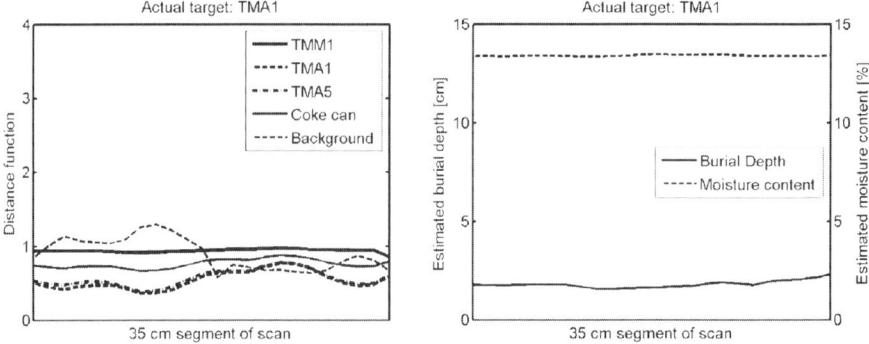

Figure 3.6. Distance function for target templates (left plot) and estimated burial depth and soil moister content for the TMA1 mine (right plot).

The classification results of the low-metal mines shown in Figs. 3.6 and 3.7 are computed from backscattered signals from targets buried beneath the surface of the road. As can be seen in the left plots of both Figs. 3.6 and 3.7, the distance function corresponding to the TMA1 and TMA5 mines are everywhere very close. Indeed, the distance function for any of these two low-metal targets turned out to be everywhere hard to distinguish from each other, particularly so when any of these targets is the actual one. From the right plots of Figs. 3.6 and 3.7, the estimated burial depth of both the TMA1 and the TMA5 is seen to be about 2 cm and the estimated moisture content about 13%. The actual moisture content was not known in any case, but was presumably closer to the assumed value of 5% used when generating the target templates, because the target's contrast was quite low both in the earth-layer and the road.

As a comparison, previously obtained measurements are used for classification of a few targets buried in an indoor sandbox at the Linköping site of FOI.

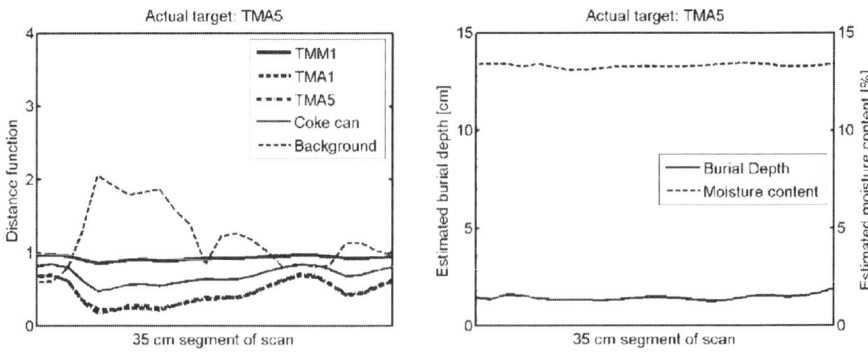

Figure 3.7. Distance function for target templates (left plot) and estimated burial depth and soil moister content for the TMA5 mine (right plot).

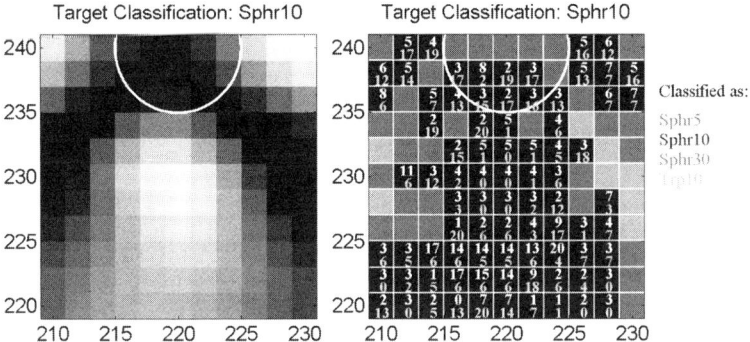

Figure 3.8. Signal strength from a target in sandbox (left) and classification result (right) when an area of the sandbox is scanned by the HUMUS GPR system. Top circle outlines the extent of the target. Top figures in black squares (correct classification) denote estimated depth (cm) and bottom figures estimated moisture content (%).

The targets here are three metal targets, spheres of diameter 5 cm, 10 cm, and 30 cm buried at a depth of 5 cm, and a dielectric target, a dummy antipersonnel mine buried at the depth of 2 cm. The antenna is in this case suspended on a computer-controlled positioning system that automatically lets the antenna traverse the surface of the sand at a selected height above the ground (here 5 cm) and perform the preprogrammed radar measurements, illuminating the target from above. Fig. 3.8 shows the classification result for one target, viz., the 10-cm metal sphere. The figures to the left and at the bottom of the two plots are coordinates in cm of the positions at the surface of the sand. A measurement is made at the center of each square of 2 cm width. Fig. 3.8, left plot, displays the signal strength of the echo returned to the radar. The right plot of Fig. 3.8 displays the classification results. The location of the target is outlined by a circle in both plots. The signal strength is displayed using shades of gray from black (very low) to white (very high) signal strength. The low target strength that occurs along two diagonals starting at the target corresponds to antenna positions where at least one of two crossing antenna elements reaches above the target.

In the right plot of Fig. 3.8, the black squares denote correct classification, dark gray squares denote any other metal target, and light gray squares denote the dielectric target. Each square that denotes correct classification contains two figures above each other. The top (bottom) figure denotes the estimated depth (moisture content). As can be seen from the two plots of Fig. 3.8, the classification of the metal sphere of diameter 10 cm together with the estimated burial depth (cm) and soil moisture content (%) are most accurate where the target strength is high, i.e., away from the two diagonals and sufficiently close to the target.

Already King and Harrington [1] pointed out that for pulses to penetrate into the earth they have to be of very low frequency. On the other hand, higher frequency content in the pulses is required to extract information from the target for accurate classification. In conclusion, it seems that the choice of frequency band used with this radar system is well suited for both soil penetration and target classification. There seems to be little point in using frequencies above a few gigahertzs for large underground targets such as antitank mines. Higher frequencies can provide more detailed information, particularly about smaller targets such as antipersonnel mines in dry soil. On the other hand, higher frequencies tend to contribute clutter from pebbles, twigs, and roots. In all cases of target classification it is important to acquire only a sufficient amount of information for accurate target classification.

3.4 Summary and Conclusions

We have investigated the possibility of classifying underground targets buried at *unknown depth* in soil of *unknown moisture content* using UWB GPR. We have implemented on a GPR system our previously developed combination of the TTM and the DEM for classifying targets using signatures generated by PWDs. The method was put to test in a field experiment were the targets included dummy antitank mines, metal and low-metal, buried underground in an earth-layer or in a road.

The response of each target at a selected burial depth and selected moisture content were used to generate a PWD signature template for the target. These templates were then used in the classification procedure. For each echo returned from a target the PWD signature was translated in order to best fit the actual template by adjusting the depth and moisture content, taken to be unknown parameters. This optimization was accomplished using a few generations of the DEM followed by a fast zooming in of parameters using a direct search method. The smallest value of the objective function, or fitness function, represented the classification. In each case, the targets were correctly classified with this method. In addition, when the proper template was used, the values of the depth and moisture parameters corresponding to the best signature fit turned out to be good and consistent predictions. Finally, we note that, although very hard to detect, mines with low-metal content buried at about 20 cm depth in a gravel road could be conveniently classified using the present classification method. In fact, a metal detector could prove too insensitive to the low-metal content in deep buried plastic mines. In such a case, this classification method could serve the combined purposes of both detecting and classifying targets.

Acknowledgements

The authors gratefully acknowledge the partial support of their respective institutions. This work was partially supported by the Office of Naval Research.

References

[1] R. W. P. King and C. W. Harrington. The transmission of electromagnetic waves and pulses into the earth. *J. Appl. Phys.*, 39:4444–4452, 1968.

[2] D. L. Moffat and R. J. Puskar. A subsurface electromagnetic pulse radar. *Geophysics*, 41:506–518, 1976.

[3] D. J. Daniels. *Surface-Penetrating Radar*. IEE, London, 1996.

[4] H. C. Strifors, S. Abrahamson, B. Brusmark, and G. C. Gaunaurd. Signature features in time–frequency of simple targets extracted by ground penetrating radar. In F. A. Sadjadi (Ed.) *Automatic Object Recognition V, Proceedings of SPIE*, vol. 2485, pp 175–186, 1995.

[5] H. C. Strifors, B. Brusmark, S. Abrahamson, A. Gustafsson, and G. C. Gaunaurd. Analysis in the joint time–frequency domain of signatures extracted from targets underground buried underground. In F. A. Sadjadi (Ed.) *Automatic Object Recognition VI, Proceedings of SPIE*, vol. 2756, pp. 152–163, 1996.

[6] G. C. Gaunaurd and H. C. Strifors. Signal analysis by means of time-frequency (Wigner-type) distributions–Applications to sonar and radar echoes. *Proc. IEEE*, 84:1231–1248, 1996.

[7] L. Cohen. *Time–Frequency Analysis*. Prentice Hall, Englewood Cliffs, NJ, 1995.

[8] R. G. Baraniuk and D. L. Jones. Signal-dependent time–frequency analysis using a radially Gaussian kernel. *Signal Process.*, 32:263–284, 1993.

[9] G. C. Gaunaurd, H. C. Strifors, and A. Sullivan. Comparison of time, frequency and time–frequency analysis of backscattered echoes from radar targets buried in realistic half-spaces. In E. L. Mokole, M. Kragalott, and K. R. Gerlach (Eds.) *Ultra-Wideband, Short-Pulse Electromagnetics.vol. 6*. Kluwer, New York, pp. 481–492, 2003.

[10] H. C. Strifors, A. Friedmann, S. Abrahamson, and G. C. Gaunaurd. Comparison of the relative merits for target recognition by ultra-wideband radar based on emitted impulse or step-frequency wave. In F. A. Sadjadi (Ed.) *Automatic Target Recognition X, Proceedings of SPIE*, vol. 4050, pp. 2–11, 2000.

[11] H. C. Strifors, G. C. Gaunaurd, and A. Sullivan. Influence of soil properties on time–frequency signatures of conducting and dielectric targets buried underground. In F. A. Sadjadi (Ed.) *Automatic Target Recognition XII, Proc. SPIE*, vol. 4726, pp. 15–25, 2002.

[12] H. C. Strifors, G. C. Gaunaurd, and A. Sullivan. Reception and processing of electromagnetic pulses after propagation through dispersive and dissipative media. In F. A. Sadjadi (Ed.) *Automatic Target Recognition XIII, Proceedings of SPIE*, vol. 5094, pp. 208–219, 2003.

[13] H. C. Strifors, G. C. Gaunaurd, and A. Sullivan. Simultaneous classification of underground targets and determination of burial depth and soil moisture content. In F. A. Sadjadi (Ed.) *Automatic Target Recognition XIV, Proceedings of SPIE*, vol. 5426, pp. 256–263, 2004.

[14] R. Storn and K. Price. Differential evolution—a fast and efficient heuristic for global optimization over continous spaces. *J. Global Optim.*, 11:341–359, 1997.

[15] K. V. Price. An introduction to differential evolution. In D. Corne, M. Dorigo, and F. Glover (Eds.) *New Ideas in Optimization*. McGraw-Hill, London, pp. 79–108, 1999.

[16] J. C. Lagarias, J. A. Reeds, M. H. Wright, and P. E. Wright. Convergence properties of the Nelder-Mead simplex method in low dimensions. *SIAM J Optim.*, 9:112–147, 1998.

[17] C. de Boor. *A Practical Guide to Splines*. Springer, New York, 2001.

[18] A. V. Oppenheim and R. W. Schafer. *Digital Signal Processing*. Prentice-Hall, Englewood Cliffs, NJ, 1975.

4

A Weighted Zak Transform, Its Properties, and Applications to Signal Processing

Izidor Gertner[1] and George A. Geri[2]

[1]Computer Science Department
City College of The City University of New York
138th Street & Convent Avenue, New York, NY 10031
[2]Link Simulation & Training, 6030 S. Kent Street, Mesa, AZ 85212

Abstract: We have previously used the weighted Zak transform (WZT) to represent complex, real-world imagery in the space/spatial frequency domain. Also, the WZT has previously been used to construct complete, orthonormal sets of functions on a von Neumann–Gabor lattice. In this chapter, we investigate the properties of the weighting function of the WZT. We then use the WZT to construct an ambiguity function that vanishes at all points on a lattice, except for the origin, thus providing the narrowest possible discretely defined function. We then prove that the ambiguity function falls off exponentially between the lattice points. Finally, we show how to obtain basis waveforms using the WZT, which are suitable for analyzing and synthesizing complex signals and images.

4.1 Introduction

The weighted Zak transform (WZT) has been widely used in constructing complete orthonormal sets of functions on a von Neumann–Gabor lattice [4, 5]. Furthermore, we have used the WZT to represent images in the space/spatial frequency domain [8].

The WZT can be used in computing [4, 5, 8] the Gabor coefficients, c_{mn}, in expansions of the form:

$$Z_f(\Omega, \tau) = Z_g(\Omega, \tau) \sum_m \sum_n c_{m2n} \cdot e^{\iota 2\pi \Omega n} e^{\iota \pi \tau m}$$

$$+ Z_g\left(\Omega, \tau + \frac{T}{2}\right) \sum_m \sum_n c_{m2n+1} \cdot e^{\iota 2\pi \Omega n} e^{\iota 2\pi \tau m}, \qquad (4.1)$$

where $c_{mn} = \langle f, \tilde{g}_{mn} \rangle = \langle Z_f, Z_{\tilde{g}_{mn}} \rangle$ [8].

We have previously applied the Zak Transform (ZT) to the computation of expansion coefficients for complex images [10]. In that case, the images were decomposed in a combined space/spatial frequency domain using Hermite functions. In related work, we used for image analysis another window function obtained by weighting the Zak transform of a Gaussian [8]. The resulting basis functions were smooth and localized, and hence suitable for the analysis of complex, real-world imagery. The basis functions described here are similar to those described in [8], and hence may also be relevant to the analysis of complex imagery consisting of localized targets on complex real-world-backgrounds. In this chapter, we consider the properties of the WZT and applications to signal processing. In signal processing one attempts to design signals with prescribed ambiguity functions (AFs), in particular, AFs that are strongly localized at the origin in the time–frequency plane [11–14]. First, we note the relation between the discrete Husimi distribution and the ambiguity function on the lattice of the Gaussian analyzing window and the test signal, which follows from [9]. We then use the WZT to construct an ambiguity function that vanishes at all points on a lattice, except for the origin, thus providing the narrowest possible discretely defined function. We then prove that the ambiguity function falls off exponentially between the lattice points. Finally, we show how to obtain basis waveforms, using the WZT, which are suitable for analyzing and synthesizing complex signals and images.

4.2 Background and Definitions

The Zak Transform (ZT) for the signal $f(t)$ is defined as follows [1, 2]:

$$Z_f(\Omega, \tau) = \sqrt{T} \sum_n e^{2\pi \iota \Omega n T} f(\tau - nT).$$ (4.2)

Correspondingly, the inverse transform gives the signal, $f(t)$:

$$f(t) = \sqrt{T} \int_0^{\frac{1}{T}} Z(\Omega, \tau) d\Omega.$$ (4.3)

Define the weighted Zak Transform as

$$z_f(\Omega, \tau) = \frac{Z_f(\Omega, \tau)}{\sqrt{S_f(\Omega, \tau)}}.$$ (4.4)

We call $\tilde{f}$ the inverse of the weighted Zak transform, $z_f(\Omega, \tau)$, as is used in

Eq. (4.2) and where the weighting function is defined as

$$S_f(\Omega, \tau) = \frac{1}{2}\left(|Z_f(\Omega, \tau)|^2 + \left|Z_f\left(\Omega, \tau + \frac{T}{2}\right)\right|^2\right)$$

$$= S_f\left(\Omega, \tau + \frac{T}{2}\right) \qquad (4.5)$$

$$= S_f\left(\Omega + \frac{1}{T}, \tau\right)$$

The weighting function is real, positive, and periodic in both variables (Ω, τ). For Gaussian the weighting function is a product of two Jacobi theta functions [5].

4.3 The Weighted Zak Transform and the Ambiguity Function

Wigner, Ambiguity Functions, and Discrete Husimi Distribution. When the analyzing window is Gaussian, the coefficients $|h_{mb,na}(f)|^2$ can be expressed in terms of the discrete Wigner function [9]:

$$|h_{mb,na}(f)|^2 = \iint W_{gg}(na - t, mb - \vartheta) \cdot W_{ff}(t, \vartheta)dtd\vartheta \qquad (4.6)$$

which is also called a discrete Husimi distribution [9]. The Discrete Fourier Transform (DFT) of $|h_{mb,na}(f)|^2$ can be computed as a product of the corresponding Ambiguity functions $A_{gg} \cdot A_{ff}$ on the lattice.

Let us elaborate on this and say the following. Denote by $H(x, p)$ the Husimi function for the signal f, and by $K(\nu, t)$ the 2-D Fourier transform of $H(x, p)$. Then from Eq. (44) of [9] it follows that for **any** ν and t, $K(\nu, t) = A_{g,g}(\nu, t) \cdot A_{f,f}(\nu, t)$.

Therefore, this equality holds also on the lattice. Then, using a well-known connection [2, 3] between the $|Z_g(\Omega, \tau)|^2$ and $|Z_f(\Omega, \tau)|^2$ and their corresponding Ambiguity functions (AFs) on the lattice, one gets the result: $K(\nu, t)$ on the lattice can be computed from the squares of the absolute values of the ZTs. We see here again that it is not neccessary to know the phase of the ZT in order to find $K(\nu, t)$ on the lattice.

Both the A_{fg} and the Wigner function play an important role in signal processing [3]. Where the cross AF, $A_{fg}(v, t)$, in the $\nu - t$ plane for the signals $f(t)$ and $g(t)$ is defined as follows [2]:

$$A_{f,g}(\nu, t) = \int_{-\infty}^{\infty} e^{-2\pi\iota\nu z} f^*\left(z - \frac{1}{2}t\right) g\left(z + \frac{1}{2}t\right) dz, \qquad (4.7)$$

where $f^*(t)$ is the complex conjugate of the signal $f(t)$. In the computations of the coefficients of Eq. (4.6) one is often interested in the values of the AF on a discrete lattice in the $\nu - t$ plane. For this purpose, there is an important relation between the cross AF, $A\left(n\frac{1}{T}, mT\right)$, on a discrete lattice and the Zak transform, $Z(\Omega, \tau)$. This relation is as follows [2, 3]:

$$Z_f^*(\Omega, \tau)Z_g(\Omega, \tau) = \sum_{m,n}(-1)^{mn} A_{f,g}\left(\frac{n}{T}, mT\right) e^{-(2\pi \iota \Omega Tm + 2\pi \iota n\frac{\tau}{T})} \qquad (4.8)$$

for $f = g$

$$|Z_f(\Omega, \tau)|^2 = \sum_{m,n}(-1)^{mn} A_{ff}\left(\frac{n}{T}, mT\right) e^{-(2\pi \iota \Omega Tm + 2\pi \iota n\frac{\tau}{T})} \qquad (4.9)$$

The cross Wigner function for two signals $f(t)$ and $g(t)$ can be defined as follows [2]:

$$W_{f,g}(t, \nu) = \int e^{-2\pi \iota \nu z} f^*\left(t - \frac{1}{2}z\right) g\left(t + \frac{1}{2}z\right) dz. \qquad (4.10)$$

One can easily check that $W_{f,g}(t, \nu)$ is the two-dimensional Fourier transform of $A_{f,g}(t, \nu)$. We have [2]

$$W_{f,g}(t, \nu) = \int\int A_{f,g}(u, \nu)e^{2\pi \iota ut - 2\pi \iota \nu \nu} du d\nu \qquad (4.11)$$

and

$$A_{f,g}(\nu, t) = 2W_{f,g}(t, \nu)\left(\frac{t}{2}, \frac{\nu}{2}\right), \qquad (4.12)$$

where $\bar{f}(t) = f(-t)$.

The Weighted Zak Transform and the Ambiguity Function. An interesting connection can be established between the WZT, $z(\Omega, \tau)$, and the AF, if we use Eq. (4.9) rewrite it for $\tau + \frac{T}{2}$, and then add the two forms of the equation. We find (summing only over even $n = 2s$, since the sum vanishes for odd $n = 2s + 1$):

$$|Z_f(\Omega, \tau)|^2 + \left|Z_f\left(\Omega, \tau + \frac{T}{2}\right)\right|^2 = 2\sum_{m,s} A_{ff}\left(\frac{2s}{T}, mT\right) e^{-(2\pi t\iota \Omega Tm + 4\pi ts\frac{\tau}{T})}. \qquad (4.13)$$

By using the definitions in Eqs. (4.4) and (4.5), Eq. (4.13) for the WZT, z, becomes (omitting subscript f).

$$|z(\Omega, \tau)|^2 + \left|z\left(\Omega, \tau + \frac{T}{2}\right)\right|^2 = 1 = 2\sum_{m,s} A^z\left(\frac{2s}{T}, mT\right) e^{-(2\pi \iota \Omega T'm + 4\pi \iota s\frac{\tau}{T})}. \qquad (4.14)$$

It therefore follows that for the normalized $z(\Omega, \tau)$, the AF in Eq. (4.14) will be $\frac{1}{2}$ at $A^z(0, 0)$ and will be zero at all other even-frequency points of the von

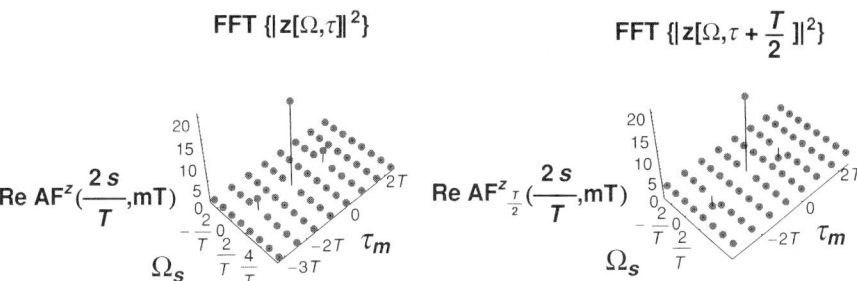

Figure 4.1. The $AF^z\left(\frac{2s}{T}, mT\right)$ and $AF^z_{\frac{T}{2}}\left(\frac{2s}{T}, mT\right)$ for $T = 2\lambda\sqrt{\alpha\pi} = 1.25331, \lambda = 0.5, \alpha = 0.5$, of the Eq. (4.14a).

Neumann– Gabor lattice $\left\{\frac{2s}{T}, mT\right\}$ in the time-frequency plane. $A^z\left(\frac{2s}{T}, mT\right)$ in Eq. (4.14) is computed in the following way by taking the 2D DFT of $\left|z(\Omega, \tau)\right|^2 + \left|z\left(\Omega, \tau + \frac{T}{2}\right)\right|^2$. Denote the DFT of the first term AF^z and the DFT of the second term $AF^z_{\frac{T}{2}}$, then,

$$A^z\left(\frac{2s}{T}, mT\right) = AF^z\left(\frac{2s}{T}, mT\right) + AF^z_{\frac{T}{2}}\left(\frac{2s}{T}, mT\right), \qquad (4.14a)$$

where the sum vanishes when $2s$ is replaced by an odd integer. This is shown in Figures 4.1 and 4.2 for a Gaussian signal.

4.4 The Weighting Function

The Zak Transform, $Z_0(\Omega, \tau)$, for Gaussian $f(t) = \frac{1}{4\sqrt{\pi\lambda^2}}e^{-\frac{t^2}{2\lambda^2}}$, from Eq. (4.2) is

$$Z_0(\Omega, \tau) = 4\sqrt{\frac{T^2}{\pi\lambda^2}}e^{\frac{-\tau^2}{2\lambda^2}}\left(1 + 2\sum_{n=1}^{\infty}\cos\left(2n\left(\pi\Omega T - \iota\frac{T^2}{2\lambda^2}\frac{\tau}{T}\right)\right)\cdot e^{-\frac{T^2}{2\lambda^2}n^2}\right). \qquad (4.15)$$

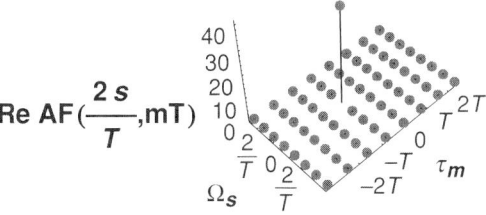

Figure 4.2. Ambiguity function on von Neumann–Gabor lattice.

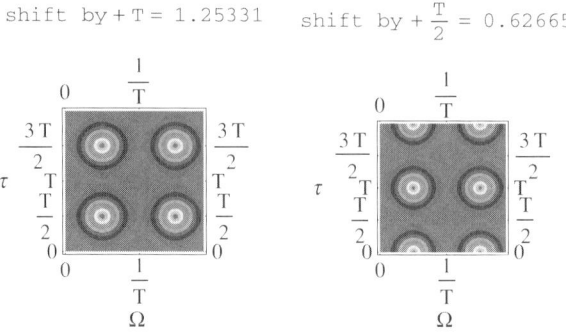

Figure 4.3. The absolute value $|Z_0(\Omega, \tau)|$ for $\lambda = 0.5$ and $\alpha = 0.5$ (see color insert).

The expression in the outer parentheses is a Jacobi Theta 3 function [6]:

$$Z_0(\Omega, \tau) = 4\sqrt{\frac{T^2}{\pi\lambda^2}}\, e^{-\frac{\tau^2}{2\lambda^2}}\, \Theta_3\left(\pi\Omega T - \iota\frac{T^2}{2\lambda^2}\frac{\tau}{T}\bigg| e^{-\frac{T^2}{2\lambda^2}}\right).$$

For computational convenience, denote $\alpha = \frac{T^2}{4\pi\lambda^2}, p = \iota \cdot 2 \cdot \alpha$, and $q = e^{\iota\pi p}$. Then Eq. (4.15) in standard notation is

$$Z_0(\Omega, \tau) = 4\sqrt{4\alpha}\, e^{-2\alpha\pi\frac{\tau^2}{T}}, \Theta_3\left(\pi\Omega T - \iota 2\pi\alpha\frac{\tau}{T}\bigg| q\right) \qquad (4.16)$$

which has a simple zero at $\zeta = \left(\pi\Omega T - \iota 2\pi\alpha\frac{\tau}{T}\right) = \frac{\pi(1-p)}{2}$ for $\Omega = \frac{1}{2T}$ and $\tau = \frac{T}{2}$, and is periodic in Ω with period $\frac{1}{T}$ (Figure 4.3).

The absolute value of the WZT of the Gaussian is shown in Fig. 4.4

The weighting function $S_j(\Omega, \tau)$ of Eq. (4.5) for Gaussian $f(t)$ is shown in Fig. 4.5 and 4.6.

Alternatively, the $S(\Omega, \tau)$ function can be written in terms of Θ_3 functions [5]. For the Gaussian, $f(t)$, the AF is $A\left(\frac{2s}{T}, mT\right) = e^{-\pi\frac{T^2}{4\pi\lambda^2}m^2}e^{-\frac{4\pi^2\lambda^2}{T^2}s^2}$, on the lattice points $\left(\frac{2s}{T}, mT\right)$. Using this AF and Eq. (4.13) one can show that the weighting function is a product of two Θ_3 functions as follows [5]:

$$|Z(\Omega, \tau)|^2 + \left|Z\left(\Omega, \tau + \frac{T}{2}\right)\right|^2 = 2\sum_m e^{-\pi\alpha m^2}e^{-2\pi\iota\Omega Tm}\sum_s e^{-\frac{\pi s^2}{\alpha}}e^{-4\pi\iota s\frac{\tau}{T}}$$

$$= 2\Theta_3(\pi\Omega T|e^{-\pi\alpha})\Theta_3\left(2\pi\frac{\tau}{T}\bigg| e^{-\frac{\pi}{\alpha}}\right) \qquad (4.17)$$

$$S(\Omega, \tau) = \frac{1}{2}\left(|Z(\Omega, \tau)|^2 + \left|Z\left(\Omega, \tau + \frac{T}{2}\right)\right|^2\right)$$

$$= \Theta_3(\pi\Omega T|e^{-\pi\alpha})\Theta_3\left(2\pi\frac{\tau}{T}\bigg| e^{-\pi\frac{1}{\alpha}}\right) \qquad (4.18)$$

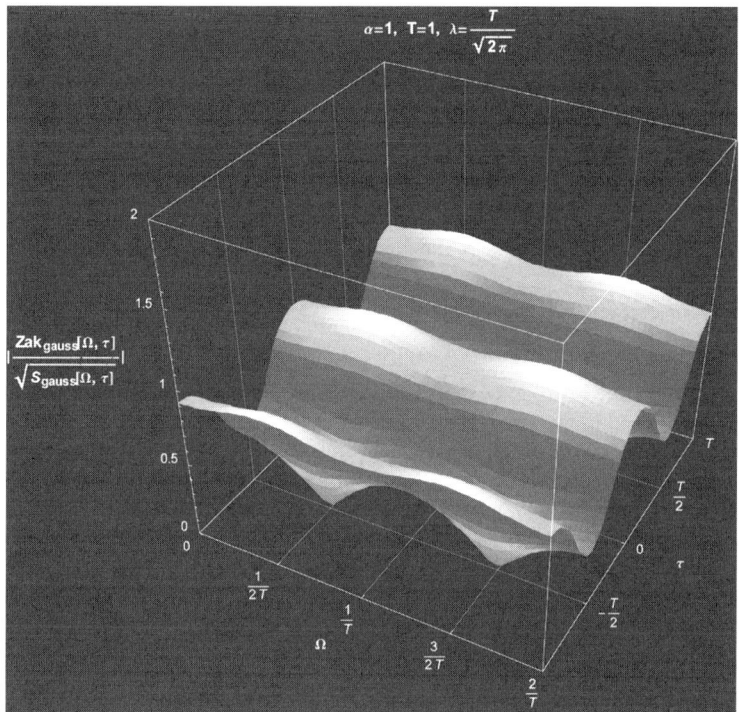

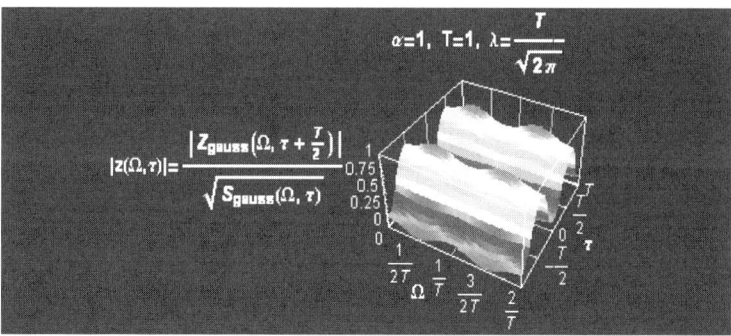

Figure 4.4. The absolute value of the weighted $|z(\Omega, \tau)| = \frac{|Z(\Omega, \tau)|}{\sqrt{S(\Omega, \tau)}}$ for the Gaussian function (see color insert).

Figure 4.6. shows plots of $S_f(\Omega, \tau)$, which is a product of two Jacobi Theta functions (see Eq. 4.17). one of which is a function of Ω and the other of τ. When T is small, the Theta function depending on Ω oscillates, while the Theta function depending on τ will be almost constant and close to 1. This can be clearly seen in the first three plots of Figure 4.6(a). For large T, the situation is be reversed: The Theta function of Ω is almost constant and is

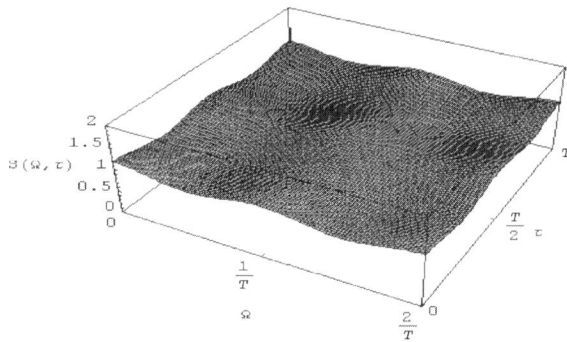

Figure 4.5. $S_f(\Omega, \tau)$ for Gaussian $f(t), \alpha = 1$.

close to 1, while the Theta function of τ oscillates. This can be seen on all the plots of Figs. 4.6(b) and 4.6(c). Qualitatively, the plots of Fig. 4.6(b) and 4.6(c) can be obtained rotating the plots of Fig. 4.4 by 90 degrees. Note that each plot of Fig. 4.6(b) and 4.6(c) have different T s and is normalized to fit the uniform display.

4.5 The Inverse of the Weighted Zak Transform

The inverse ZT of the WZT $z(\Omega, \tau)$ is obtained using Eq. 4.3 and is depicted in Figs. 4.7, and 4.8 for various periods T and a fixed Gaussian parameter $\lambda = 0.5$. Figure 4.8 compares the inverse of the WZT $z(\Omega, \tau)$ with the Gaussian.

Decay Rate of the Expansion Coefficients

The AF on a lattice that was constructed using the WZT is shown in the Fig. 4.2

Keeping in mind the results shown in Eq. 4.16 for any normalized $z_0(\Omega, \tau)$, it is clear also that $z_0(\Omega, \tau)$ [see Eq. (4.14) will lead to an AF, $A\left(\frac{2s}{T}, mT\right)$, [see Eq. (4.16) that will vanish at all points of the lattice $\left(\frac{2s}{T}, mT\right)$ in the time-frequency plane, with the exception, of the point $s = m = 0$ at the origin. However, for the Gaussian function the AF, $A\left(\frac{2s}{T}, mT\right)$, will in addition fall off exponentially for increasing absolute values of $|s|$ and $|m|$ of the integers s, m. The exponential fall-off can be shown in the following way. For the Gaussian function (the ground state of the harmonic oscillator) we have for

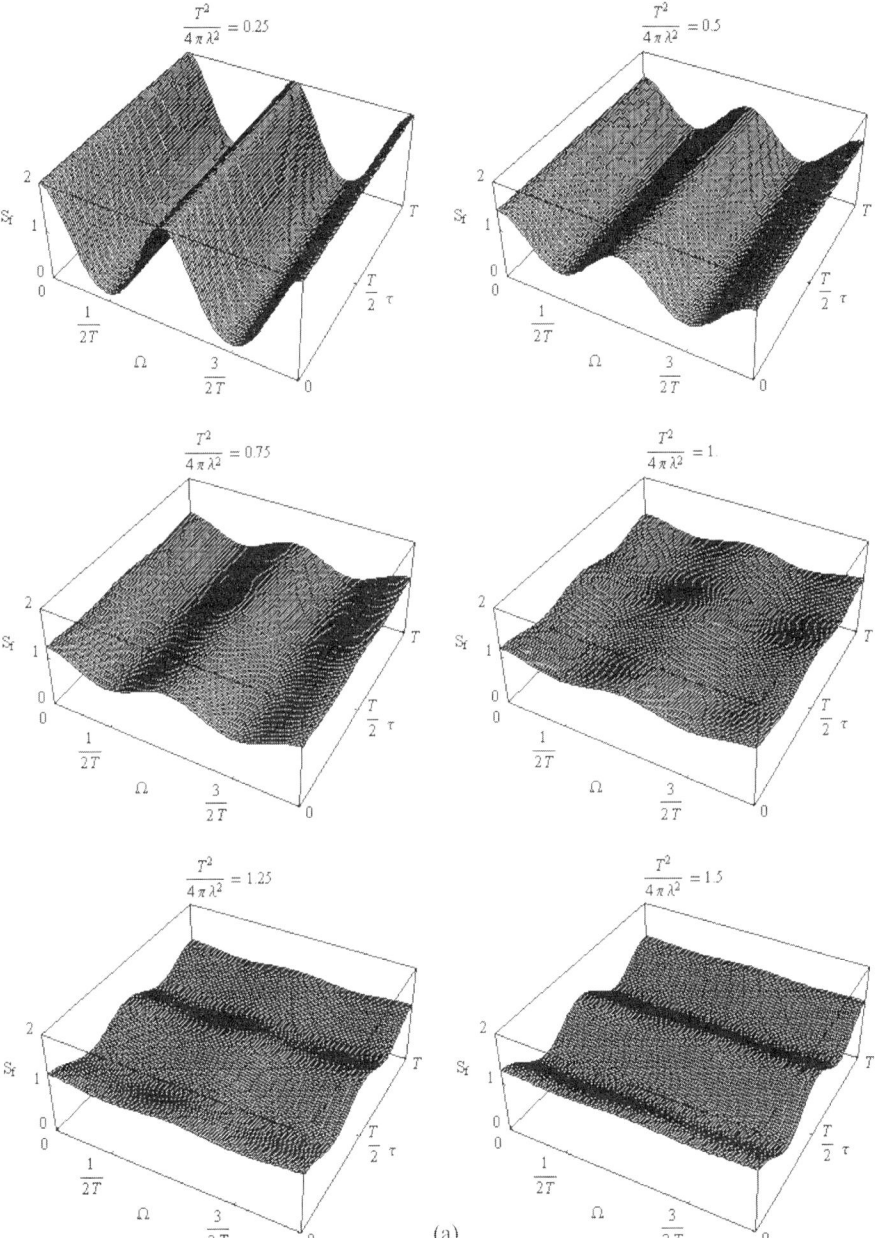

Figure 4.6. The weighting function, $S_f(\Omega, \tau)$, for various T and fixed $\lambda = 0.5$.

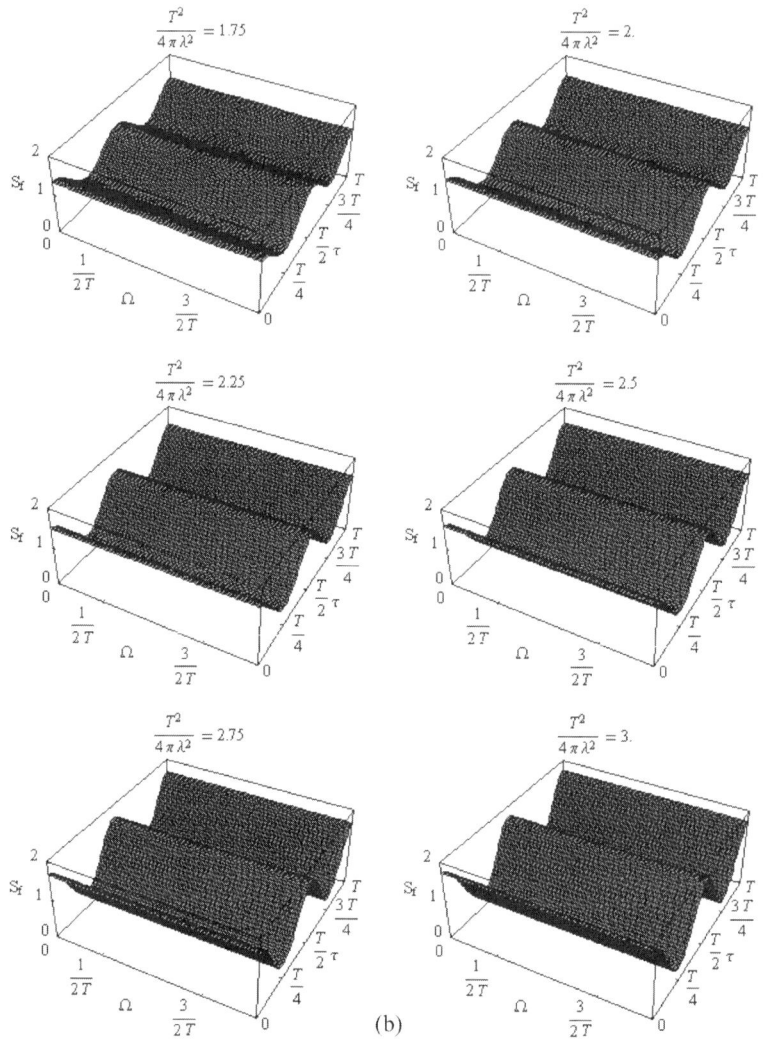

Figure 4.6. (*Continued*)

the parameter $p = \iota$ (See Eq. (4.18) for $\frac{T^2}{4\pi\lambda^2} = 1$) for normalized $|z_0(\Omega, \tau)|^2$:

$$|z_0(\Omega, \tau)|^2 = \frac{\sum_m \sum_n (-1)^{mn} e^{-\pi\left[m^2 - \left(\frac{n}{2}\right)^n\right] - 2\pi\iota\Omega Tm + \iota 2\frac{\pi}{T}\iota n}}{\Theta_3(\pi\Omega T|\iota)\Theta_3\left(\frac{2\pi\tau}{T}|\iota\right)}. \tag{4.19}$$

For real $\Omega, \tau |z_0(\Omega, \tau)|^2$ is periodic in Ω and τ with period $\frac{1}{T}$ and T, respectively. One can extend Ω and τ into the complex plane $\Omega' + \iota\Omega'', \tau' + \iota\tau''$ and write

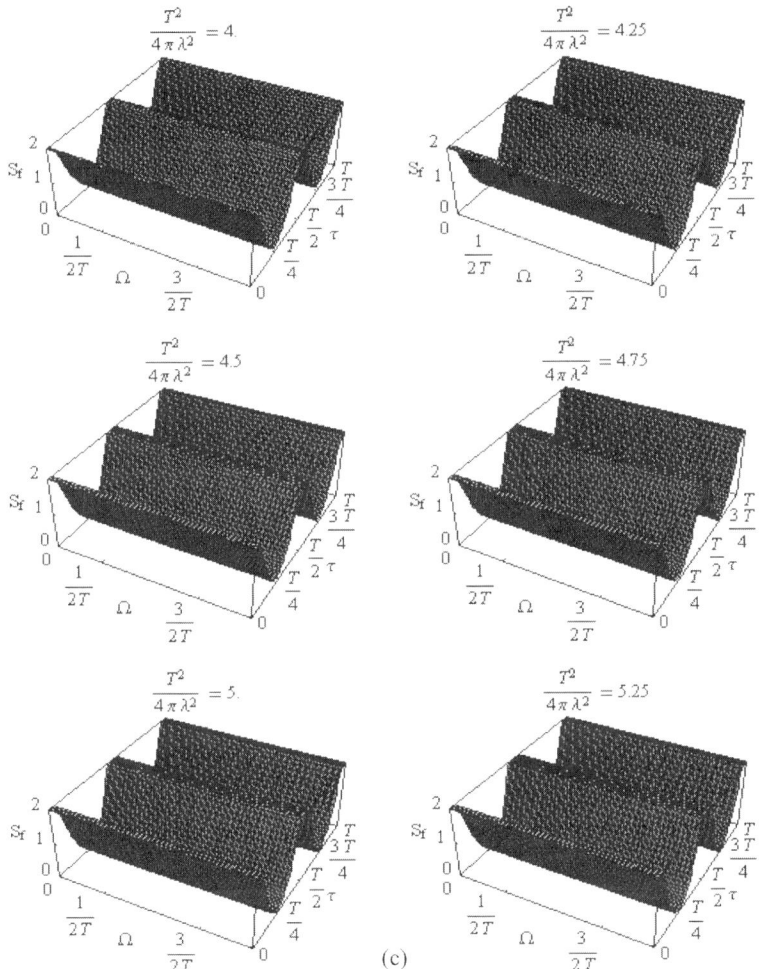

Figure 4.6. (*Continued*)

the series for $|z_0(\Omega, \tau)|^2$ in the following way:

$$|z_0(\Omega, \tau)|^2 = \sum_m \sum_n z_0(m, n) e^{2\pi\Omega''Tm - \frac{2\pi}{T}\tau''n - 2\pi\iota\Omega'Tm + \iota\frac{2\pi}{T}\tau'n}. \tag{4.20}$$

Since $|z_0(\Omega, \tau)|^2$ is analytic for all real Ω' and τ' and for complex parts $|\Omega''| < \frac{1}{2T}$, $\tau'' < \frac{T}{4}$ (the zero of $\Theta_3(\zeta|\iota)$ is at $\zeta = \frac{\pi}{2}(1 + \iota)$, and $|z_0(\Omega, \tau)|^2$ has to fall off as $e^{-\pi m - \frac{\pi}{2}n}$. This follows from a well-known theorem for Fourier series [7]. But $z_0(m, n)$ in Eq. 4.20 is proportional to the Wigner function (or Ambiguity function) that will also fall off exponentially on the lattice in phase space (time–frequency space) (Figure 4.9).

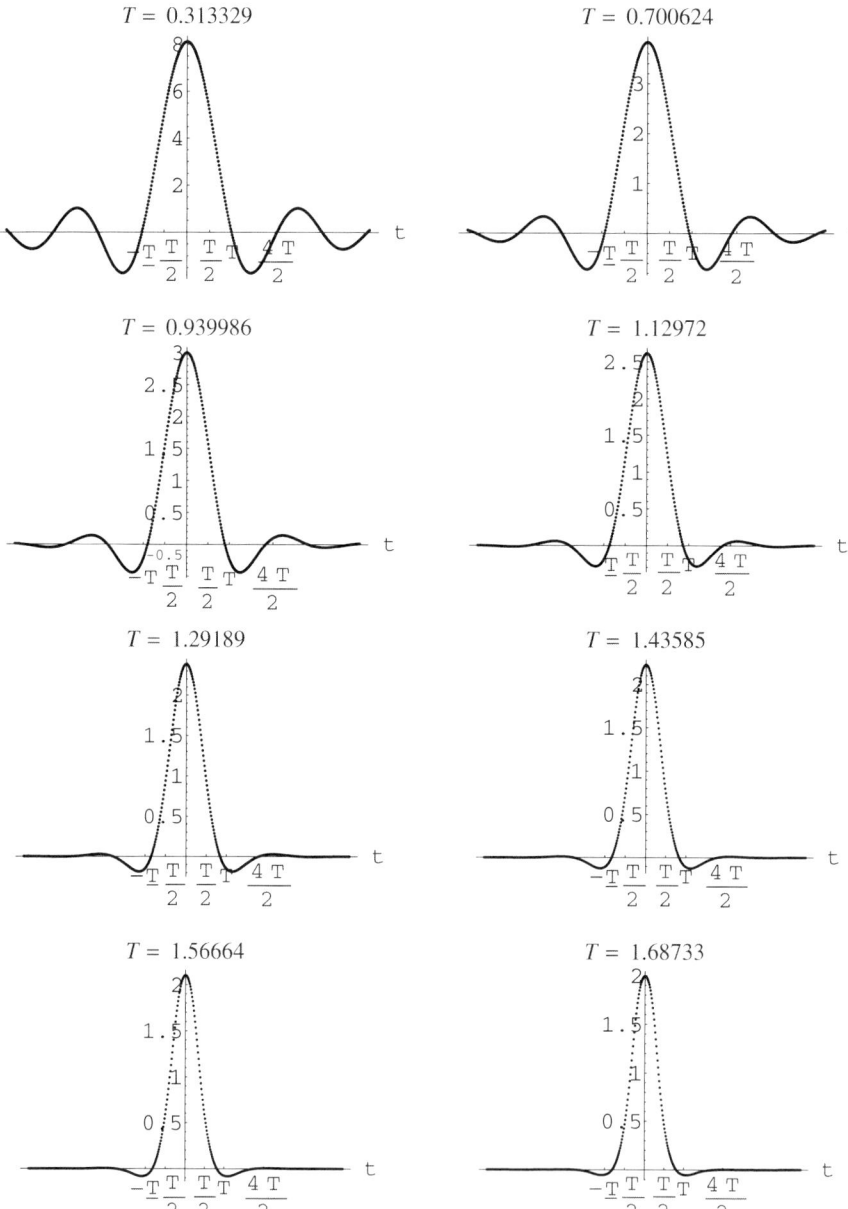

Figure 4.7. The basis function derived from the Gaussian for various T and fixed $\lambda = 0.5$.

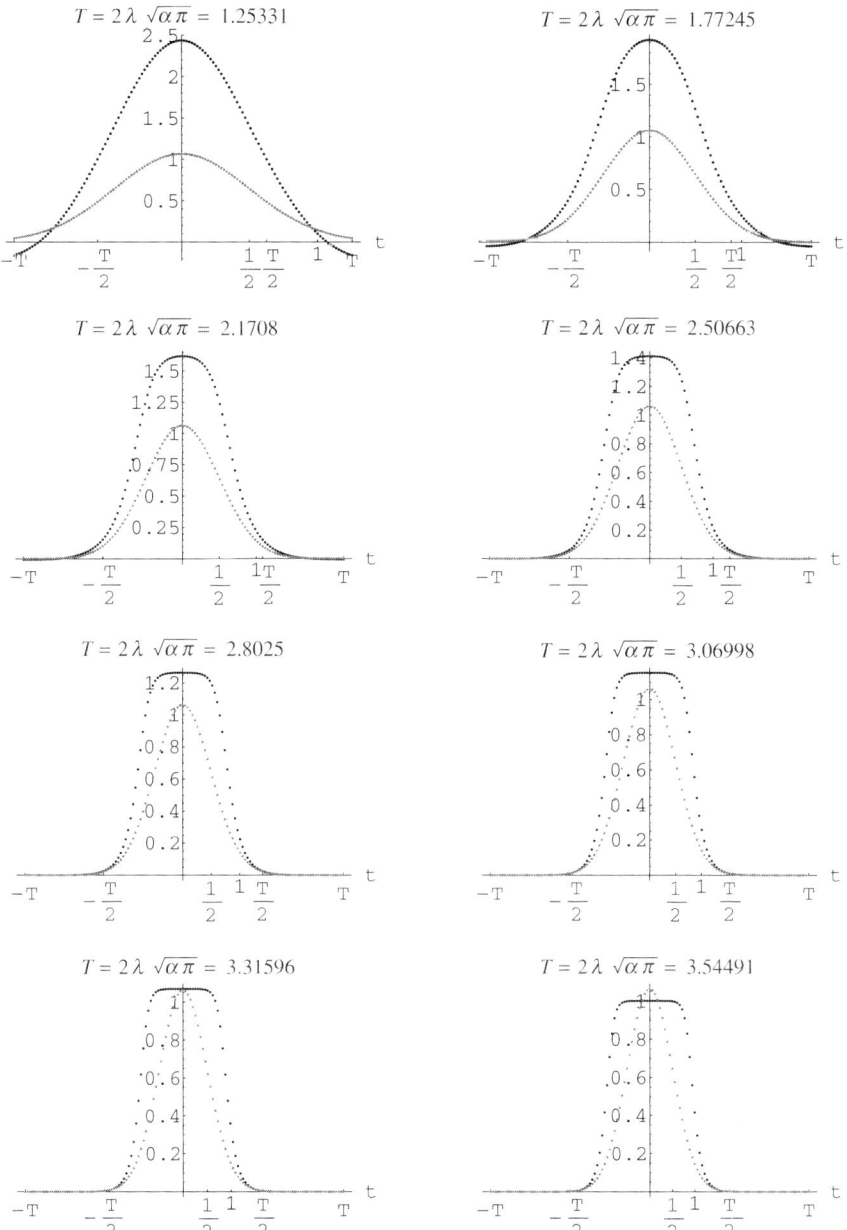

Figure 4.8. The inverse of the weighted $z(\Omega, \tau)$ for larger periods T. The dotted red plot is Gaussian $\lambda = 0.5$. Note the scale of each plot is different: for reference purpose marks of $1, \frac{1}{2}$ on t axis are shown (see color insert).

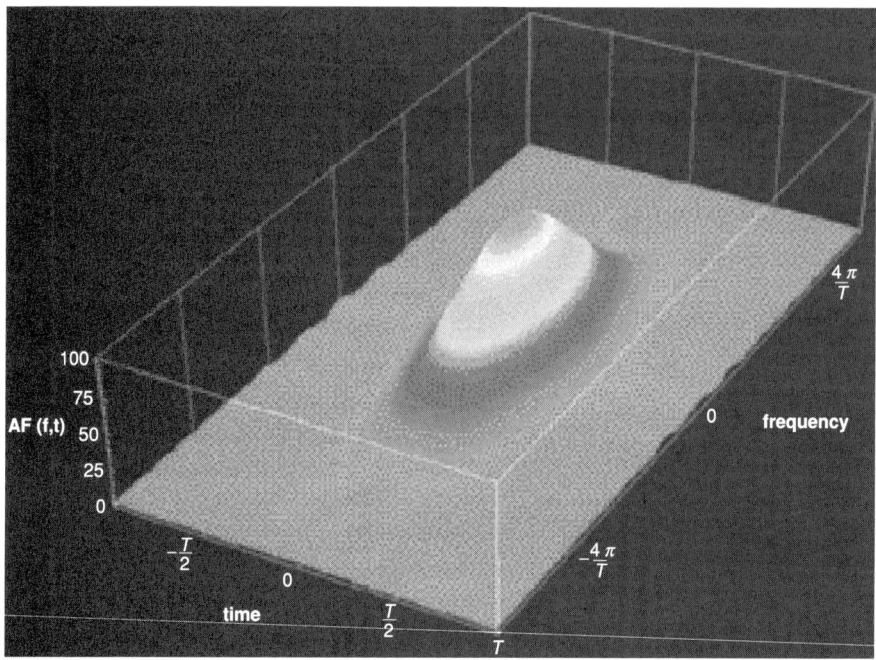

Figure 4.9. The absolute value of the Ambiguity function (see color insert).

4.6 Signal Synthesis Using the Weighted Zak Transform

For any given signal, $g(t), S(\Omega, \tau)$ is periodic in both variables with period $\frac{1}{T}$ and $\frac{T}{2}$, respectively. This function can be written as Fourier series as follows:

$$
\begin{aligned}
\frac{1}{\sqrt{S(\Omega, \tau)}} &= \sum_n a_n(\Omega) e^{\iota \frac{2\pi}{T} \tau n} \\
&= \sum_n b_n(\tau) e^{\iota 2\pi \Omega \frac{T}{2} n},
\end{aligned} \tag{4.21}
$$

where $a_n(\Omega)$ is periodic in Ω, and $b_n(\tau)$ in τ.

By taking the inverse ZT of the WZT, $\frac{Z(\Omega, \tau)}{\sqrt{S(\Omega, \tau)}}$, we can get the signal $g(t)$ corresponding to it, or its Fourier transform $G(\nu)$. We have, by using Eq. (4.21)

$$
\tilde{g}(t) = \sqrt{T} \int_0^{\frac{1}{T}} \frac{Z(\Omega, \tau)}{\sqrt{S(\Omega, \tau)}} d\Omega = \sum_n b_n(t) g(t - nT). \tag{4.22}
$$

And, correspondingly, using the connection between the Fourier Transform and the ZT:

$$\tilde{G}(\nu) = \frac{1}{\sqrt{T}} \int_0^1 \frac{Z(\nu, \tau)}{\sqrt{S(\nu, \tau)}} e^{-\iota 2\pi \nu \tau} d\tau = \sum_n a_n(\nu) G\left(\nu - \frac{n}{T}\right), \qquad (4.23)$$

where $\tilde{G}(\nu)$ is the Fourier transform of the signal $g(t)$.

Result represented by Eq. (4.22) is a generalization of formula (4.3) in Janssen's paper [2]: instead of one window, $m(t)$, we have a set of windows, $b_n(t)$. Result represented by Eq. (4.23) is, likewise, a generalization of formula (4.2) in Janssen's paper [2].

Acknowledgment

The authors thank Professor J. Zak for numerous discussions that resulted in the work presented here.

The first author was partially sponsored by ONR grant No. N000140210122. This work was also supported in part by U.S. Air Force Contract FA8650-05-D-6502 to Link Simulation and Training.

References

[1] J. Zak. *Phys. Rev. Lett.*, 19:1385 (1967).

[2] A. J. E. M. Janssen. *Philips J. Res.*, 43, 23 (1998).

[3] L. Auslander, I. C. Gertner, and Richard Tolimieri. *IEEE Trans. Sig. Proc.*, 39:825 (1991).

[4] I. Daubechies, S. Jaffard, and Jean-Lin Journe. *SIAM J. Math. Anal.*, 22, 554 (1991).

[5] J. Zak. *J. Phys. A: Math. Gen.*, 36, L553 (2003).

[6] E. T. Whittaker and G. N. Watson. *Course in Modern Analysis*. Cambridge University Press, 1950.

[7] J. Des Cloizeaux. *Phys. Rev.*, 135, A685(1964): 692, 693.

[8] I. C. Gertner, G. A. Geri, *J. Opt. Soc. Am.* A, 11(8):2215–2219, 1994.

[9] J. Zak. Discrete Weyl-Heisenberg Transforms. *J. Math. Phys.* 37(B): 3815–3823, 1996.

[10] I. Gertner and G. A. Geri. *Biol. Cybernet.*, 71: 147–151, 1994.

[11] B. Moran. *Mathematics of Radar*, 20th Century Harmonic Analysis—A Celebration, NATO Advanced Study Institute, pp. 295–328, (2000).

[12] J. P. Costas. *Proceedings of The IEEE*, 72(8):996–100, 1984.

[13] I. Gladkova. *IEEE AES*, 37 (4) 1458–1464, (2001).

[14] W. Schempp. *Harmonic Analysis on the Heisenberg Nilpotent Lie Group*, Pitman Research Notes in Mathematics Series, 147, 1984.

Using Polarization Features of Visible Light for Automatic Landmine Detection

Wim de Jong and John Schavemaker

TNO Defence, Security and Safety

5.1 Introduction

This chapter describes the usage of polarization features of visible light for automatic landmine detection. The Section 5.2 gives an introduction to landmine detection and the usage of camera systems. In Section 5.3 detection concepts and methods that use polarization features are described. Section 5.4 describes how these detection concepts have been tested and evaluated. The results of these tests are given in Section 5.5. Conclusions from these results are given in Section 5.6.

5.2 Landmine Detection with Camera Systems

Landmines as left behind after military or civil conflicts are a huge problem. The main current practice to clear mine fields is by prodding and by use of a metal detector. Prodding is very time consuming and dangerous and demands a lot of concentration from the deminer. The metal detector is used to find all metal objects. However, landmines that contain little or no metal also exist and moreover, in former battle zones a lot of metal shrapnel is left behind. The lack of sophisticated tools that are reliable makes the process of demining very slow. To improve the clearance speed in the future, many nations have put effort in landmine-detection research.

The research on sensor systems focuses on three main topics [1, 2]. First, the development of new sensors. Second, the improvement of existing sensors. The third research topic is the integration of these sensors into a sensor-fusion system. The use of one sensor is generally believed to be insufficient for landmine detection meeting the requirements of humanitarian demining

(100% detection) for the reason that a single sensor has a false-alarm rate which is too high or a detection rate which is too low. The goals of sensor-fusion are to reduce the probability of false alarms $P(\text{fa})$, to increase the probability of detection $P(\text{d})$, or to improve a combination of both.

5.2.1 Polarization Camera

One of the sensors that are considered for a multisensor landmine detection system is the thermal infrared (TIR) camera [3]. TIR cameras are able to detect small temperature differences (as low as 15 mK). Landmines often have different heat conductivity and heat capacity compared to the soil and the vegetation around them. Because of these differences in thermal properties, differences in temperature between a landmine and the background may develop when the soil is heated or cooled down. However, TIR images of landmines in natural scenes contain clutter, since other (natural) objects like trunks, holes, and rocks also may have different thermal properties compared to the background. In the visual spectrum it is well known that unpolarized light reflected from a smooth surface becomes polarized [4]. This is also true for TIR radiation. However, for TIR radiation not only the reflection, but also the emission is polarized. Since in general the surfaces of landmines are smoother than the surfaces found in a natural background, the presence of significantly polarized TIR radiation is an extra indication for landmines (or other nonnatural objects). Using a polarization setup, the performance of the TIR camera can be improved and thus have a larger contribution to the multisensor system [5–7].

When compared to a TIR polarization camera a VIS polarization camera is in general much cheaper and more robust. The drawback that buried land mines cannot be detected with a VIS polarization camera is in some applications less important. This chapter only deals with the automatic detection of landmines using a VIS polarization camera.

5.2.2 The Use of VIS Polarization in Area Reduction

Reduction of the suspected area (before close-in detection is started) is very important in mine-clearance operations, irrespective of the detection technique (prodding or using an advanced multisensor system). Area reduction is the process through which the initial area indicated as contaminated (during a general survey) is reduced to a smaller area. Area reduction consists mainly of collecting more reliable information on the extent of the hazardous area.

Mechanical methods to reduce the area for mine clearance are among others rollers and flails. So-called Pearson rollers [8], mounted on an armored commercial front-end loader, are sometimes used to rapidly reduce areas adjacent to locations where antipersonnel (AP) landmines are suspected. Since the AP mine rollers have proven to be successful within humanitarian mine clearance, The HALO Trust has developed an anti-tank (AT) mine roller unit to be mounted on a front-end loader like the AP mine roller. The AT roller system is designed to detonate antitank mines, but each detonation of an AT mine will incur a downtime for repair of the rollers, since some damage is expected on an AT mine detonation.

In order to minimize the amount of physical AT mine detonations, a camera system is developed that uses polarization properties of visual light to reach a high-detection rate of AT mines, combined with a low false-alarm rate. The camera system is fitted to the roof of the vehicle and is forward looking. The goal is to detect as many surface-laid AT mines as possible and to stop the vehicle before the roller detonates the mine. Contrary to close-in mine detection, 100% detection is not a requirement in this scenario, since mines missed by the camera will be detonated by the rollers. For some applications a very low false-alarm rate (less than 3 alarms per 100 m forward movement of the vehicle) combined with a moderate detection rate would be a valuable system. The following list presents some possible camera system concepts and the assumptions they have on the camera image modality:

– Black and white (BW), visible light, video camera;
– color, visible light, video camera;
– BW, visible light, video camera with polarization;
– Color, visible light, video camera with polarization.

All mine-detection camera systems are based on assumptions regarding the appearance of landmines for image processing and landmine detection. The above-mentioned camera concepts of a visible-light camera with or without polarization have different mine-detection capabilities.

5.3 Detection

In addition to the conventional way of detecting mines using image intensity [9] and/or color features, an additional way is created by the use of polarizers. Polarization can add significantly to the systems robustness and its detection performance [5–7], especially in the case of detecting artificial objects within a natural background. However, the performance of polarization features depends more on the operating conditions than the color or intensity features, which are more invariant to those conditions. For example, the position of the sun (both azimuth and elevation), viewing angle with respect to the sun position, and weather conditions (daylight, clouds) have an impact on the use of polarization. Extensive tests have given some insights into the application of polarization system and its effect on the system characteristics.

5.3.1 Detection Concepts

The mine-detection concept that we propose should combine features and cues from different image-processing techniques:

Color and/or intensity analysis: Detection is based on object color [10] or intensity contrast with the surrounding background [9]. The contrast threshold is defined by global or local image statistics (mean and standard deviation) of the image.

Polarization analysis: Objects are detected on the basis of their polarization contrast with the surrounding background [7].

Edge detection and grouping: Straight or partly circular edges are extracted because they can indicate artificial objects. The subsequent step groups edges into hypothetical artificial objects (e.g., using an edge-based Hough transform [11]).

For this concept, each image-processing technique results in a set of possible object detections with computed color, edge, or polarization features for which additional features (e.g., morphological) can be calculated. The resulting set of objects is put into a pattern-recognition classifier that performs the final mine detection based on all computed features and previously learned examples. In order to ensure robustness of this classifier, it is important that the image-processing techniques deliver invariant features. For example, scale invariance [12] can be found with proper camera calibration.

5.3.2 Detection Methods

For the tests that are described in Section 5.4, we have made a first partial implementation of the mine-detection concepts described in the previous paragraph. For these tests we have implemented a threshold on intensity and polarization, and a blob analysis scheme as pattern-recognition classifier.

5.3.2.1 Threshold on Intensity

The assumption that landmines have intensity values that are different from local surroundings can be exploited using intensity contrast images. As mines can be brighter or darker than the background, the intensity contrast can be a so-called positive or negative contrast with the background. The (normalized) positive intensity contrast image is defined as

$$I_{\mathrm{pos}}(x) = \frac{I(x) - \min(I)}{\max(I) - \min(I)} \tag{5.1}$$

The (normalized) negative intensity contrast image is defined as

$$I_{\mathrm{neg}}(x) = 1 - \frac{I(x) - \min(I)}{\max(I) - \min(I)} \tag{5.2}$$

On both contrast images a threshold can be applied to obtain an image with binary values indicating the detection of landmines. The applied threshold is a parameter of the detection method, and its choice of value depends on the scenario of application and required performance. The positive intensity contrast detections are defined as

$$D_{I_{\text{pos}}}(x) = \{I_{\text{pos}}(x)\}_t \qquad (5.3)$$

The negative intensity contrast detections are defined as

$$D_{I_{\text{neg}}}(x) = \{I_{\text{neg}}(x)\}_t \qquad (5.4)$$

where the threshold operator is defined as

$$\{f(x)\}_t = \begin{cases} 1 : f(x) \geq t \\ 0 : f(x) < t \end{cases} \qquad (5.5)$$

5.3.2.2 Threshold on Polarization

Landmines that have a smooth artificial surface (and/or have intensity values that are different from local surroundings) can be detected using polarization contrast. Polarization contrast can be defined using the Stokes vector [4], which is a mathematical representation of polarization. Using this representation, optical elements can be described as matrix operators on these vectors. The Stokes vector is constructed by measuring the intensity of four different polarization states:

- I_0 measures all states equally through an isotropic filter,
- I_h measures the intensity through a horizontal linear polarizer,
- I_{45} measures the intensity through a linear polarizer oriented at 45 degrees,
- I_L measures the intensity through a polarizer which is opaque for left circular polarization.

Using these measured intensities, the Stokes parameters I, Q, U, V are defined as

$$\begin{cases} I = 2I_0 \\ Q = 2I_h - 2I_0 \\ U = 2I_{45} - 2I_0 \\ V = 2I_L - 2I_0 \end{cases} \qquad (5.6)$$

Often different representations are used for the linear polarization:

$$\text{LP} = \sqrt{Q^2 + U^2} \qquad (5.7)$$

$$\text{DoLP} = \frac{\text{LP}}{I} \qquad (5.8)$$

$$\Psi = \frac{1}{2}\arctan\left(\frac{U}{Q}\right) \qquad (5.9)$$

with LP the amount of linear polarization, DoLP the degree of linear polarization, and Ψ the angle of polarization (orientation of the polarization ellipse). The polarization detections are defined as

$$D_{\text{pol}}(x) = \{f(I, Q, U)(x)\}_t \tag{5.10}$$

where $f(I, Q, U)(x)$ is a function that projects the first three elements of the Stokes Vector to a scalar. For example, the linear polarization LP (Eq. (5.7)), with Q and U being the Stokes components that are determined for each pixel. The amount of linear polarization is, in the implementation for the tests, taken as the polarization feature. The normalized polarization contrast image is defined as

$$\text{LP}_{\text{pos}}(x) = \frac{\text{LP}(x) - \min(\text{LP})}{\max(\text{LP}) - \min(\text{LP})} \tag{5.11}$$

The polarization detections are defined as

$$D_{\text{LP}_{\text{pos}}}(x) = \{\text{LP}_{\text{pos}}(x)\}_t \tag{5.12}$$

5.3.2.3 Combination of Intensity and Polarization

For implementation for the tests, we have combined intensity and polarization in the following way:

$$F(x) = \{I_{\text{pos}}(x)\}_{t_1} \vee \{I_{\text{neg}}(x) \cdot \text{LP}_{\text{pos}}(x)\}_{t_2} \tag{5.13}$$

The blobs in the resulting binary image $F(x)$ are extracted by means of a connected-component algorithm. Blobs are groups of connected pixels that represent possible landmine detections. Blobs with a size smaller than threshold t_3 are removed from image $F(x)$. As such, the complete detection procedure has three parameters: thresholds t_1, t_2, and t_3.

The reason for this combination is as follows: The combination takes a disjunction of the positive intensity and polarization contrast to use either positive intensity or polarization, whichever has more contrast at the appropriate moment. The conjunction between the negative intensity contrast and the polarization is applied for reasons of false-alarm reduction. Furthermore, it is assumed that intensity and polarization are complementary features. This assumption has been validated from the results of the tests.

5.4 Detection Tests

To test the detection concepts, detection tests have been performed. This section describes the camera hardware and the detection and performance evaluation software that have been used for the detection tests. These detection tests have been performed using a static setup, looking at a static scene. This section also describes the test procedure, including preparation of the test fields, selection of the test mines, and pre-processing of the measured data before detection evaluation. Several factors that are expected to influence the detection performance have been taken into account when setting up and performing the static tests. Factors that are expected to influence the detection results are as follows

– Light conditions (elevation and relative azimuth of the sun);
– Distance between camera and mine (number of pixels on mine);
– Mine type;
– Condition of mine surface;
– Burial depth (fraction of mine body visible for camera);
– Vegetation;
– Orientation of top surface of mine (horizontal or tilted).

5.4.1 Polarization Measurement Setup

Generally there are two different approaches used for the measurement of (infrared or visual) polarization. Either time division or spatial division is necessary to measure up to four elements of the Stokes vector using only one focal plane.

With time division, different polarization images are measured sequentially. This is usually performed by mounting a polarization filter in front of the camera and taking a sequence of images with different polarization directions. For measurements of the full (four elements) Stokes vector, a retarder (for instance, a quarter wave plate) is rotated followed by a fixed linear polarization filter. This common approach of either rotating a polarizer or a retarder is reported by the majority of literature [13–15].

Using spatial division, the different polarization states are measured simultaneously at the cost of reduced spatial resolution. For example, every four adjacent pixels of a focal plane array (FPA) are grouped. In front of each of these four pixels a different polarization filter is mounted, each with a different orientation [16, 17].

When more than one focal plane is available, an optical prism assembly, mounted behind the camera lens can be used to separate an image into three equal components. Each image is captured with a CCD. In front of each CCD element a polarization filter with a different, fixed orientation is mounted. This solution is known only for visible light since visible light CCDs are much cheaper than IR FPAs.

Our approach for the measurement setup is the use of time division and a rotating polarization filter. With this setup only linear polarization can be measured (the first three Stokes components). The measurement setup consisted of a black-and-white visible-light camera system, mounted on a tripod. The camera has a frame rate of 25 Hz and the images are grabbed from the analog video output of the camera using a frame grabber. The rotating polarization filter (Polaroid sheet) was mounted in front of the camera lens.

The filter was rotating at a constant speed. A trigger pulse from the rotation setup started the acquisition of the frame grabber in the computer. Since the camera was running at a fixed frame rate, the time between the frames and thus the rotation of the polarization filter between the different frames is fixed. A recorded sequence (about 56 images) contains at least one full rotation of the filter, which had a rotation time of just over 2 s. This camera had a fixed iris and a limited dynamic range, which appeared to be a limitation with shiny mines.

5.4.2 Test Setup

5.4.2.1 Test Fields

In order to perform measurements with different relative sun positions, four test fields were constructed, each in a different direction relative to the camera, which was placed in a central position. Three of the four fields had a stony background; the field in the west direction had a grass background. During data collection, the camera made recordings of the different measurement areas with the camera being rotated on the tripod to view the different areas. From the camera viewpoint, the four areas faced the North, South, East, and West directions (Fig. 5.1).

5.4.2.2 Test Mines

Replicas of four different types of AT mines that appear frequently in mine afflicted countries have been chosen for the tests. One type has a shiny surface (TM62P-3), while the surface of the PMN consists of mat rubber. The surface of the P2 consists of concentric rings which are expected to cause detection difficulties. Quite often only the cap of the mine is visible and not the whole

Figure 5.1. Two examples of the four measurement areas (facing North, South, East and West). The example on the left is the West measurement area that has grass as background. The other measurement areas have stones as background clutter as shown in the example on the right.

Table 5.1. Lay-out of all four measurement areas. The objects are placed on a 1-m grid. The six false alarms mentioned in this table are boulders with sizes and/or colors that are comparable to the mines that are used.

	Distance to camera	Column 1	Column 2	Column 3
Row 1	13 m	PRBM-3	TM62P-3	PRBM-3
Row 2	12 m	TM62P-3	P2	TM62P-3
Row 3	11 m	P2	PRBM-3	P2
Row 4	10 m	False alarm	False alarm	False alarm
Row 5	09 m	PMN	False alarm	PMN
Row 6	08 m	False alarm	PMN	False alarm

mine body. In the test field only the TM62P-3 mines were surface laid. From the P2 and PRBM-3 only replicas of the detonators were placed (as if the mine body was buried). The PMN AP mines were used since these mines have roughly the same diameter as the detonator of an AT mine. These PMNs were surface laid. Each measurement area contains three samples of four mine types and six false alarms. The mines and false alarms were placed in a rectangular area with a width of 3 m and a length of 6 m. Eighteen objects (3 × 6) were placed on a square grid. Table 5.1 shows the layout of the measurement areas.

Specular reflectivity and thus polarization contrast depends on the smoothness of the mine surface. Measurements were performed with clean mine surfaces and with mine surfaces with some sand on them.

During most measurements, the mines were placed horizontally. However, some measurements were performed with tilted mines, with tilt angles in different directions relative to the camera.

5.4.2.3 Measurement Procedure

Every 30 minutes during data gathering, recordings were taken of each field. The time difference between the recordings of the four fields is less than 1 min. Most recordings took place between sunrise and sunset. Only a few recordings were taken during the twilight period.

5.4.2.4 Data Pre-Processing

The detection algorithms as described in Section 5.3.2 use Stokes images as input. The recorded data consists of image sequences with different orientations of the rotating polarization filter for the different images in the sequence. Several pre-processing steps are needed to convert the raw data to calibrated Stokes images. (1). Determination of rotation frequency of polarization filter. (2). Correction for intrinsic polarization of camera. (3). Determination of offset angle of filter. (4). Determination of three Stokes parameters for each pixel.

5.4.3 Performance Evaluation Tools

The performance of the camera system was evaluated using the ROC curve. In an ROC curve the detection rate is plotted against the false-alarm rate for adjustable optimization parameters. Each working point on the ROC corresponds with a set of values for the thresholds t_1, t_2, and t_3 (Eq. (5.13)). The detection rate is defined as the fraction of detected landmines.

The corresponding number of false alarms per unit area is calculated using the following method: The camera system will be fitted on top of a vehicle that moves forward, and we count multiple false alarms that are on the same horizontal line of the measurement area as one false alarm. This is because the vehicle must stop only once if one or more false alarms are detected on the same horizontal line in front of the vehicle.

In order to evaluate the performance of the results of the tests, a tool was developed to construct the ground truth of the recorded sequences. Figure 5.2 shows the graphical user interface for the tool. In the left window of the tool a visualization of the recorded polarization sequence is shown. Within that window we can mark the region of interest by drawing a boundary box, which appears as a shaded grey area on the right. Detections and/or false

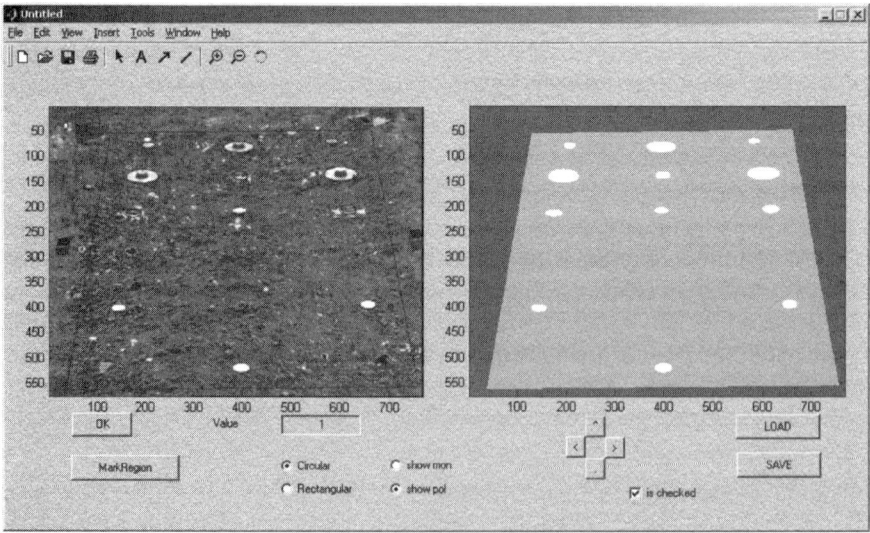

Figure 5.2. Performance evaluation tool to make ground truth of measurement areas. The left panel shows the calculated LP image (Eq. (5.7)). The light-grey area, which is roughly 3 m wide and 6 m deep, in the right panel is used in the detection step. The white circles give the ground truth.

alarms outside that box are not taken into account for the results of that sequence. The ground-truth mine positions can be entered by placing a circular or rectangular object on the mines on the left. The size of the mine can be altered and the ground-truth mines appear in white on the right.

5.5 Test Results

This section presents detection results on recordings made in the late summer period on the test field described in the previous section.

In the following subsections the detection results for different example data sets are presented. The examples are given to show the advantage of using polarization and to assess the influence of several factors (as mentioned in the previous section) to the detection performance as well as to validate the landmine-detection concepts.

5.5.1 Polarization Versus Intensity

In this subsection we look at the difference in detection results obtained on a data set that was recorded under ideal recording conditions: completely clear sky during the whole day, with a shortwave irradiance that continuously increases from 300 W/m^2 at 9:00 till 750 W/m^2 at 14:00 local time. Figure 5.3 shows two ROC curves obtained from the complete data set (all hours, all

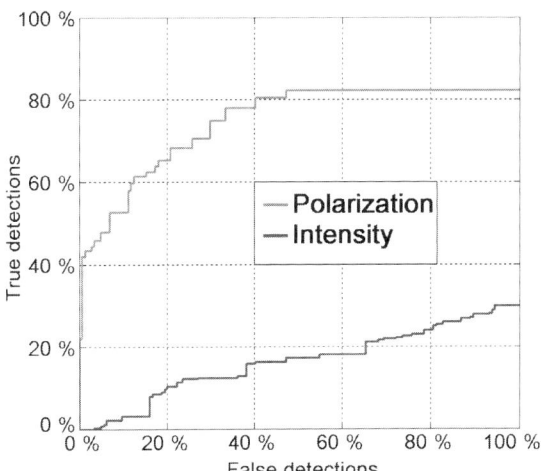

Figure 5.3. Two ROC curves obtained from one data set that show a clear difference in detection performance when using polarization features. In the bottom curve only intensity features are used for detection. In the top curve only polarization features are used (see color insert).

mines, all viewing directions). In the bottom curve only positive intensity contrast has been used for detection. In the top curve only the amount of linear polarization has been used. From this figure we may conclude that polarization is the most important feature for detection. Although detection based on intensity only has very low performance, the combination of polarization and intensity features generally results in a better ROC. For this specific data set, fusion of both features gives an increase in true detections of not more than 5%.

5.5.2 Position of the Sun

In this subsection we assess the influence of the relative position of the sun on the performance of the system. Again a data set has been used that was recorded under ideal conditions: completely clear sky during the whole day, with a shortwave irradiance that continuously increases from 250 W/m^2 at 9:00 till 800 W/m^2 at 14:00 local time. For each individual measurement hour and each individual mine type an ROC was calculated. This results in 4 ROC curves per measurement hour. Subsequently in each of those ROC curves the percentage of true detections at a false-alarm rate of 3% was chosen. Figure 5.4(a) shows this percentage of true detections versus the time of measurement for the four mine types. ROC curves were also calculated for each viewing direction (and all mines types combined, with the exclusion of the P2 mine type). Figure 5.4(b) shows the percentage of true detections versus the time of measurement for the four viewing directions. The horizontal dashed lines in the two graphs are the detection results for the whole measurement period (instead of the result of 1 hour) per mine type (in Fig. 5.4(a)) and per viewing direction (in Fig. 5.4(b)), also with a false-alarm rate of 3%. From Fig. 5.4(a) we may conclude that detection results per mine type do not differ much, except for the P2 mine, that shows a very low performance. This effect can be explained by the fact that the P2 mine has no flat top surface whereas the other mines have. As a result this specific mine type cannot be detected very well using polarization features.

Furthermore, Fig. 5.4(a) clearly shows the dependency between the recording hour and the detection rate. For all mines (except the P2 mine type) the detection graph shows similar behavior. At 9:00 to 10:00 hours we have top performance, at 12:00 hours the graph shows a local minimum, and after 12:00 hours there is increased performance again, with the exception of 15:00 hours. This indicates that there is a relationship between the relative position of the sun and the detection results, given that the recording conditions were ideal.

From Fig. 5.4(b) we may conclude that three out of four viewing directions perform better than 50% detection with a false-alarm rate of 3% when we look at the detection results for the whole measurement period (horizontal dashed lines). However, looking at the results per hour, there are some

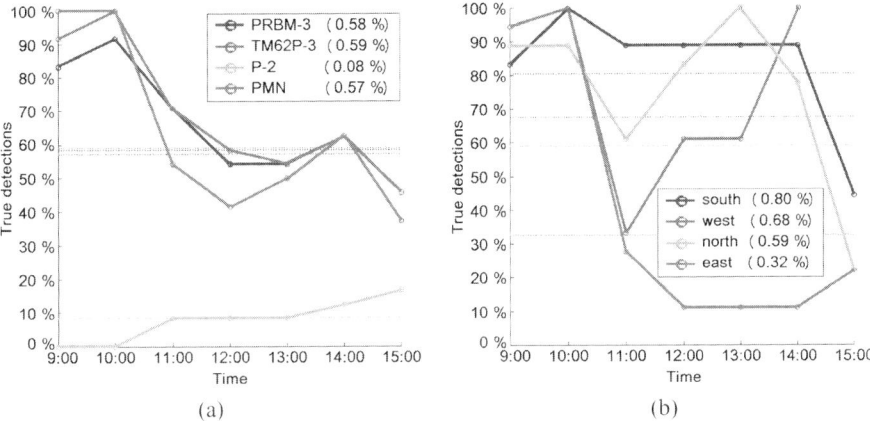

Figure 5.4. (a) Detection rate vs. time of measurement for the four mine types. (b) Detection rate vs. time of measurement for the four viewing directions. The results for all mine types with the exclusion of the P2 mine type are shown. In both figures a false-alarm rate of 3% was chosen. The horizontal dashed lines in the figures are the detection results for the whole measurement period (instead of the results of 1 hour) per mine type (a) or per viewing direction (b) (see color insert).

significant variations. This indicates that different parameter settings are used for different hours which do not closely match the global settings for the whole period. From a detailed inspection of the recorded images, we may conclude that a viewing direction directly toward the sun results in the highest polarization contrast of the mines. When the viewing direction is along with the sun, the polarization contrast of the mines is lowest.

5.5.3 Influence of Clouds

In this section we look at the detection results for a data set that was recorded under cloudy conditions. The shortwave irradiance changes between 200 W/m^2 and 800 W/m^2 on a time scale of only a few minutes. With the detection results of this data set we assess the influence of cloudy weather on the performance of the system.

Figure. 5.5(a) presents the detection results for the four mine types. With the exception of the P2 mine the detection results are good. Furthermore, the variation in the results is significant only in the morning and at the beginning of the evening. When we compare the results with the results for the four mine types recorded under a completely clear sky (Fig. 5.4(a)) we can see that there is no drop in performance during the middle of the day. Also the results for the whole period (dashed horizontal lines) are significantly higher.

Figure. 5.5(b) presents the detection results for the four viewing directions. For all viewing directions the results for the whole period are above the

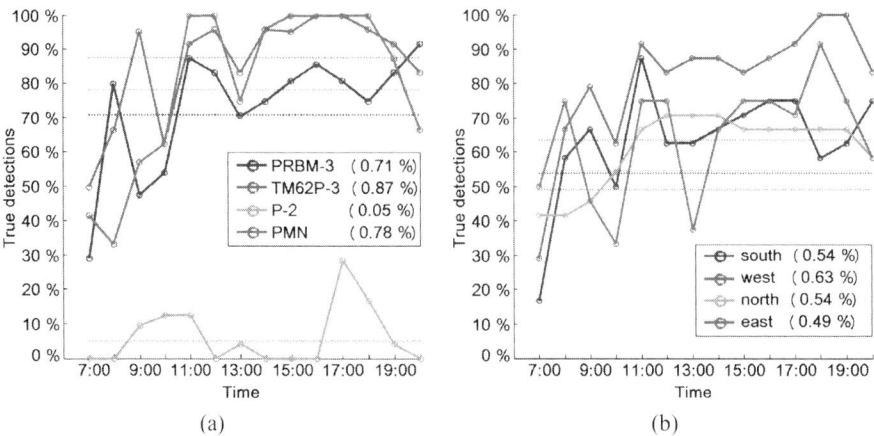

Figure 5.5. (a) Detection rate vs. time of measurement for the four different mine types. (b) Detection rate vs. time of measurement for the four viewing directions. A false-alarm rate of 3% was chosen (see color insert).

required 50% detection rate. Note that these results include also the P2 mine which explains the somewhat lower performance in Fig. 5.5(b) when compared to the dashed lines in Fig. 5.5(a) and Fig. 5.5(b). Furthermore, the variation in the results is significant only in the morning and at the beginning of the evening, with the exception of the East direction. When we compare these results with the results for the four viewing directions recorded under a completely clear sky (Fig. 5.4) we can see that there is far less variation between the four viewing directions. This probably stems from the fact that under cloudy conditions a more uniform light source, which varies less with viewing direction, illuminates the mines. Under the clear sky conditions the sun as a point source gives the highest contribution to the illumination.

5.5.4 Surface Orientation

This section describes the results of tests performed using the camera system on tilted mines. The tests were conducted to assess the influence of the mine orientation on polarization results when compared with mines that are not tilted The tests were divided into three parts:

– Mines tilted away from the camera;
– Mines tilted toward the camera; and
– Mines tilted to the side.

In all three situations a tilt angle of 10 degrees was used. Only the mines in the 1st and 3rd column of the four measurement areas were tilted (see Table 5.1). The mines in the inner column remained horizontal to serve as a reference point.

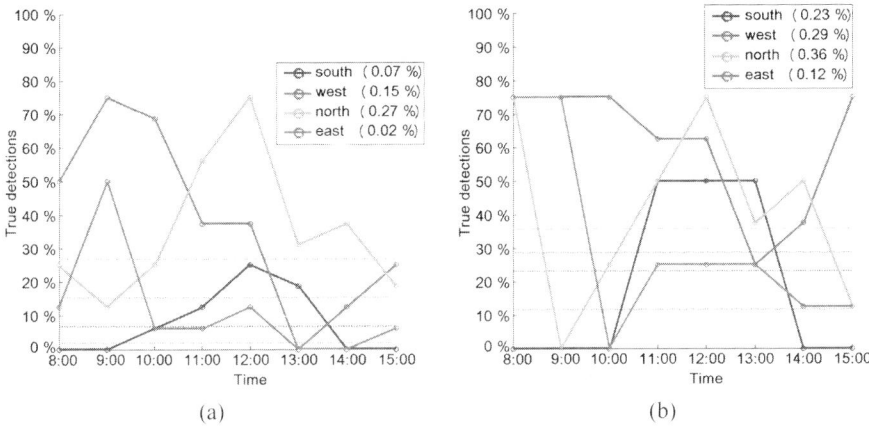

Figure 5.6. Detection rate vs. time of measurement for the four viewing directions. A false-alarm rate of 3% was chosen. (a) Results for mines tilted away from the camera. (b) Results for the nontilted mines (see color insert).

5.5.4.1 Results for Mines Tilted Away From Camera

Figure 5.6 presents the detection results for mines tilted away from the camera (Fig. 5.6(a)) and nontilted mines (Fig. 5.6(b)) for the four viewing directions. Comparison of the dashed lines shows that for each viewing direction the detection results of the mines tilted away from the camera are at least 10% lower than the detection results of the horizontal mines.

5.5.4.2 Results for Mines Tilted Toward the Camera

Figure. 5.7 presents the detection results for mines tilted toward the camera (Fig. 5.7(a)) and nontilted mines (Fig. 5.7(b)) for the four viewing directions. Comparison of the dashed lines shows that detection results for mines tilted toward the camera only deteriorated for the North viewing direction when compared with the horizontal mines of the same data set. The other viewing directions showed improved performance. The East viewing direction performs poorly for tilted and nontilted mines because of shadows in the morning imagery and background clutter in the recordings later on the day.

5.5.4.3 Results for Mines Tilted Along the Viewing Direction

Figure 5.8 presents the detection results for mines tilted along the viewing direction (Fig. 5.8(a)) and nontilted mines (Fig. 5.8(b)) for the four viewing directions. Comparison of the dashed lines shows that except for the north direction detection results deteriorate for tilted mines when compared with

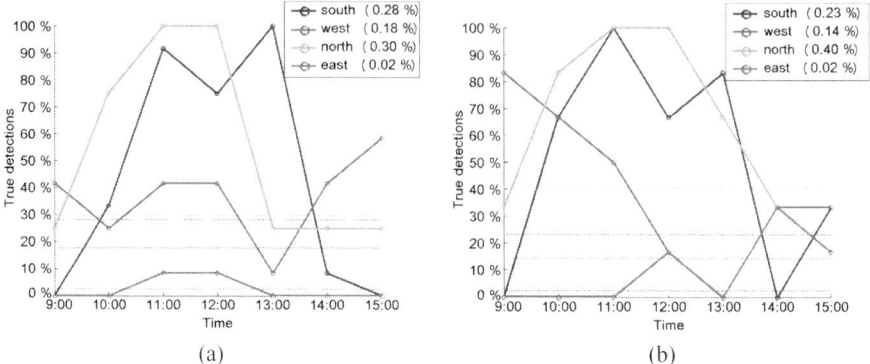

(a)   (b)

Figure 5.7. Detection rate vs. time of measurement for the four viewing directions. A false-alarm rate of 3% was chosen. (a) Results for mines tilted toward the camera. (b) Results for the nontilted mines (see color insert).

the nontilted mines. However, the results per hour of the west direction do not differ much.

5.6 Conclusions

In this chapter we have described the usage of polarization features of visible light for automatic landmine detection.

Given the results of the conducted experiments, the following conclusions can be drawn:

– Combination of intensity and polarization gives the best performance of an automatic landmine detection system.

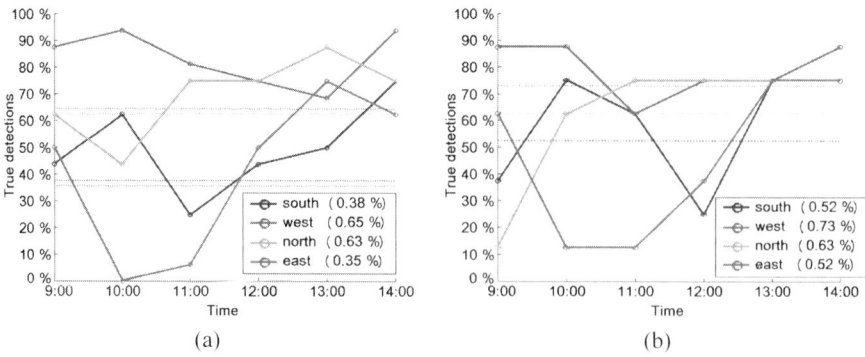

(a)   (b)

Figure 5.8. Detection rate vs. time of measurement for the four viewing directions. A false-alarm rate of 3% was chosen. (a) Results for mines tilted along the viewing direction. (b) Results for the nontilted mines (see color insert).

– Performance depends on viewing direction, relative to the position of the sun. Looking toward the sun (i.e., when the sun is in a sector of ±45 degrees when looking in the forward motion direction) seems to be the best viewing direction. This is illustrated by the results for some of the data sets from the West viewing direction (see, for example, the magenta curve in Fig. 5.4(b)). Note that for the West viewing direction the background clutter is far less (grass field) when compared with the other directions, which have debris as background clutter.

– The East viewing direction performs less well because of the position of the sun relative to the viewing direction during most measuring hours.

– Performance depends on the time of day and shows large variations, which may be partly due to weather conditions, shadows, and a fixed camera iris. A camera with a larger dynamic range could solve part of these variations.

– Performance depends on the type of mine: detection results do not differ much for the mines used in the test, with the exception of the P2 mine that shows a very low performance. This can be explained by the fact that the P2 mine has no flat top surface and, as such, cannot be detected very well using polarization.

– Detection of tilted mines proves to be more difficult, except for mines tilted toward the camera.

References

[1] B. Gros and C. Bruschini. Sensor technologies for the detection of antiperson-nel mines: A survey of current research and system developments. In *Proceedings of the International Symposium on Measurement and Control in Robotics (ISMCR'96)*, pp. 509–518, Brussels, Belgium, May 1996.

[2] F. Cremer, K. Schutte, J. G. M. Schavemaker, and E. den Breejen. A comparison of decision-level sensor-fusion methods for anti-personnel landmine detection. *Inform. Fusion* 2: 187–208, Sept. 2001.

[3] J. G. Schavemaker, E. den Breejen, K. W. Benoist, K. Schutte, P. Tettelaar, M. de Bijl, P. J. Fritz. LOTUS field demonstration of integrated multisensor mine detection system in Bosnia. In *Proc. SPIE Vol. 5089, Detection and Remediation Technologies for Mines and Minelike Targets VIII*, R. S. Harmon, J. T. Broach, and J. John H. Holloway, eds., Orlando, FL, April 2003.

[4] E. Hecht. *Optics*, Addison-Wesley Publishing Company, Reading (MA), USA, ed., 1987.

[5] F. Cremer, W. de Jong, and K. Schutte. Infrared polarisation measurements and modelling applied to surface laid anti-personnel landmines. *Opt. Eng.* 41: 1021–1032, May 2002.

[6] F. Cremer, W. de Jong, K. Schutte, A. Yarovoy, V. Kovalenko, and R. Bloemenkamp. Feature level fusion of polarimetric infrared and gpr data for land-mine detection. In *International Conference on Requirements and Technologies for the Detection, Removal and Neutralization of Landmines and UXO*, H. Sahli, A. Bottoms, and J. Cornelis, eds., pp. 638–642, Brussels, Belgium, Sept. 2003.

[7] F. Cremer. *Polarimetric Infrared and Sensor Fusion for the Detection of Land-mines.* PhD dissertation, Technische Universiteit Delft, Delft, Nov. 2003.

[8] *Technical Note 09.50/01; Guide to mechanical mine clearance/ground prepara-tion using commercial tractors and front loaders.* http://www. mineactionstan-dards.org/tnma_list.htm, 2002. Version 1.0.

[9] J. G. M. Schavemaker, F. Cremer, K. Schutte, and E. den Breejen. Infrared pro-cessing and sensor fusion for anti-personnel land-mine detection. In *Proceedings of IEEE Student Branch Eindhoven: Symposium Imaging*, pp. 61–71, (Eindhoven, the Netherlands), May 2000.

[10] M. Swain and D. Ballard. Color indexing. *International. Journal of Computer Vision* 7: 11–32, Jan. 1991.

[11] W. A. C. M. Messelink, K. Schutte, A. M. Vossepoel, F. Cremer, J. G. M. Schavemaker, and E. den Breejen. Feature-based detection of landmines in in-frared images. In *Proc. SPIE Vol. 4742, Detection and Remediation Technolo-gies for Mines and Minelike Targets VII*, J. T. Broach, R. S. Harmon, and G. J. Dobeck, eds., pp. 108–119, Orlando, FL, April 2002.

[12] C. F. Olson. Adaptive-scale filtering and feature detection using range data. *IEEE Trans. on Pattern. Anal. Mach. Intell.* 22, Sept. 2000.

[13] T. Rogne, S. Stewart, and M. Metzler. Infrared polarimetry: what, why, how and the way ahead. In *Proc. of Third NATO IRIS Joint Symposium*, pp. 357–368, Quebec, Canada, Oct. 1998.

[14] K. P. Bishop, H. D. McIntire, M. P. Fetrow, and L. McMackin. Multi-spectral polarimeter imaging in the visible to near IR. In *Proc. SPIE Vol. 3699, Tar-gets and Backgrounds: Characterization and Representation V*, W. R. Watkins, D. Clement, and W. R. Reynolds, eds., pp. 49–57, Orlando, FL, April 1999.

[15] M. H. Smith. Optimizing a dual-rotating-retarder Mueller matrix polarimeter. In *Proc. SPIE Vol. 4481, Polarization and Remote Sensing IV*, W. G. Egan and M. J. Duggin, eds., pp. 31–36, San Diego, CA, July 2001.

[16] F. A. Sadjadi and C. S. L. Chun. Application of a passive polarimetric infrared sensor in improved detection and classification of targets. *Int. J. Infrared Mil-limeter Waves* 19(11) pp. 1541–1559, 1998.

[17] G. P. Nordin, J. T. Meier, P. C. Deguzman, and M. W. Jones. Micropolarizer array for infrared imaging polarimetry. *J. Opt. Soc. Am. A* 16:1168–1174, May 1999.

The Physics of Polarization-Sensitive Optical Imaging

Cornell S. L. Chun

Physics Innovations Inc.
P.O. Box 2171
Inver Grove Heights, MN 55076–8171

6.1 Introduction

This volume addresses the fundamental physical bases of sensing and information extraction in state-of-the-art automatic target recognition systems. The physical basis of sensing light and its properties of intensity, coherence, and wavelength are discussed elsewhere in this book. In this chapter, another property of light, polarization, will be described. All light, natural or man-made, is polarized to some degree. Light from a target contains information in its polarization state. Often this information relates to the three-dimensional shape of the target and is independent of the intensity and wavelengths radiated. Intensity and wavelengths can vary with the illumination source, the temperature of the target, and the surface coating on the target. However, for most cases of interest the three-dimensional shape of the target will remain unchanged.

In order to use this information in automatic target recognition systems, three technical capabilities must be developed: (1) an understanding of how light polarization is altered when scattered from or transmitted through objects, (2) development of imaging sensors to capture the polarization of light from targets, and (3) development of algorithms which use the captured polarization information to aid in automatic target recognition. The latter two capabilities are currently being discussed in the literature and will not be covered here. This chapter will describe how objects alter the polarization of light and what polarization states should be expected for light from targets when viewed using an polarimetric imaging sensor.

First a description of the polarization state of incoherent light will be given in terms of the Stokes vector. When light interacts with matter, the transformation from the input to output polarization states is given in terms of the Mueller matrix. A description will then be given of a common occurrence that changes the polarization, the reflection and transmission at a planar air/dielectric interface. Reflection and transmission are described by the Fresnel equations. Finally, examples will be given where polarization information captured during the imaging of objects can be used to aid automatic target recognition.

6.2 Stokes Vectors and Mueller Matrices

A beam of incoherent radiation emitted or reflected from a target's surface can be completely described at a given wavelength by the four Stokes parameters I, Q, U, and V [1]. The first Stokes parameter I is a measure of the total intensity of radiation. The second parameter Q measures the amount of linear polarization in the horizontal direction. The third parameter U measures the amount of linear polarization in a plane rotated $45°$ from the horizontal. The fourth parameter V is associated with the circular polarization.

The first three Stokes parameters can be transformed into degree of linear polarization P and angle of polarization ϕ using the relations [2],

$$P = \frac{\sqrt{Q^2 + U^2}}{I}$$

$$\phi = \frac{1}{2} \cdot \arctan\left(\frac{U}{Q}\right) \tag{6.1}$$

In Section 6.3, P and ϕ will be shown to be related to the orientation of the surface element of an object radiating the light.

The four Stokes parameters together form the Stokes vector $\mathbf{S} = (I, Q, U, V)$. Upon reflection from or transmission through an object, the change in the input Stokes vector $\mathbf{S}_{in}$ is described by a 4×4 matrix $\mathbf{M}$ called the Mueller matrix [1]. The output Stokes vector $\mathbf{S}_{out}$ is equal to $\mathbf{M} \cdot \mathbf{S}_{in}$.

6.3 Target Shape Information from Polarization Imaging

6.3.1 Reflection at an Air/Dielectric Interface

6.3.1.1 Polarization from Reflection—Theory

The relationship between the polarization characteristics of light and surface orientation can be demonstrated using the common experience of polarized reflection. When sunlight is incident at a grazing angle on the horizontal surface of water the portion of the light which is reflected is partially linearly polarized with a horizontal plane of polarization. (Polarizers in sunglasses are oriented to attenuate horizontally polarized light.) Similarly, sunlight which

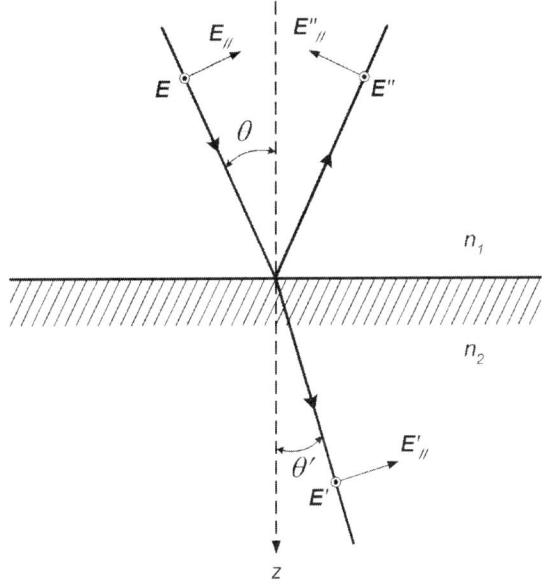

Figure 6.1. The reflection and transmission of light at an interface.

is reflected from a vertical glass window on the side of a building is vertically linearly polarized.

A theoretical description of the polarization of light upon reflection from a planar interface is given by the Fresnel equations. The derivation described here will follow the notation in [3]. Fig. 6.1 shows the reflection and transmission of light at a boundary between media of index of refraction n_1 and n_2. The z-axis is normal to the interface. Negative z is in the n_1 region, and positive z in the n_2 region. The x and y axes are in perpendicular and parallel to the scattering plane, respectively. The electric field vectors are defined as shown in the figure.

When there is a planar interface between air $n_1 = 1$ and an absorbing dielectric with a complex index of refraction $n_2 = n - i\kappa$, the reflectivity becomes

$$\rho_{||}(\theta) = \frac{(n \cos\theta - 1)^2 + (\kappa \cos\theta)^2}{(n \cos\theta + 1)^2 + (\kappa \cos\theta)^2} \tag{6.2a}$$

and

$$\rho_{\perp}(\theta) = \frac{(n - \cos\theta)^2 + \kappa^2}{(n + \cos\theta)^2 + \kappa^2} \tag{6.2b}$$

for incident light linearly polarized parallel and perpendicular to the plane of incidence, respectively [4]. The angle of incidence is θ, i.e., the angle between the incident light and the normal to the interface. When incident light is

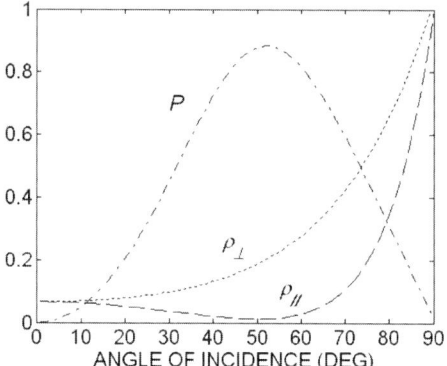

Figure 6.2. Variation with angle of the reflectivity and degree of polarization.

unpolarized, the reflected light has a degree of polarization given by

$$P(\theta) = \frac{\rho_\perp(\theta) - \rho_\parallel(\theta)}{\rho_\perp(\theta) + \rho_\parallel(\theta)}. \qquad (6.3)$$

If the incident light is unpolarized and the sensor is not sensitive to polarization, then the measured reflectivity is

$$\rho(\theta) = \frac{\rho_\perp(\theta) + \rho_\parallel(\theta)}{2} \qquad (6.4)$$

$\rho_\parallel$, $\rho_\perp$, and P are plotted in Fig. 6.2 for the case where the complex index of refraction is $1.5 + i0.15$, which is representative of paint in the wavelength range, 3 to 5 μm, the mid-wave infrared region [5].

Figure 6.2 shows that $\rho_\perp > \rho_\parallel$ for all angles θ except at normal incidence. When the incident light is unpolarized, the reflected light is linearly polarized perpendicular to the scattering plane. This corresponds to the common observation of polarized reflection of the sun off planar surfaces as mentioned above. The scattering plane contains the normal vector to the surface element. Furthermore, the maximum degree of polarization P occurs when viewing the surface at a grazing angle, similar to the Brewster angle. These are constraints on the direction of the normal vector within an image pixel which can be used to aid in recognition of the target from its three-dimensional shape.

6.3.1.2 Polarization from Reflection—Examples

This section discusses an example of how sensing the polarization of reflected light can aid target detection and recognition. Section 6.3.1.1 described how light, which is reflected from a planar surface, is linearly polarized perpendicular to the scattering plane. An example of polarized reflection giving cues of the three-dimensional shape of target is shown in Fig. 6.3 [6]. In these images, large values of the degree of polarization P occur in regions where the smooth

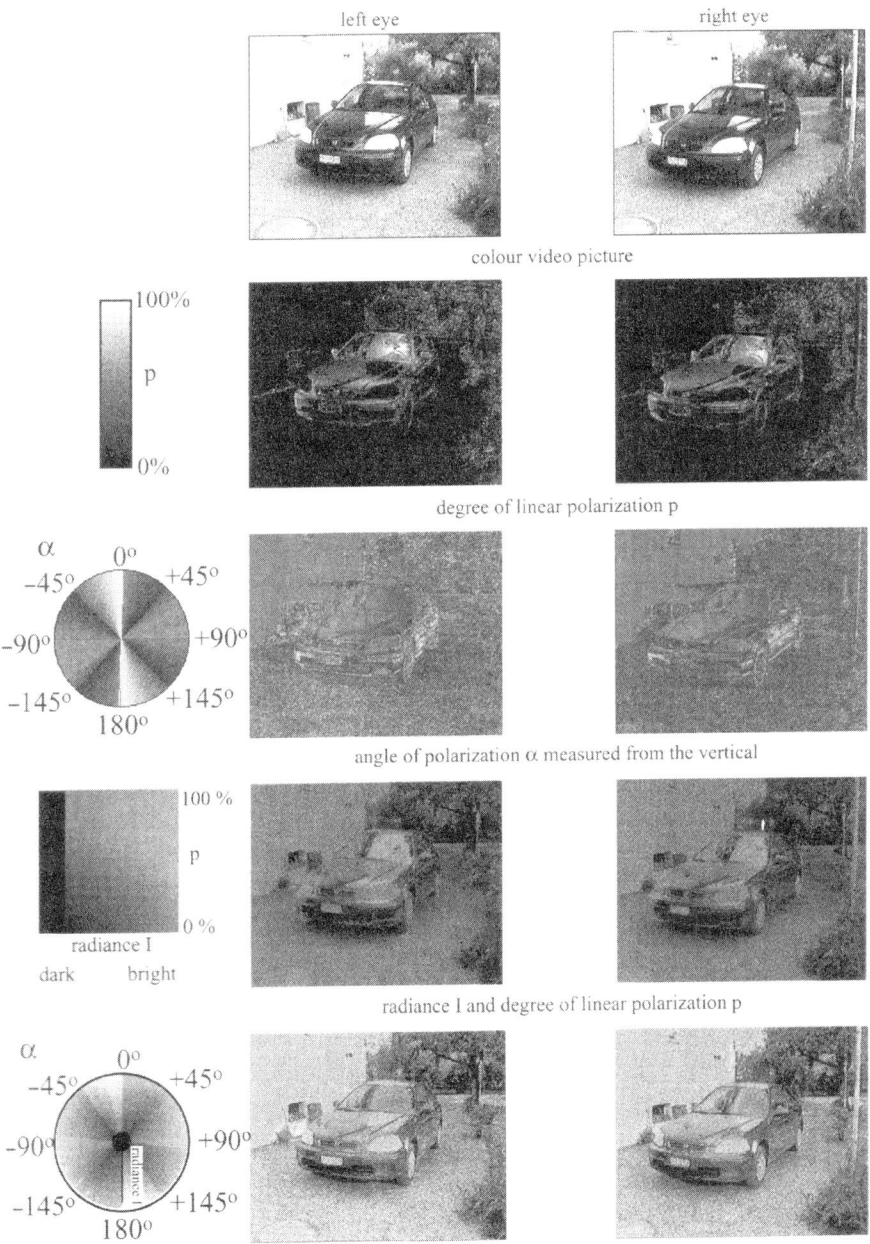

Figure 6.3. Polarization images of a scene with reflected light (see color insert). [Reprinted with permission of IS&T: The Society for Imaging Science and Technology sole copyright owners of The Journal of Imaging Science and Technology.]

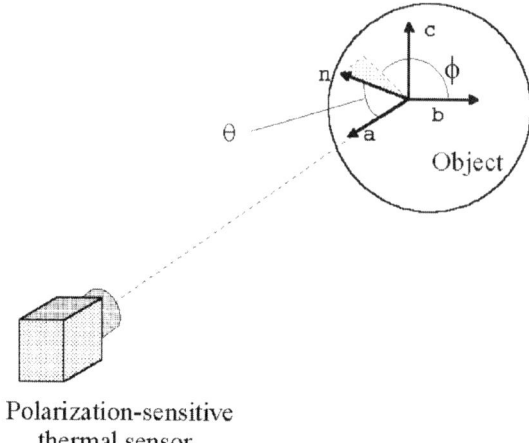

Polarization-sensitive
thermal sensor

Figure 6.4. Thermal radiation from a surface on an object is received by a polarization-sensitive sensor.

surface of the automobile is oriented at a grazing angle to the camera. Furthermore, in regions where P is large, the angle α of the polarization plane is perpendicular to the normal vector of the surface.

6.3.2 Thermal Emission

6.3.2.1 Polarization from Thermal Emission—Theory

The previous section described how unpolarized light, when incident at a grazing angle on the surface of water, is reflected with horizontal linear polarization (Fig. 6.2). The portion of the light, which is transmitted into the water is linearly polarized with a vertical plane of polarization. By the same mechanism, if light originates below the surface and passes through the surface, the transmitted light would be linearly polarized with a vertical polarization plane. The polarization plane is identical to the scattering plane and contains the normal vector to the surface. The transmitted light when exiting at a more grazing angle to the surface will have a greater degree of polarization.

Thermal emission is the light originating below the surface. Referring to Fig. 6.4, when thermal radiation is received by a polarization-sensitive sensor from a surface element on an object, the plane of polarization gives the azimuth angle ϕ and the degree of polarization P can be related to the polar angle θ. Together θ and ϕ determine the normal vector $\mathbf{n}$ to the surface element.

The relationship between P and θ can be found using the Fresnel equations [Eqs. (6.2)] for light reflection at a planar interface and Kirchhoff's law relating the emitting and absorbing properties of a body, $\varepsilon(\theta) = 1 - \rho(\theta)$ [4].

Define the plane of emission as the plane containing the normal vector $\mathbf{n}$ to the surface element and the vector $\mathbf{a}$ from the surface element to the sensor (Fig. 6.4). Using these equations, the emissivity $\varepsilon_{||}$ for radiation linearly polarized parallel to the plane of emission and the emissivity $\varepsilon_\perp$ for radiation linearly polarized perpendicular to the plane of emission are given by [4]

$$\varepsilon_{||}(\theta) = \frac{4n\,\cos\theta}{[(n^2+\kappa^2)\cos^2\theta]+[2n\,\cos\theta]+1} \tag{6.5a}$$

$$\varepsilon_\perp(\theta) = \frac{4n\,\cos\theta}{\cos^2\theta+[2n\,\cos\theta]+n^2+\kappa^2}, \tag{6.5b}$$

where θ is the angle of emission, i.e., the angle from the surface normal $\mathbf{n}$ to the emission direction $\mathbf{a}$. The thermal radiation is assumed to be traveling from an absorbing material with complex index of refraction $n_2 = n - i\kappa$ through the surface into air with $n_1 = 1$. The degree of polarization P is given by

$$P = \frac{\varepsilon_{||}(\theta) - \varepsilon_\perp(\theta)}{\varepsilon_{||}(\theta) + \varepsilon_\perp(\theta)} \tag{6.6}$$

If the sensor is not sensitive to polarization, then the measured emissivity will be

$$\varepsilon(\theta) = \frac{\varepsilon_{||}(\theta) + \varepsilon_\perp(\theta)}{2} \tag{6.7}$$

Using the same complex index of refraction as in the Section 6.3.1.1, $\varepsilon_{||}$, $\varepsilon_\perp$, and P can be calculated and are shown in Fig. 6.5. Examples of P and θ images of objects captured using polarized thermal emission sensing are shown in the next section.

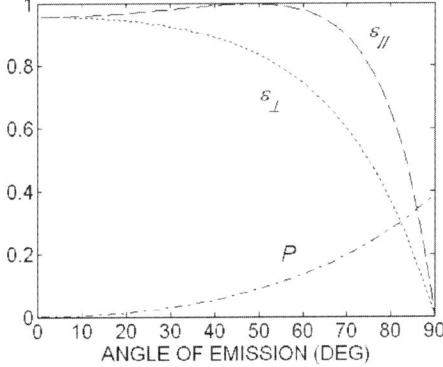

Figure 6.5. Variation with angle of the directional emissivity and degree of polarization.

6.3.2.2 Polarization from Thermal Emission—Examples

Section 6.3.2.1 described how thermally emitted light, which originates below the surface of an object and passes through the surface, will be linearly polarized with a vertical polarization plane. The polarization plane contains the normal vector to the surface (Fig. 6.4). The transmitted light when exiting at a more grazing angle to the surface will have a greater degree of polarization P (Fig. 6.5).

Figure 6.6 shows polarimetric images of a cube captured using a polarization-sensitive thermal imaging sensor [7]. The cube is made of solid metal painted a flat black and is heated internally. The I image corresponds closely with the image expected from intensity, i.e., a forward-looking infrared (FLIR) image. From the above discussion on polarized emission, the degree of polarization P should be larger when viewing a surface at a more grazing angle (Fig. 6.5). The top of the cube in the P image in Fig. 6.6 is brighter than the sides which face the viewer. (In the P image, the bright pixels that surround the cube are artifacts of the imaging technique [7].) In the image of the angle of the plane of polarization ϕ, the gray levels are chosen as follows. Where the surface emits light with a polarization plane such that the electric field vector E points in the 3 o'clock direction, the pixel is white. Where the E vector points in the 12 o'clock direction, the pixel is mid-gray. Where the E vector points to 9 o'clock, the pixel is black. The above discussion described how, for thermal emission, the plane of polarization contains the normal vector to the surface. The ϕ image clearly shows that the right-hand side of the cube, where the normal vector **n** points to the right, is white, the upper surface, where **n** points up, is mid-gray, and the left-hand side, where **n** points to the left, is black. The two parameters, P and ϕ, are directly related to the two angles which determine surface orientation. From the P and ϕ images in Fig. 6.6 the three-dimensional shape of the target can be derived.

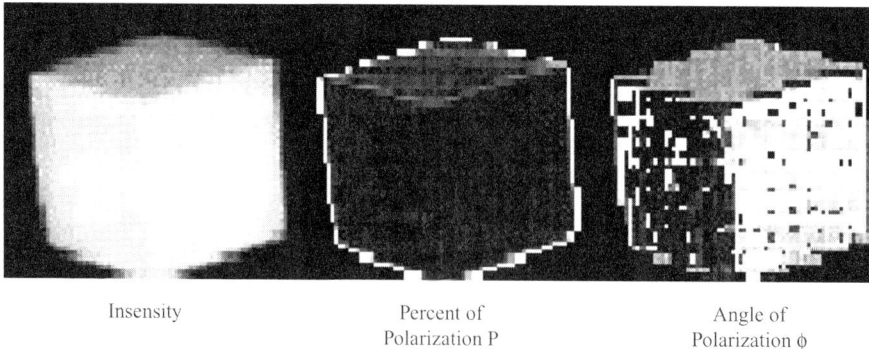

Insensity   Percent of   Angle of
                             Polarization P   Polarization φ

Figure 6.6. I, P, and ϕ images of a heated cube.

Figure 6.7. High-resolution polarization images of a scale model of an aircraft.

Another set of polarimetric images with greater spatial resolution and a target with more facets is shown in Fig. 6.7 [8]. The target is a die-cast metal scale model of an F-117 stealth aircraft. The model is painted a flat black and heated above the temperature of the background. In this figure, the three-dimensional shape and orientation of the target is readily apparent. In the ϕ image, surfaces on the aircraft facing towards the viewer's upper right are dark gray, and surfaces towards the upper left are light gray.

6.3.3 Circular Polarization

6.3.3.1 Advantages of Imaging Circular Polarization

The previous section describes how thermal emission generates linearly polarized radiation. When linearly polarized light is reflected from an air/dielectric interface, circularly polarized light is generated [1]. It is surprising then, that imaging systems that detect the circular polarization of light have not been developed for automatic target identification. The sign and magnitude of the circular polarization can often reveal the spatial orientation, material, and surface roughness of the target surface [9]. In addition, since very few natural materials can generate circular polarized light, such a sensor would be able to detect man-made targets in a background free of clutter.

6.3.3.2 Polarization Ray Tracing

The section describes the simulation, by ray tracing, of the polarization images of an idealized terrestrial scene. In this scene, shown in Fig. 6.8, the sensor is located 1 m above the surface of the ground. The ground has a planar surface. The target is a sphere of diameter 0.308 m and the center of the sphere is 1 m above the surface of the ground. The ground and sphere are dielectric with a complex index of refraction of $1.5 + i0.15$, which is representative of paint in the mid-wave infrared wavelength range, 3 to 5 μm [5]. The surfaces of the sphere and ground are smooth and transmission and reflection are described

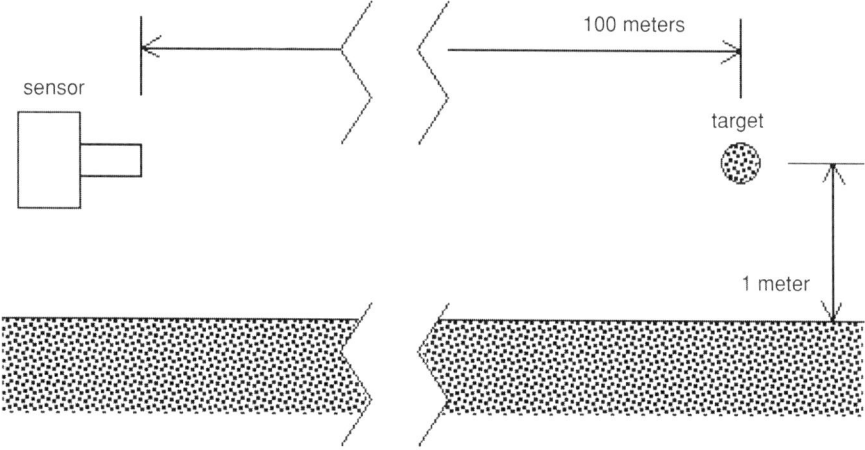

Figure 6.8. Idealized terrestrial scene for ray tracing images: Sphere above ground.

by Fresnel equations. All surfaces are at the temperature 300 K. The ray tracing simulates the effects of polarized thermal emission and the effects on intensity and polarization components as light is reflected and refracted at interfaces.

Using polarization ray-tracing, polarimetric images of size 64 × 64 pixels of the target were generated. Images of the Stokes parameters I, Q, U, and V are shown in Fig. 6.9(a). From these Stokes parameters, the degree of linear polarization (DOLP) and degree of circular polarization (DOCP) [Fig. 6.9(b) and (c)] can be calculated using

$$\text{DOLP} = \frac{\sqrt{Q^2 + U^2}}{I} \quad \text{and} \quad \text{DOCP} = \frac{V}{I} \tag{6.8}$$

These polarimetric images show that there are distinctive regions on the target sphere, referred to as Regions 1, 2, and 3 in Fig. 6.9(d). In Region 1, the upper half of the sphere, thermal radiation is emitted which the DOLP image shows as being linearly polarized. The Stokes I image, or intensity image, shows that the upper half of the sphere contributes less radiation than the lower half of the sphere. The lower half can be divided into two regions. Region 2 is the area in the lower half of the sphere away from the limb, and Region 3 is the area near the limb. Radiation from both areas is dominated by reflection of thermal emission from the ground plane. Radiation, coming from Region 2, originated as thermal emission from the ground in front of the sphere, i.e., the ground between the sensor and the target sphere. From the point of view of the sensor viewing the sphere, light is linearly polarized over most of this area. In this area light is also circularly polarized, especially near the center of the sphere.

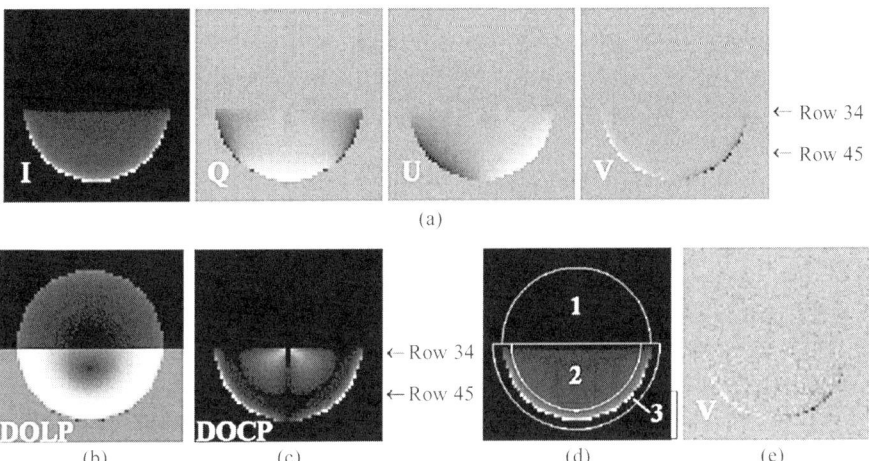

(a)

(b)   (c)   (d)   (e)

Figure 6.9. (a) Stokes parameter images of a sphere above a ground plane (Fig. 6.8). (b) Degree of linear polarization. (c) Degree of circular polarization. (d) On the target sphere image, three distinctive regions of polarization characteristics. (e) Stokes parameter V image at output of wide-band MWIR imaging sensor.

Light from Region 3, near the limb of the sphere, has significantly higher intensity and higher degree of circular polarization than the other two regions. Radiation, coming from this region, originated as thermal emission from the ground behind the sphere. This radiation reflects off the sphere at a grazing angle. Because the surface of the sphere is curved, a larger area on the ground contributes to light received by a pixel viewing near the limb than by pixels viewing Region 2. This emission from a larger ground area results in a higher intensity I for a pixels in Region 3. Furthermore, the grazing angle results in a larger conversion, upon reflection, of the linearly polarized ground emission into circularly polarized radiation.

Figure 6.10(a) is a plot of degree of circular polarization across two rows of pixels, rows 34 and 45 of the V image in Fig. 6.9(a). Row 34, plotted as ($\times$), is just below the equator of the sphere and includes pixels with the largest magnitude of V in Region 2. The plot of Row 34 shows that DOCP is largest near the center of the row with a value approximately 1.5%. Row 45, plotted as ($\lozenge$) is half way from the equator to the south pole of the sphere and includes pixels with the largest magnitude of V in Region 3. The plot of Row 45 shows DOCP is largest near the limb with a value approximately 2.5%.

6.3.3.3 Simulation of Sensor Output Images

For a polarimetric imaging sensor operating in the mid-wave infrared [9], the sensitivity was determined by simulating the image for circular

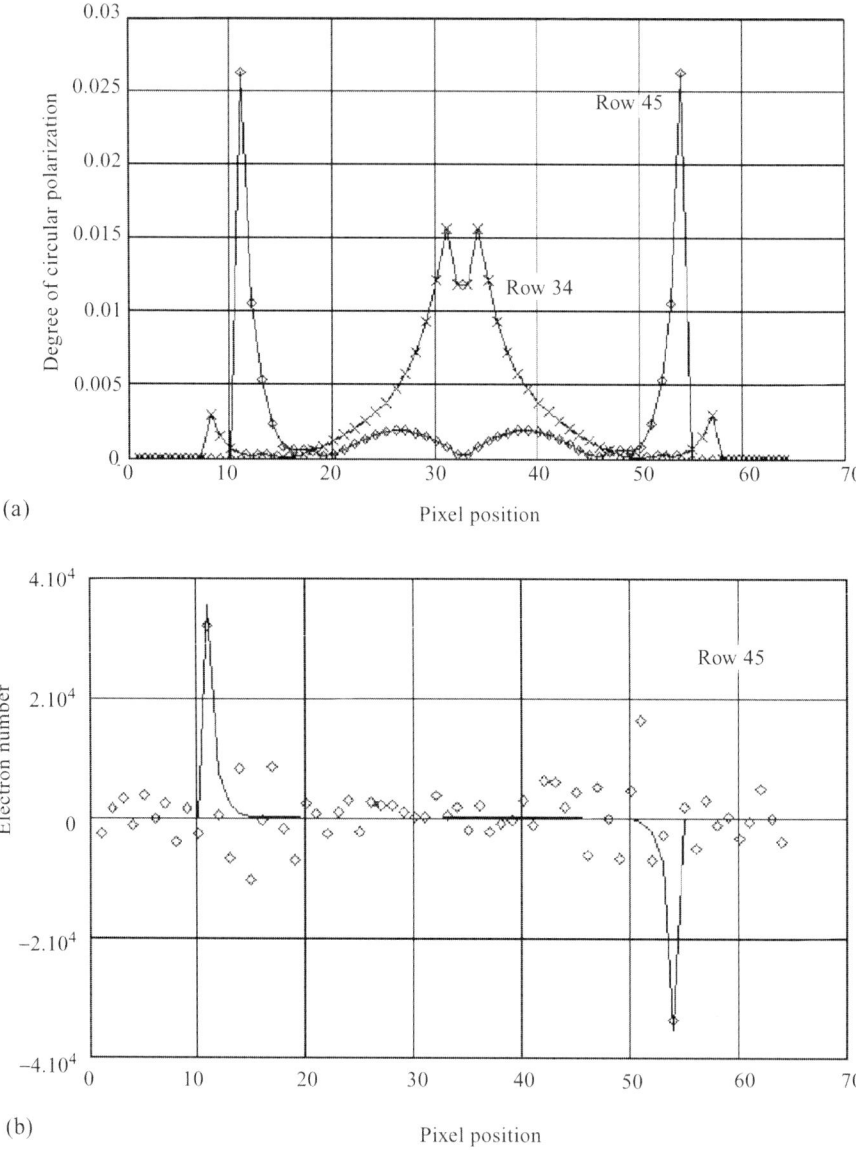

(a)

(b)

Figure 6.10. (a) Degree of circular polarization in rows 34 and 45 in the DOCP image. (b) Values of pixels in row 45 of stokes parameter V image (Fig. 6.9(e)) from a wide-band MWIR imaging sensor.

polarization that would appear at the output of the sensor. The output V image is shown in Fig. 6.9(e) and contains sensor noise. A plot of the value of circular polarization parameter V across row 45 is shown in Fig. 6.10(b). V is in units of electron numbers, which is the number of photoelectrons accumulated during an integration time at a pixel. In Fig. 6.10(b) the solid line is the true value of V, and the symbol ($\Diamond$) is the value of V measured by the sensor. Row 45 is half way from the equator to the south pole of the sphere and includes pixels with the largest magnitude of V in Region 3. The plot of Row 45 shows that the measured data reproduces V for pixels near the limb where V is largest. However, away from the limb, the noise is too large to determine V. By combining these observations with the DOCP plot in Fig. 6.10(a), we can conclude that the wide-band sensor is able to resolve Stokes parameter V when the degree of circular polarization is approximately 2% or greater.

6.4 Summary

This chapter describes the polarization state of light from targets. Polarization information often gives clues to the three-dimensional shape of the target and is independent of the intensity and wavelengths of light radiated from the target. Light which is polarized by a reflection from a surface element has a polarization state which give clues to the orientation of the surface. Thermally emitted light also has a polarization state which reveals the surface orientation. Polarization imaging gives information which can lead to signification improvements in automatic target recognition systems.

References

[1] D. Goldstein. *Polarized Light*, 2nd edn. Dekker, New York, 2003.
[2] C. S. L. Chun, D. L. Fleming, and E. J. Torok. Polarization-sensitive, thermal imaging. In F. A. Sadjadi (Ed.) *Automatic Object Recognition IV, Proceedings of SPIE*, vol. 2234, pp. 275–286, 1994.
[3] D. Clarke and J. F. Grainger. *Polarized Light and Optical Measurement*. Pergamon Press, Oxford, 1971.
[4] R. Siegel and J. R. Howell. *Thermal Radiation Heat Transfer*, 4th edn. Taylor & Francis, New York, 2002.
[5] T. W. Nee and S. M. F. Nee. Infrared polarization signatures for targets. In W. R. Watkins and D. Clement (Eds.) *Targets and Backgrounds: Characterization and Representation, Proceedings of SPIE*, vol. 2469, pp. 231–241, 1995.
[6] F. Mizera, B. Bernáth, G. Kriska, and G. Horvath. Stereo videopolarimetry: Measuring and visualizing polarization patterns in three dimensions. *J. Imaging Sci. Technol.*, 45:393–399, 2001.
[7] C. S. L. Chun, D. L. Fleming, W. A. Harvey, and E. J. Torok. Polarization-sensitive thermal imaging sensor. In B. F. Andresen and M. Strojnik (Eds.) *Infrared Technology XXI, Proceedings of SPIE*, vol. 2552, pp. 438–444, 1995.

[8] F. A. Sadjadi and C. S. L. Chun. Remote sensing using passive infrared Stokes parameters. *Opt. Eng.*, 43:2283–2291, 2004.

[9] C. S. L. Chun. Microscale waveplates for polarimetric infrared imaging. In B. F. Andresen and G. F. Fulop (Eds.) *Infrared Technology and Applications XXIX, Proceedings of SPIE*, vol. 5074, pp. 286–297, 2003.

7

Dispersion, Its Effects, and Compensation

Leon Cohen[1] and Patrick Loughlin[2]

[1]City University of New York, 695 Park Avenue, New York, NY 10021, USA
[2]University of Pittsburgh, 348 Benedum Hall, Pittsburgh, PA 15261, USA

7.1 Introduction

As a pulse propagates in a dispersive medium, many of the fundamental characteristics of the pulse change and therefore different observers at different locations/times will see a different pulse. Of particular importance, for example, are the duration and bandwidth of the pulse but these quantities change as the pulse evolves. It is, of course, important for different observers or sensors to be able to report that it is the same pulse with a common origin. That is, one has to compensate for dispersion if a recognizer is based on the characteristics of the signal at the source. Channel models that simulate such propagation effects can be effectively used for this purpose provided that the model is accurate, which requires detailed information about the propagation channel. In the absence of such information, or when it is inaccurate, other approaches are needed.

Our principal aim is to define moments of a wave that may be used as features for classification. We discuss various moment-like features of a propagating wave. In the first set of moments, namely spatial moments at a given time, we give explicit formulations that quantify the effects of dispersion. Accordingly, one can then compensate for the effects of dispersion on these moments. We also consider the dual case of temporal moments at a particular location. We then consider another class of mixed moments, which are invariant to dispersion and hence may be useful as features for dispersive propagation, particularly in situations where knowledge of the propagation environment is limited. These moments are based on a joint position-wavenumber (or analogously a joint time-frequency) approach to wave propagation, which we have found to be particularly illuminating.

To illustrate the effects of dispersion, we show in Figure. 7.1 spectrograms (squared-magnitude of the short-time Fourier transform) of an acoustic

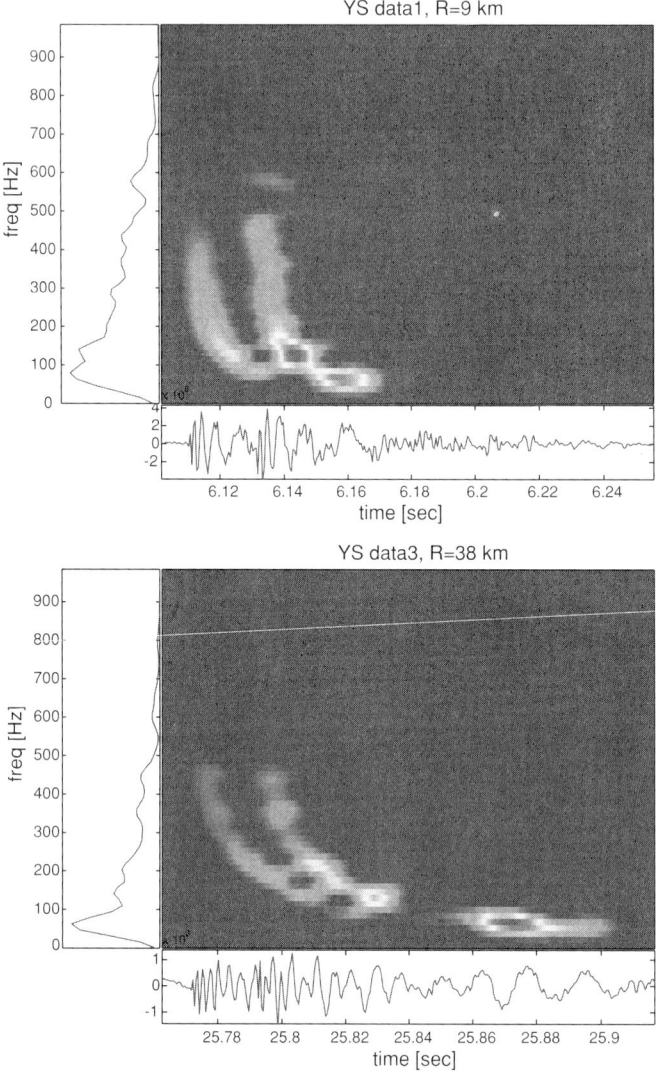

Figure 7.1. Spectrograph (main panel) of an underwater sound pressure wave measured at a distance of 9 km (top) and 38 km (bottom) from the initial source. Note the changes in the wave, which become more pronounced as it propagates, owing to dispersion (see color insert). [Data courtesy of P. Dahl, Applied Physics Laboratory, University of Washington]

wave recorded underwater at two different locations from the initial source. Notice in the spectrogram (main color panel) the dramatic change in the time-frequency trajectory (which is related to the "group delay") of the wave as it propagates from 9 km to 38 km. Below the main panel we plot the wave itself, where one can also observe that it has changed dramatically. Note how the duration of the wave has increased with propagation, because of dispersion. Accordingly, the duration is likely not a good feature to use for classification, unless one can quantify the dispersion effects and account for them. We do so here, as well as consider other moment-like features that are invariant to dispersion.

7.2 Basic Ideas and Definitions

A propagating pulse, $u(x,t)$, is a field; that is, it is a function of both space and time.[1] The method of generation of a pulse can be broken into two cases. In the first case one disturbs the medium at a given time and follows the evolution. An example of such a case is when one plucks a string and then observes the evolution of each point along the string at each time. That is we are given $u(x,0)$ and determine $u(x,t)$. The other case is where at a given position we produce a wave as a function of time, for example as in active radar, sonar, or the fiber optics case. Then we observe the resulting wave over time at different positions. That is, we are given $u(0,t)$ and obtain $u(x,t)$. We differentiate these two cases as

$$\text{Initially given}: u(x,0) \qquad \text{(Case A)} \qquad (7.1)$$
$$: u(0,t) \qquad \text{(Case B)} \qquad (7.2)$$

While one can develop an approach that combines the two cases, we have found that it is easier and more revealing to consider these two cases separately. From a mathematical point of view the two cases are almost identical with, of course, different interpretation of the quantities. This will be discussed in detail later on. Hence for the sake of clarity we will develop the two cases separately. We first discuss case A.

Obtaining $u(x,t)$ with initial condition $u(x,0)$ is a textbook problem when the wave equation is linear with constant coefficients. One defines the initial spatial spectrum, $S(k,0)$, from the initial pulse by

$$S(k,0) = \frac{1}{\sqrt{2\pi}} \int u(x,0)\, e^{-ikx} dx, \qquad (7.3)$$

where k is spatial frequency or the wavenumber. The general solution is then [8, 16]

$$u(x,t) = \frac{1}{\sqrt{2\pi}} \int S(k,0)\, e^{ikx - i\omega(k)t}\, dk \qquad (7.4)$$

---

[1] For simplicity of notation and presentation, we consider only one spatial dimension x.

per mode, where $\omega = \omega(k)$ is the dispersion relation, which couples spatial frequency k and radial frequency ω. If one defines

$$S(k,t) = S(k,0)\, e^{-i\omega(k)t} \tag{7.5}$$

then

$$u(x,t) = \frac{1}{\sqrt{2\pi}} \int S(k,t)\, e^{ikx}\, dk \tag{7.6}$$

$$S(k,t) = \frac{1}{\sqrt{2\pi}} \int u(x,t)\, e^{-ikx}\, dx. \tag{7.7}$$

That is, $u(x,t)$ and $S(k,t)$ form Fourier transform pairs between k and x for all times t. There can be many solutions to the dispersion relation and each solution is called a mode; the general solution is then the sum of the modes. Depending on whether $\omega(k)$ is complex or real we will have damping (i.e., losses) or not. In this chapter we will deal only with the case of no damping (lossless propagation), for which $\omega(k)$ is real. As usual we define the group velocity, $v(k)$, by

$$v(k) = \omega'(k) = \frac{d\omega(k)}{dk} \tag{7.8}$$

It is convenient to express the wave and its spectrum in terms of their respective amplitude and phase as

$$u(x,t) = A(x,t)e^{i\varphi(x,t)} \tag{7.9}$$

$$S(k,t) = B(k,t)e^{i\psi(k,t)}. \tag{7.10}$$

Note that

$$S(k,0) = B(k,0)e^{i\psi(k,0)} \tag{7.11}$$

and hence by virtue of Eq. (7.5) we have that

$$S(k,t) = B(k,0)e^{i\psi(k,0)}e^{-i\omega(k)t} \tag{7.12}$$

and therefore

$$B(k,t) = B(k,0) \quad \text{(for real dispersion relation)} \tag{7.13}$$

$$\psi(k,t) = \psi(k,0) - \omega(k)\,t \tag{7.14}$$

These relations will be useful later.

7.3 Spatial Moments of a Pulse

Moments have been widely used to characterize signals and images and as features for classification in a variety of applications. We consider here spatial moments of the wave at a given time, and investigate the effects of dispersion on these moments. We subsequently define a new kind of moment that is unaffected by dispersion; hence, these moments do not change with propagation and may therefore serve as useful features for classification.

The global spatial moments of a pulse are

$$\langle x^n \rangle_t = \int x^n \, |u(x,t)|^2 \, dx. \tag{7.15}$$

Because $u(x,t)$ and $S(k,t)$ form a Fourier transform pair, it is also the case that one can advantageously calculate the moments from the spectrum by way of

$$\langle x^n \rangle_t = \int S^*(k,t) \, \mathcal{X}^n \, S(k,t) \, dk, \tag{7.16}$$

where $\mathcal{X}$ is the position operator in the k representation

$$\mathcal{X} = i \frac{\partial}{\partial k}. \tag{7.17}$$

It is typically easier to use Eq. (7.16) than Eq. (7.15) since the differentiation indicated by $\mathcal{X}^n \, S(k,t)$ can easily be carried out.

For the first moment one obtains [2, 3]

$$\langle x \rangle_t = \langle x \rangle_0 + V t \tag{7.18}$$

where

$$V = \int v(k) \, |S(k,0)|^2 \, dk. \tag{7.19}$$

Note that the usual definition of group velocity, Eq. (7.8), enters naturally in Eq. (7.19). The second moment is [2, 3]

$$\langle x^2 \rangle_t = \langle x^2 \rangle_0 + t \langle v\mathcal{X} + \mathcal{X}v \rangle_0 + t^2 \langle v^2 \rangle. \tag{7.20}$$

We now explain the various quantities appearing in Eq. (7.20). We define $\langle v\mathcal{X} + \mathcal{X}v \rangle_t$ for an arbitrary time as

$$\langle v\mathcal{X} + \mathcal{X}v \rangle_t = \langle [v,x]_+ \rangle_t \tag{7.21}$$

$$= \int S^*(k,t) \, [v(k)\mathcal{X} + \mathcal{X}v(k)] \, S(k,t) \, dk \tag{7.22}$$

$$= \int S^*(k,t) \left[v(k)i\frac{\partial}{\partial k} + i\frac{\partial}{\partial k}v(k) \right] S(k,t) \, dk \tag{7.23}$$

$$= i \int S^*(k,t) \left[2v(k)\frac{\partial S(k,t)}{\partial k} + v'(k)S(k,t) \right] dk \tag{7.24}$$

$$= \int u^*(x,t) \, [v(\mathcal{K})x + xv(\mathcal{K})] \, u(x,t) \, dx \tag{7.25}$$

where

$$K = -i\frac{\partial}{\partial x}. \tag{7.26}$$

As to the quantity $\langle v^2 \rangle$ it is obtained from

$$\langle v^2 \rangle = \int v^2(k) \, |S(k,0)|^2 \, dk. \tag{7.27}$$

The spread at a particular time defined by

$$\sigma_{x|t}^2 = \langle x^2 \rangle_t - \langle x \rangle_t^2 \tag{7.28}$$

works out to be

$$\sigma_{x|t}^2 = \sigma_{x|0}^2 + 2\,t\,\mathrm{Cov}_{xv|0} + t^2\sigma_v^2, \tag{7.29}$$

where

$$\sigma_v^2 = \langle v^2 \rangle - \langle v \rangle^2 \tag{7.30}$$

$$= \int (v(k) - V)^2 \, |S(k,0)|^2 \, dk \tag{7.31}$$

$$\mathrm{Cov}_{xv|0} = \frac{1}{2}\langle v\mathcal{X} + \mathcal{X}v \rangle_0 - \langle v \rangle_0 \langle x \rangle_0. \tag{7.32}$$

The covariance is a mixed moment that couples two variables. We define the first mixed moment of position and group velocity by [2, 3, 6]

$$\langle xv \rangle_t = \frac{1}{2}\langle v\mathcal{X} + \mathcal{X}v \rangle_t \tag{7.33}$$

$$= \frac{1}{2}\int S^*(k,t)\,(v\mathcal{X} + \mathcal{X}v)\,S(k,t)\,dk. \tag{7.34}$$

This works out to

$$\langle xv \rangle_t = -\int v(k)\,\frac{\partial \psi(k,t)}{\partial k}\,|S(k,t)|^2\,dk, \tag{7.35}$$

where $\psi(k,t)$ is the phase of the spatial spectrum defined by Eq. (7.10). By Eq. (7.14) we also have that

$$\frac{\partial \psi(k,t)}{\partial k} = \frac{\partial \psi(k,0)}{\partial k} - v(k)t \tag{7.36}$$

and substituting this into Eq. (7.35) gives

$$\langle xv \rangle_t = -\int v(k)\,\frac{\partial \psi(k,t)}{\partial k}\,|S(k,t)|^2\,dk \tag{7.37}$$

$$= -\int v(k)\left[\frac{\partial \psi(k,0)}{\partial k} - v(k)t\right]|S(k,t)|^2\,dk \tag{7.38}$$

$$= -\int v(k)\,\frac{\partial \psi(k,0)}{\partial k}\,|S(k,t)|^2\,dk + \int v^2(k)t\,|S(k,t)|^2\,dk. \tag{7.39}$$

which is

$$\langle x\,v \rangle_t = \langle x\,v \rangle_0 + \langle v^2 \rangle t \tag{7.40}$$

It is easy to calculate that the covariance evolves as [2, 3, 6],

$$\mathrm{Cov}_{xv|t} = \langle xv \rangle_t - \langle x \rangle_t \langle v \rangle_t \tag{7.41}$$

$$= \mathrm{Cov}_{xv|0} + \sigma_v^2 t. \tag{7.42}$$

We note that the covariance, $\mathrm{Cov}_{xv|t}$, may be negative at any one time but must eventually become positive as time increases.

The correlation coefficient between x and v is [2, 3, 6]

$$\rho_{xv|t} = \frac{\mathrm{Cov}_{xv|t}}{\sigma_{x|t}\sigma_{v|t}} = \frac{\mathrm{Cov}_{xv|0} + t\sigma_v^2}{\sigma_v\sqrt{\sigma_{x|0}^2 + 2\,t\,\mathrm{Cov}_{xv|0} + t^2\sigma_v^2}} \tag{7.43}$$

and for large times the correlation goes as

$$\rho_{xv|t} \curvearrowright \left[1 + \frac{2\,\mathrm{Cov}_{xv|0}^2 - \sigma_{x|0}^2\sigma_v^2}{2t^2\sigma_v^4} \cdots \right]. \tag{7.44}$$

It follows then that

$$\rho_{xv|t} \to 1 \quad \text{as} \quad t \to \infty, \tag{7.45}$$

which shows that there is perfect correlation between group velocity and position for large times. For small times, the correlation goes as

$$\rho_{xv|t} \curvearrowright \rho_{xv|0} + t\left(1 - \rho_{xv|0}^2\right)\frac{\sigma_v}{\sigma_{x|0}} \cdots. \tag{7.46}$$

The fact that we have exact equations for the spatial moments allows us to compensate for the evolution of a pulse. For example, suppose that a recognizer uses the spread of a signal as a classification feature; but if the recognizer is programmed with $\sigma_{x|0}^2$ and the signal is observed at a time t, the spread measured will be $\sigma_{x|t}^2$ and the pulse would hence not be correctly recognized. The problem then becomes how to compensate for dispersion and obtain $\sigma_{x|0}^2$ to be fed into the classifier. The above exact results show how to do that and what is needed to do it. In particular, one requires knowledge of the initial covariance between position and group velocity, $\mathrm{Cov}_{xv|0}$, Eq. (7.32); the spread σ_v of the group velocity as defined by Eq. (7.31); and the propagation time t.

Alternately if one can make multiple measurements of the spatial spread at different times, it is possible to solve for the initial covariance and spread of the group velocity from the explicit formulation for the spatial spread. Specifically, suppose that we measure the spread $\sigma_{x|t}^2$ at three different times

$t = \{t_1, t_2, t_3\}$. Taking the differences $\sigma^2_{x|t_3} - \sigma^2_{x|t_2}$ and $\sigma^2_{x|t_2} - \sigma^2_{x|t_1}$ and using Eq. (7.29) yields a system of two equations with two unknowns, namely $\mathrm{Cov}_{xv|0}$ and σ^2_v, for which we can then solve. With these quantities thus obtained, we can determine the initial spread $\sigma^2_{x|0}$.

7.4 Dispersion-Invariant Moments

The moments considered above change as the wave propagates, e.g., $\sigma^2_{x|t} \neq \sigma^2_{x|0}$, and hence to be useful as features for classification, one must first compensate for the effects of dispersion, as discussed above. Another alternative, particularly appealing for the case where limited knowledge of the propagation channel or initial pulse is available, is to use moments that do not change with propagation. One example of such moments is the spectral (spatial frequency) moments,

$$\langle k^n \rangle_t = \int k^n \, |S(k,t)|^2 \, dk. \tag{7.47}$$

For real dispersion relation we have that $|S(k,t)| = |S(k,0)|$ per mode, by Eq. (7.5), and therefore $\langle k^n \rangle_t$ is independent of time, $\langle k^n \rangle_t = \langle k^n \rangle_0$, even though $u(x,t) \neq u(x,0)$ when there is dispersion.

We give another class of moments that are similar to spatial moments but, like the spatial frequency moments, they do not change with dispersion. Hence these moments complement the spectral moments and provide additional (spatial) information that may be useful for classification.

To begin, let $\langle g(x) \rangle_t$ denote a spatial moment at a given t, defined by

$$\langle g(x) \rangle_t = \int g(x) \, |u(x,t)|^2 \, dx = \int u^*(x,t)g(x)u(x,t)dx \tag{7.48}$$

$$= \int S^*(k,t) \, g(\mathcal{X}) \, S(k,t) \, dk, \tag{7.49}$$

where, as mentioned previously, the latter equation follows by the properties of the Fourier transform relation between $u(x,t)$ and $S(k,t)$. So, for example, the duration, which is a central moment, is obtained by taking $g(x) = (x - \langle x \rangle_t)^2$, by which we have

$$\sigma^2_{x|t} = \int u^*(x,t)(x - \langle x \rangle_t)^2 u(x,t)dx \tag{7.50}$$

$$= \int S^*(k,t) \, (\mathcal{X} - \langle x \rangle_t)^2 \, S(k,t) \, dk. \tag{7.51}$$

Consider now a new set of moments similar to the above but rather than taking them centered around the mean position for a given time, $\langle x \rangle_t$, we take the moments to be centered about the spatial group delay, defined by

$$x_g(k,t) = -\psi'(k,t) = -\frac{\partial \psi(k,t)}{\partial k}, \tag{7.52}$$

where $\psi(k,t)$ is the spatial phase as per Eqs. (7.10) and (7.14). We define the "spatial moments centered about the spatial group delay" by [12]

$$A_n(t) = \int S^*(k,t) \left(\mathcal{X} - x_g(k,t) \right)^n S(k,t) \, dk \tag{7.53}$$

$$= \int S^*(k,t) \left(i\frac{\partial}{\partial k} + \psi'(k,t) \right)^n S(k,t) \, dk. \tag{7.54}$$

The spatial group delay $x_g(k,t)$ can be considered as a *local* moment akin to the spatial mean for a particular wavenumber k. It is related to the ordinary spatial mean by averaging over k, specifically,

$$\int x_g(k,t) \left| S(k,t) \right|^2 dk = \langle x \rangle_t, \tag{7.55}$$

which follows from Eqs. (7.48) and (7.49) for $n = 1$ and $g(x) = x$, with $\mathcal{X}$ given by Eq. (7.17).

Thus, the moments $A_n(t)$ are similar to the standard duration $\sigma_{x|t}$ (for $n = 2$) and other ordinary central spatial moments, but with the key difference being that the centroid is dependent on k. The significance of this difference is that with dispersion, different frequencies propagate at different velocities, and hence the mean position of the wave is different for different k. Taking this into account by using a frequency-dependent central moment as defined above compensates for the effects of dispersion. Further physical meaning of these moments and an alternate formulation will be discussed shortly but first we show that indeed they do not change with propagation, i.e., that $A_n(t)$ are independent of time,

$$A_n(t) = A_n(0) \tag{7.56}$$

for each mode.

To show that these moments do not change as the wave propagates, consider first

$$\left(i\frac{\partial}{\partial k} + \psi'(k,t) \right)^n S(k,t) = \left(i\frac{\partial}{\partial k} + \psi'(k,t) \right)^n B(k,t) e^{i\psi(k,t)} \tag{7.57}$$

$$= \left(i\frac{\partial}{\partial k} + \psi'(k,t) \right)^{n-1} iB'(k,t) e^{i\psi(k,t)}. \tag{7.58}$$

Therefore, repeating n times we have that

$$\left(i \frac{\partial}{\partial k} + \psi'(k,t) \right)^n S(k,t) = i^n B^{(n)}(k,t) e^{i\psi(k,t)}, \tag{7.59}$$

where $B^{(n)}(k,t)$ denotes the nth derivative with respect to k. Substituting this into Eq. (7.54) and recalling Eq. (7.13), we obtain,

$$A_n(t) = \int S^*(k,t) \left(\mathcal{X} - x_g(k,t) \right)^n S(k,t) \, dk \tag{7.60}$$

$$= \int B(k,t) e^{-i\psi(k,t)} i^n B^{(n)}(k,t) e^{i\psi(k,t)} \, dk \tag{7.61}$$

$$= \int i^n B(k,t) B^{(n)}(k,t) \, dk \tag{7.62}$$

$$= \int i^n B(k,0) B^{(n)}(k,0) \, dk \tag{7.63}$$

$$= A_n(0). \tag{7.64}$$

Accordingly, even though the wave changes as it propagates when there is dispersion, i.e., $u(x,t) \neq u(x,0)$, the moments $A_n(t)$, like the spectral moments $\langle k^n \rangle_t$ given earlier, are the same per mode for $u(x,t)$ and $u(x,0)$.

To gain further physical insight for these new moments, we show that they can be calculated in the wave domain, just as with ordinary spatial moments. In particular it follows from Eqs. (7.48) and (7.49) that

$$A_n(t) = \int i^n B(k,t) B^{(n)}(k,t) \, dk \tag{7.65}$$

$$= \int x^n |u_A(x,t)|^2 \, dx, \tag{7.66}$$

where

$$u_A(x,t) = \frac{1}{\sqrt{2\pi}} \int B(k,t) \, e^{ikx} \, dk. \tag{7.67}$$

Thus the moments $A_n(t)$ are the spatial moments of the pulse $u_A(x,t)$ which is a pulse constructed from only the spectral amplitude $B(k,t)$ of the wave $u(x,t)$. Not only does this provide another physical interpretation of the moments $A_n(t)$, but it also gives us another way to compute them, namely from $u_A(x,t)$, which may be computationally simpler.

7.5 Example

We illustrate the above with an exactly solvable example. We take a quadratic dispersion relation

$$\omega(k) = ck + \gamma k^2/2 \tag{7.68}$$

and for the initial pulse we take,

$$u(x,0) = (\alpha/\pi)^{1/4} \, e^{-\alpha x^2/2 + i\beta x^2/2 + ik_0 x} \tag{7.69}$$

$$(\alpha/\pi)^{1/4} \, e^{-\eta \alpha x^2/2 + ik_0 x}, \tag{7.70}$$

where for convenience we let

$$\eta = \alpha - i\beta. \tag{7.71}$$

The initial spectrum is

$$S(k,0) = \frac{1}{\sqrt{2\pi}} \int u(x,0) \, e^{-ikx} dx \tag{7.72}$$

$$= \frac{(\alpha/\pi)^{1/4}}{\sqrt{\eta}} \, \exp\left[-\frac{(k-k_0)^2}{2\eta} \right] \tag{7.73}$$

$$= \frac{(\alpha/\pi)^{1/4}}{\sqrt{\alpha - i\beta}} \, \exp\left[-\frac{\alpha(k-k_0)^2}{2(\alpha^2+\beta^2)} - i\frac{\beta(k-k_0)^2}{2(\alpha^2+\beta^2)} \right] \tag{7.74}$$

and therefore the time-dependent spectrum is

$$S(k,t) = S(k,0) \, e^{-i\omega(k)t} \tag{7.75}$$

$$= \frac{(\alpha/\pi)^{1/4}}{\sqrt{\eta}} \, \exp\left[-\frac{(k-k_0)^2}{2\eta} - i(ck + \gamma k^2/2)t \right] \tag{7.76}$$

$$= \frac{(\alpha/\pi)^{1/4}}{\sqrt{\alpha - i\beta}} \, \exp\left[-\frac{\alpha(k-k_0)^2}{2(\alpha^2+\beta^2)} - i\frac{\beta(k-k_0)^2}{2(\alpha^2+\beta^2)} - i(ck + \gamma k^2/2)t \right]. \tag{7.77}$$

The squared amplitude spectrum and spectral phase (up to a constant) are

$$|S(k,t)|^2 = \frac{(\alpha/\pi)^{1/2}}{\sqrt{\alpha^2+\beta^2}} \, \exp\left[-\frac{\alpha(k-k_0)^2}{(\alpha^2+\beta^2)} \right] \tag{7.78}$$

$$\psi(k,t) = -\frac{\beta(k-k_0)^2}{2(\alpha^2+\beta^2)} - (ck + \gamma k^2/2)t. \tag{7.79}$$

Note that $|S(k,t)|^2 = |S(k,0)|^2$ as expected for real dispersion relation $\omega(k)$. At the initial time, $t = 0$, the means and standard deviations of x and k are given by

$$\langle x \rangle_0 = 0; \qquad \langle k \rangle_0 = k_0 \tag{7.80}$$

$$\sigma_{x|0}^2 = \frac{1}{2\alpha}; \qquad \sigma_{k|0}^2 = \frac{\alpha^2+\beta^2}{2\alpha}. \tag{7.81}$$

The exact solution is given by

$$
u(x,t) = \frac{(\alpha/\pi)^{1/4}}{\sqrt{1+i\gamma\eta t}} \exp\left[-\frac{\eta}{2}\frac{(x-ct-k_0\gamma t)^2}{1+i\gamma\eta t} + ik_0(x-ct) - i\gamma k_0^2 t/2\right]
$$

$$
= \frac{1}{(2\pi\sigma_{x|t}^2)^{1/4}} \exp\left[-\frac{(x-ct-k_0\gamma t)^2}{4\sigma_{x|t}^2}\right] \tag{7.82}
$$

$$
\times \exp\left[i\frac{(x-ct-k_0\gamma t)^2\{\beta+\gamma(\alpha^2+\beta^2)t\}}{4\alpha\sigma_{x|t}^2} + ik_0(x-ct) - i\gamma k_0^2 t/2 - i\delta\right] \tag{7.83}
$$

where $\sigma_{x|t}^2$ is given below and

$$
\delta = \frac{1}{2}\arctan\frac{\gamma\alpha t}{1+\gamma\beta t}. \tag{7.84}
$$

The average goes as

$$
\langle x\rangle_t = (c+\gamma k_0)\,t. \tag{7.85}
$$

Calculating the spread $\sigma_{x|t}^2$, using Eq. (7.29), we first obtain

$$
\mathrm{Cov}_{xv} = \frac{\gamma\beta}{2\alpha} \tag{7.86}
$$

$$
\langle v\rangle = c+\gamma k_0 \tag{7.87}
$$

$$
\langle v^2\rangle = \gamma^2\frac{\alpha^2+\beta^2}{2\alpha} - (c+\gamma k_0)^2 \tag{7.88}
$$

$$
\sigma_v^2 = \gamma^2\frac{\alpha^2+\beta^2}{2\alpha}. \tag{7.89}
$$

Substituting these values into Eq. (7.29), we have

$$
\sigma_{x|t}^2 = \sigma_{x|0}^2 + 2t\frac{\gamma\beta}{2\alpha} + t^2\gamma^2\frac{\alpha^2+\beta^2}{2\alpha}. \tag{7.90}
$$

Since $\sigma_{x|0}^2 = 1/(2\alpha)$ we can rewrite it as

$$
\sigma_{x|t}^2 = \sigma_{x|0}^2\left[1+2\beta\gamma t+\gamma^2(\alpha^2+\beta^2)t^2\right]. \tag{7.91}
$$

7.5.1 Dispersion-Invariant Moments

We now consider the spatial moments centered about the group delay for the above example. We do so directly from Eq. (7.54), even though we know from our general proof that it has to be independent of time and there are simpler ways to calculate the moments. The derivative of the spectral phase is given by

$$\psi'(k,t) = -\frac{\beta(k-k_0)}{(\alpha^2+\beta^2)} - (c+\gamma k)t \tag{7.92}$$

and therefore we have to calculate

$$A_n(t) = \int S^*(k,t) \left(i\frac{\partial}{\partial k} - \frac{\beta(k-k_0)}{(\alpha^2+\beta^2)} - (c+\gamma k)t \right)^n S(k,t)\, dk. \tag{7.93}$$

That this is independent of time is certainly not obvious. To see how the time dependence drops out, let us calculate the first and second moments, for which we need

$$\left(i\frac{\partial}{\partial k} - \frac{\beta(k-k_0)}{(\alpha^2+\beta^2)} - (c+\gamma k)t \right) S(k,t) = \left(i\frac{\partial}{\partial k} - \frac{\beta(k-k_0)}{(\alpha^2+\beta^2)} - (c+\gamma k)t \right) \tag{7.94}$$

$$\frac{(\alpha/\pi)^{1/4}}{\sqrt{\alpha-i\beta}} \exp\left[-\frac{\alpha(k-k_0)^2}{2(\alpha^2+\beta^2)} - i\frac{\beta(k-k_0)^2}{2(\alpha^2+\beta^2)} - i(ck+\gamma k^2/2)t \right]$$

$$= -i\frac{\alpha(k-k_0)}{(\alpha^2+\beta^2)} S(k,t). \tag{7.95}$$

Therefore

$$A_1(t) = \int S^*(k,t) \left(i\frac{\partial}{\partial k} - \frac{\beta(k-k_0)}{(\alpha^2+\beta^2)} - (c+\gamma k)t \right) S(k,t)\, dk \tag{7.96}$$

$$= \int \left[-i\frac{\alpha(k-k_0)}{(\alpha^2+\beta^2)} \right] |S(k,0)|^2\, dk \tag{7.97}$$

which is indeed independent of time. For this case the moment is zero, since $|S(k,0)|^2$ is a Gaussian with mean $\langle k \rangle = k_0$. (In fact, $A_1(t) = 0$ in general.)

Now consider the second moment. We have

$$A_2(t) = \int S^*(k,t) \left(i\frac{\partial}{\partial k} - \frac{\beta(k-k_0)}{(\alpha^2+\beta^2)} - (c+\gamma k)t \right)^2 S(k,t)\, dk \tag{7.98}$$

$$= \int \left| \left(i\frac{\partial}{\partial k} - \frac{\beta(k-k_0)}{(\alpha^2+\beta^2)} - (c+\gamma k)t \right) S(k,t) \right|^2 dk, \tag{7.99}$$

where the last step follows by the usual rules of operator algebra because the operator is Hermitian. Using Eq. (7.95), we have that

$$A_2(t) = \int \left| \left[-i\frac{\alpha(k-k_0)}{(\alpha^2+\beta^2)} \right] S(k,t) \right|^2 dk \tag{7.100}$$

$$= \frac{\alpha^2}{(\alpha^2+\beta^2)^2} \int (k-k_0)^2 |S(k,t)|^2\, dk \tag{7.101}$$

$$= \frac{\alpha^2}{(\alpha^2+\beta^2)^2} \frac{(\alpha/\pi)^{1/2}}{\sqrt{\alpha^2+\beta^2}} \int (k-k_0)^2 \exp\left[-\frac{\alpha(k-k_0)^2}{(\alpha^2+\beta^2)} \right] dk \tag{7.102}$$

$$= \frac{\alpha}{2(\alpha^2+\beta^2)} \tag{7.103}$$

and we see that it is independent of time. To check further, we calculate using Eq. (7.63)

$$A_2(t) = \int i^2 \, B(k,0) B^{(2)}(k,0) \, dk \qquad (7.104)$$

$$= -\int B(k,0) \frac{\partial^2}{\partial k^2} B(k,0) \, dk \qquad (7.105)$$

$$= \int \left(\frac{\partial}{\partial k} B(k,0) \right)^2 dk \qquad (7.106)$$

$$= \frac{\alpha^2}{(\alpha^2 + \beta^2)^2} \int (k - k_0)^2 B^2(k,0) \, dk, \qquad (7.107)$$

which is the same as Eq. (7.103). (Recall $B^2(k,0) = |S(k,0)|^2$ which is a Guassian for this example and hence the final integral is the spread $\sigma_{x|0}^2$ of the initial pulse, which is $1/2\alpha$).

7.6 Wigner Approximation for Wave Propagation, and Approximately Invariant Moments

The moments considered above are global moments, in that they are averages over all space. Suppose that we are interested in *local* moments, such as averages for a particular spatial frequency of interest. One way to obtain such moments is to filter the wave about the frequency of interest, and then calculate the moments above for the filtered wave [12]. Another way to obtain local moments is to consider a phase-space distribution of the wave, such as the Wigner distribution, and obtain moments from that [11, 12]. We consider this approach here and define local mixed moments analogous to the A_n above, and show that they, too, are (approximately) invariant to dispersion.

The spatial/spatial-frequency Wigner distribution of the wave $u(x,t)$ is defined by [2, 9]

$$W(x, k; t) = \frac{1}{2\pi} \int u\left(x + \frac{\lambda}{2}, t\right) u^*\left(x - \frac{\lambda}{2}, t\right) e^{-ik\lambda} \, d\lambda. \qquad (7.108)$$

The local Wigner central spatial moments for a particular t and k are given by

$$\langle (x - \langle x \rangle_{t,k})^n \rangle_{t,k} = \frac{1}{\int W(x, k; t) \, dx} \int (x - \langle x \rangle_{t,k})^n \, W(x, k; t) \, dx \qquad (7.109)$$

$$= \frac{1}{B^2(k,t)} \int x^n \, W(x + \langle x \rangle_{t,k}, k; t) \, dx, \qquad (7.110)$$

where [2]

$$\int W(x, k; t)\, dx = B^2(k, t) \tag{7.111}$$

$$\langle x \rangle_{t,k} = \frac{1}{\int W(x, k; t) dx} \int x\, W(x, k; t)\, dx \tag{7.112}$$

$$= -\psi'(k, t) \tag{7.113}$$

$$= -\psi'(k, 0) + v(k)t. \tag{7.114}$$

It can be shown that the Wigner distribution at t is approximately related to the Wigner distribution of the initial pulse $u(x, 0)$ by [5, 13]

$$W(x, k; t) \approx W(x - v(k)t, k; 0) \tag{7.115}$$

for real dispersion relation $\omega(k)$, where $v(k)$ is the group velocity defined by Eq. (7.8). Plugging Eqs. (7.115) and (7.114) into Eq. (7.110), it follows that

$$\langle (x - \langle x \rangle_{t,k})^n \rangle_{t,k} \approx \frac{1}{B^2(k, t)} \int x^n W(x - \psi'(k, 0), k; 0)\, dx \tag{7.116}$$

$$= \frac{1}{B^2(k, 0)} \int x^n W(x - \psi'(k, 0), k; 0)\, dx \tag{7.117}$$

$$= \langle (x - \langle x \rangle_{0,k})^n \rangle_{0,k}. \tag{7.118}$$

Hence the local central spatial moments of the Wigner distribution are approximately dispersion-invariant; i.e., they are independent of t. Indeed the invariance is exact for the first two moments, which we work out explicitly; by Eq. (7.109), and using Eq. (7.114), we have, for $n = 1$,

$$(x - \langle x \rangle_{t,k}))_{t,k} = \frac{1}{B^2(k, t)} \int (x + \psi'(k, 0) - v(k)t) W(x, k; t)\, dx \tag{7.119}$$

$$= -\psi'(k, t) + \psi'(k, 0) - v(k)t = 0. \tag{7.120}$$

For $n = 2$ we obtain

$$(x - \langle x \rangle_{t,k})^2 \rangle_{t,k} = \frac{1}{B^2(k, t)} \int (x - \langle x \rangle_{k,t})^2 W(x, k; t)\, dx \tag{7.121}$$

$$= \frac{1}{2} \left[\left(\frac{B'(k, t)}{B(k, t)} \right)^2 - \frac{B''(k, t)}{B(k, t)} \right] \tag{7.122}$$

$$= \frac{1}{2} \left[\left(\frac{B'(k, 0)}{B(k, 0)} \right)^2 - \frac{B''(k, 0)}{B(k, 0)} \right], \tag{7.123}$$

where the last step follows from Eq. (7.13). Hence, the first two local central spatial moments of the Wigner distribution are exactly invariant to dispersion. Higher moments are approximately invariant, as given by Eq. (7.117).

Relation to global moments. Averaging the local moments over k gives global moments. For the Wigner distribution moments above, this is accomplished by

$$\langle (x - \langle x \rangle_{t,k})^n \rangle_t = \int \langle (x - \langle x \rangle_{t,k})^n \rangle_{t,k} \, B^2(k,t) \, dk. \tag{7.124}$$

While these global moments are not the same as those defined by Eq. (7.54), they are dispersion-invariant for $n = 1, 2$ and approximately so for $n \geq 3$.

7.7 Dual Case: Initially Given $u(0, t)$

As mentioned at the outset, the propagation of waves can be thought of in terms of two cases: (A) given $u(x, 0)$ we solve for $u(x, t)$; this is the case considered above; and (B) given $u(0, t)$ we solve for $u(x, t)$. We now consider this case here. The mathematics and results are analogous to the first case, except for a sign change by convention in the Fourier transforms and associated operators.

7.7.1 Preliminaries

Given $u(0, t)$, one defines the Fourier spectrum by

$$F(0, \omega) = \frac{1}{\sqrt{2\pi}} \int u(0, t) e^{i\omega t} dt. \tag{7.125}$$

Note the sign difference in the exponent, by convention, with respect to the case of $u(x, 0)$ and $S(k, 0)$ considered previously. The general solution is then

$$u(x, t) = \frac{1}{\sqrt{2\pi}} \int F(0, \omega) e^{ik(\omega)x - i\omega t} \, d\omega \tag{7.126}$$

per mode, where $k(\omega)$ is the dispersion relation written in terms of k as a function of ω. There can be many solutions to the dispersion relation, and each solution is called a mode; the general solution is then the sum of the modes. As with the previous case, we will deal only with the case of real dispersion relation $k(\omega)$ here. Analogous to the group velocity and the case of $u(x, 0)$, the derivative of the dispersion relation $k(\omega)$ arises for the case of $u(0, t)$. We call this the "group slowness" (or the "unit transit time"), defined as

$$\tau_g(\omega) = k'(\omega) = \frac{dk(\omega)}{d\omega}. \tag{7.127}$$

If one defines

$$F(x, \omega) = F(0, \omega) e^{ik(\omega)x} \tag{7.128}$$

then we have the Fourier pair

$$u(x, t) = \frac{1}{\sqrt{2\pi}} \int F(x, \omega) e^{-i\omega t} d\omega \tag{7.129}$$

$$F(x, \omega) = \frac{1}{\sqrt{2\pi}} \int u(x, t) e^{i\omega t} dt \tag{7.130}$$

for all x. As done previously, it is convenient to express the wave and its spectrum in terms of their respective amplitude and phase,

$$u(x,t) = A(x,t)e^{i\varphi(x,t)} \tag{7.131}$$

$$F(x,\omega) = B(x,\omega)e^{i\psi(x,\omega)}. \tag{7.132}$$

Note that

$$F(0,\omega) = B(0,\omega)e^{i\psi(0,\omega)} \tag{7.133}$$

and hence we have that

$$F(x,\omega) = B(0,\omega)e^{i\psi(0,\omega)}e^{ik(\omega)x} \tag{7.134}$$

and therefore

$$B(x,\omega) = B(0,\omega) \quad \text{(for real dispersion relation } k(\omega)) \tag{7.135}$$

$$\psi(x,\omega) = \psi(0,\omega) + k(\omega)x. \tag{7.136}$$

7.7.2 Dispersion-Invariant Temporal Moments

We give a new kind of central temporal moment analogous to the central spatial moments considered previously. We begin by considering the duration of the pulse at a particular location x, which is given by

$$\sigma^2_{x|t} = \int u^*(x,t)\,(t - \langle t \rangle_x)^2\, u(x,t)\,dt \tag{7.137}$$

$$= \int F^*(x,\omega)\left(-i\frac{\partial}{\partial\omega} - \langle t \rangle_x\right)^2 F(x,\omega)\,d\omega, \tag{7.138}$$

where we have again made use of the properties of the Fourier transform relation between $u(x,t)$ and $F(x,\omega)$ to express the duration equivalently in terms of $F(x,\omega)$.[2] Then, as with the local central spatial moments, we define the "moments centered about the group delay" by

$$A_n(x) = \int F^*(x,\omega)\left(-i\frac{\partial}{\partial\omega} - t_g(x,\omega)\right)^n F(x,\omega)\,d\omega$$

$$= \int F^*(x,\omega)\left(-i\frac{\partial}{\partial\omega} - \psi'(x,\omega)\right)^n F(x,\omega)\,d\omega, \tag{7.139}$$

where $t_g(x,\omega)$ is the *group delay* of the wave, defined by[3]

$$t_g(x,\omega) = \psi'(x,\omega) = \frac{\partial}{\partial\omega}\psi(x,\omega) \tag{7.140}$$

$$= \psi'(0,\omega) + k'(\omega)x. \tag{7.141}$$

---

[2] The negative sign in front of $-i\frac{\partial}{\partial\omega}$ is due to the definition of the Fourier transform as per Eq. (7.129). Also, we note that analogous moments were first defined by K. Davidson for signals in *Instantaneous Moments of A Signal* (Ph.D. dissertation, Univ. of Pittsburgh, 2000), in which he also derived the analogous identity given in Eq. (7.59).

[3] Consistent with the sign change in the Fourier transform, we take the group delay to be $\psi'(x,\omega)$ rather than $-\psi'(x,\omega)$.

Note, then, that these moments are similar to those given by Eq. (7.138), except that rather than being centered about the average time $\langle t \rangle_x$, they are centered about the group delay $t_g(x, \omega)$ of the wave. The main result is that the moments, $A_n(x)$, do not depend on position. In particular,

$$A_n(x) = \int F^*(x, \omega) \left(-i \frac{\partial}{\partial \omega} - \psi'(x, \omega) \right)^n F(x, \omega) \, d\omega \qquad (7.142)$$

$$= \frac{1}{i^n} \int B(x, \omega) \, e^{-i\psi(x, \omega)} B^{(n)}(x, \omega) \, e^{i\psi(x, \omega)} \, d\omega \qquad (7.143)$$

$$= \frac{1}{i^n} \int B(x, \omega) B^{(n)}(x, \omega) \, d\omega \qquad (7.144)$$

$$= \frac{1}{i^n} \int B(0, \omega) B^{(n)}(0, \omega) \, d\omega = A_n(0), \qquad (7.145)$$

where the last step follows for real $k(\omega)$. Thus we have shown that indeed these moments are independent of position x, $A_n(x) = A_n(0)$. We also note that for odd n, these moments are zero, which is easy to show using integration by parts, and given that $B(0, \omega) = 0$ for $|\omega| \to \infty$.

Analogous to the case with the spatial moments $A_n(t)$, we can express the moments $A_n(x)$ in the time domain, in terms of the wave $u_A(x, t)$ obtained from the spectral amplitude $B(x, \omega)$ of the wave $u(x, t)$ [12],

$$A_n(x) = \frac{1}{i^n} \int B(x, \omega) B^{(n)}(x, \omega) \, d\omega \qquad (7.146)$$

$$= \int t^n \, |u_A(x, t)|^2 \, dt \qquad (7.147)$$

where

$$u_A(x, t) = \frac{1}{\sqrt{2\pi}} \int B(x, \omega) \, e^{-i\omega t} \, d\omega. \qquad (7.148)$$

7.7.3 Wigner Approximation and Approximately Invariant Moments

We now give local central temporal moments analogous to the local central spatial moments, using a phase-space/Wigner distribution approach as in Section 7.6. Because of the convention in the sign difference in the Fourier transform relation for $F(x, \omega)$ versus $S(k, t)$, we adopt the same convention in defining the Wigner distribution $W(t, \omega; x)$ versus $W(x, k; t)$. Doing so will preserve the analogy between the results for the two cases. We define the Wigner distribution in t, ω of $u(x, t)$ at a particular x by

$$W(t, \omega; x) = \frac{1}{2\pi} \int u \left(x, t + \frac{\tau}{2} \right) u^* \left(x, t - \frac{\tau}{2} \right) e^{i\omega \tau} \, d\tau. \qquad (7.149)$$

This can be expressed equivalently in terms of $F(x, \omega)$ as

$$W(t, \omega \,;\, x) = \frac{1}{2\pi} \int F\left(x, \omega - \frac{\theta}{2}\right) F^*\left(x, \omega + \frac{\theta}{2}\right) e^{i\theta t} \, d\theta \qquad (7.150)$$

$$= \frac{1}{2\pi} \int F\left(0, \omega - \frac{\theta}{2}\right) F^*\left(0, \omega + \frac{\theta}{2}\right)$$

$$e^{i\theta t} e^{-ix\left[k^*\left(\omega + \frac{\theta}{2}\right) - k\left(\omega - \frac{\theta}{2}\right)\right]} d\theta, \qquad (7.151)$$

which is readily shown by substituting in the Fourier relation for $u(x, t)$ in terms of $F(x, \omega)$ into the expression for $W(t, \omega; x)$ above.

At $x = 0$, we have

$$W(t, \omega;\, 0) = \frac{1}{2\pi} \int F\left(0, \omega - \frac{\theta}{2}\right) F^*\left(0, \omega + \frac{\theta}{2}\right) e^{i\theta t} \, d\theta \qquad (7.152)$$

therefore

$$F\left(0, \omega - \frac{\theta}{2}\right) F^*\left(0, \omega + \frac{\theta}{2}\right) = \int W(t', \omega;\, 0) \, e^{-i\theta.t'} \, dt'. \qquad (7.153)$$

It follows that

$$W(t, \omega;\, x) = \frac{1}{2\pi} \int \int W(t', \omega;\, 0) \, e^{-i\theta(t'-t)} e^{-ix\left[k^*\left(\omega + \frac{\theta}{2}\right) - k\left(\omega - \frac{\theta}{2}\right)\right]} dt' d\theta, \qquad (7.154)$$

which relates the Wigner distribution of the pulse at position x to the Wigner distribution of the initial pulse at $x = 0$.

Expanding the exponent in a power series in θ, namely

$$k\left(\omega \pm \frac{\theta}{2}\right) = k(\omega) + \left(\pm \frac{\theta}{2}\right) k'(\omega) + \frac{1}{2!} \left(\pm \frac{\theta}{2}\right)^2 k''(\omega) + \cdots \qquad (7.155)$$

and keeping just the first two terms, we obtain an approximation for the Wigner distribution,

$$W(t, \omega;\, x) \approx \frac{1}{2\pi} \int \int W(t', \omega;\, 0) \, e^{-i\theta\left(t'-t+\tau_g(\omega)x\right)} dt' d\theta \qquad (7.156)$$

$$= \int W(t', \omega;\, 0) \, \delta\left(t' - t + \tau_g(\omega) x\right) dt' \qquad (7.157)$$

or

$$W(t, \omega;\, x) \approx W\left(t - \tau_g(\omega) x, \omega;\, 0\right). \qquad (7.158)$$

Note the exact analogy of this approximation with that for the spatial Wigner distribution $W(x, k;\, t)$ given in Eq. (7.115).

Because of this analogy, all of the results from the previous case transcribe readily to this case. In particular, the local central temporal moments of the

Wigner distribution for a given x and ω are given by

$$\langle (t - \langle t \rangle_{x,\omega})^n \rangle_{x,\omega} = \frac{1}{\int W(t,\omega;\, x)dt} \int (t - \langle t \rangle_{x,\omega})^n \, W(t,\omega;\, x) \, dt \quad (7.159)$$

$$= \frac{1}{B^2(x,\omega)} \int t^n \, W(t + \langle t \rangle_{x,\omega}, \omega;\, x) \, dt, \quad (7.160)$$

where[4]

$$\int W(t,\omega;\, x) \, dt = B^2(x,\omega) \quad (7.161)$$

$$\langle t \rangle_{x,\omega} = \frac{1}{\int W(t,\omega;\, x) \, dt} \int t \, W(t,\omega;\, x) \, dt \quad (7.162)$$

$$= \psi'(x,\omega) \quad (7.163)$$

$$= \psi'(0,\omega) + \tau_g(\omega)x. \quad (7.164)$$

By the approximation above for the Wigner distribution at x in terms of the initial Wigner distribution, these central moments, like the local central spatial moments considered previously, are (approximately) dispersion-invariant, in that they do not depend on x:

$$\langle (t - \langle t \rangle_{x,\omega})^n \rangle_{x,\omega} \approx \langle (t - \langle t \rangle_{0,\omega})^n \rangle_{0,\omega} \quad (7.165)$$

This result is exact for $n = 1, 2$.

7.8 Conclusion

Moments of a pulse are basic characterizers of a pulse. The results presented show how one can compensate for the fact that the spatial and temporal moments change because of dispersion. In addition, one can obtain dispersion-invariant central moments, as shown, by taking moments centered about the group delay. We gave two different approaches for doing this, one based on a phase-space (Wigner distribution) approach, which gives approximately invariant local central moments; and a second using an operator approach which gives exactly invariant central moments, for real dispersion relations.

It is important to appreciate that the results above are per mode. In general each mode travels with a different group velocity; that is, each mode has a different dispersion relation. Accordingly, the spectral and A_n moments of a wave composed of multiple modes will change with propagation distance/time,

---

[4] Because of the sign change in the exponential of the Wigner distribution as defined in Eq. (7.149) vs. the usual definition, there is a sign change in the first conditional moment obtained, i.e., Eq. (7.163).

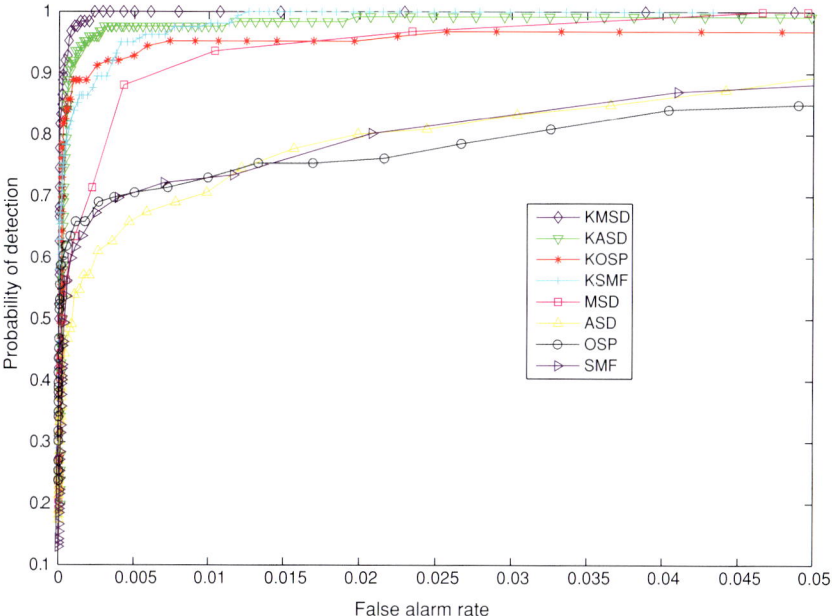

Figure 1.6. ROC curves obtained by conventional detectors and the corresponding kernel versions for the DR-II image.

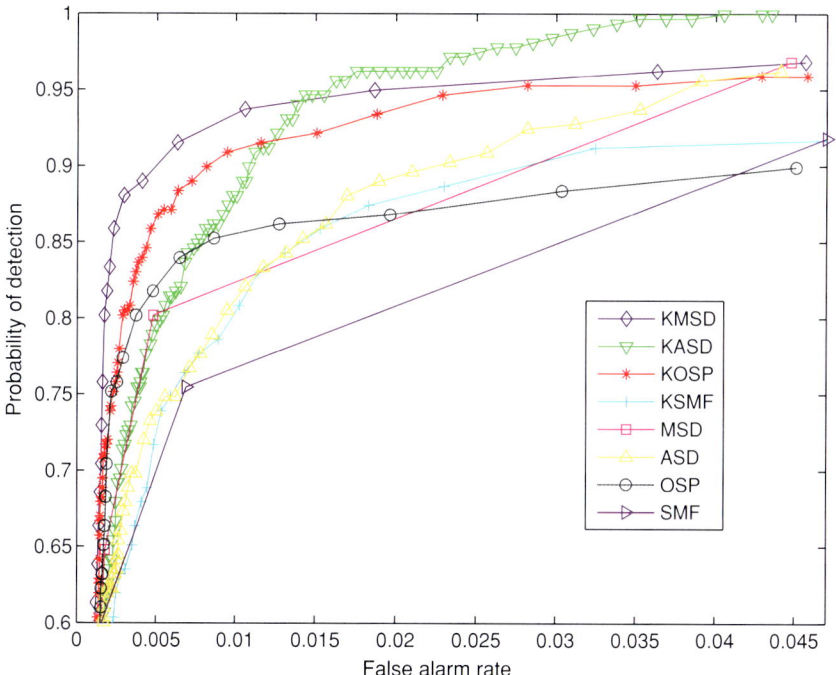

Figure 1.7. ROC curves obtained by conventional detectors and the corresponding kernel versions for the DR-II image.

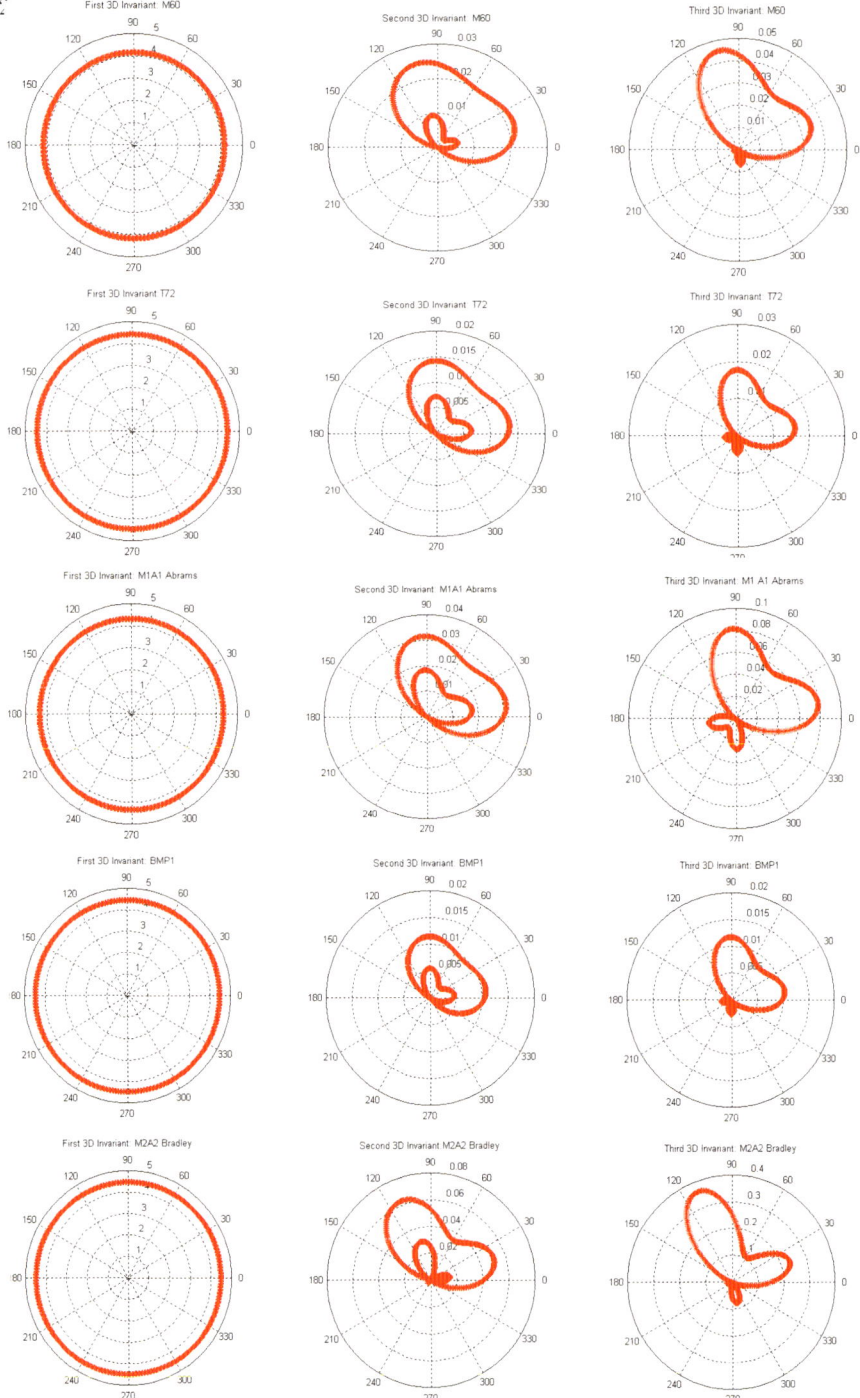

Figure 2.5. Plot of invariant values for five different targets as the targets rotate around z-axis from $0°$ to $360°$.

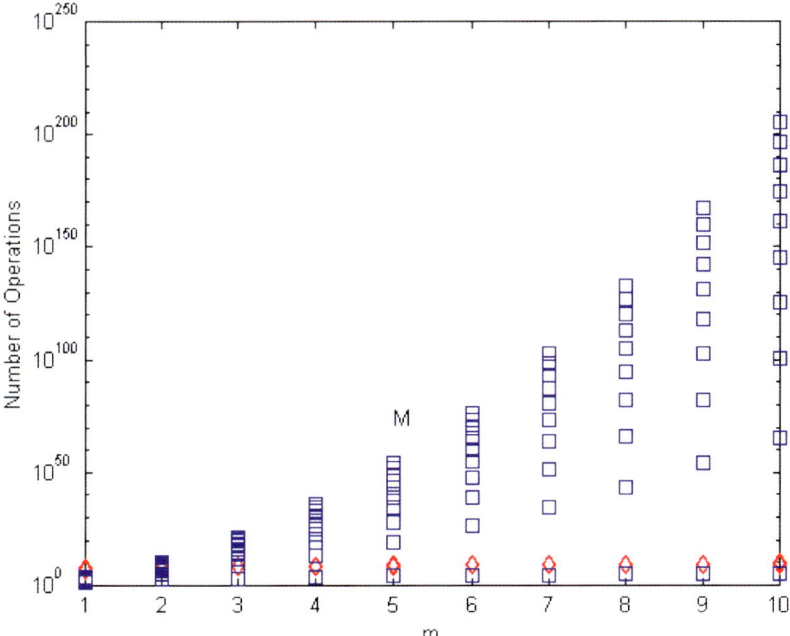

Figure 2.9. The 2-D plots of the number of operations (on a log scale) as functions of the template size and material classes for the presented approach and a typical noninvariant approach.

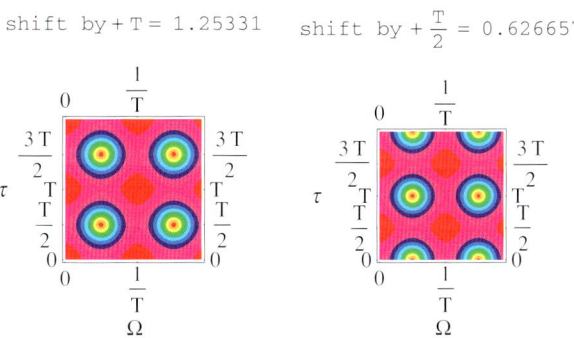

Figure 4.3. The absolute value $|Z_0(\Omega, \tau)|$ for $\lambda = 0.5$ and $\alpha = 0.5$.

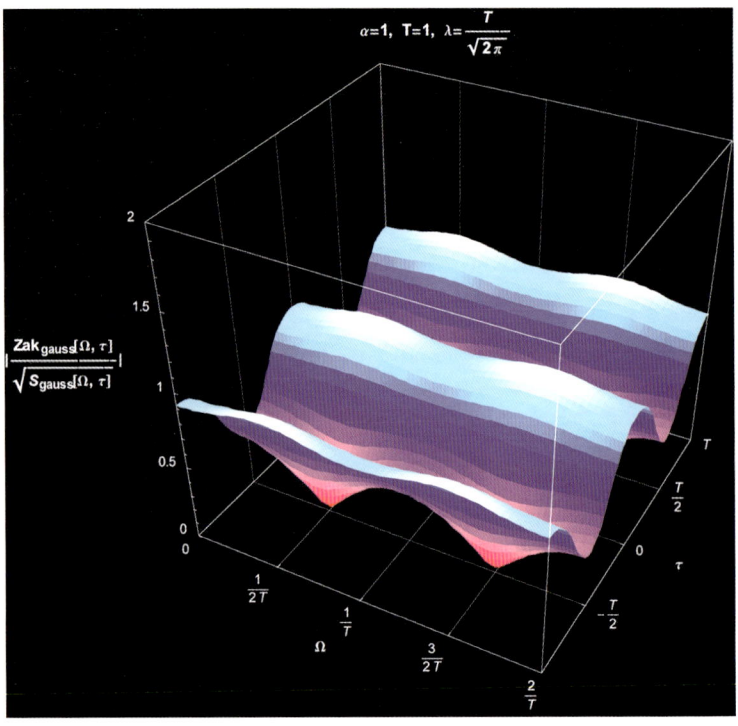

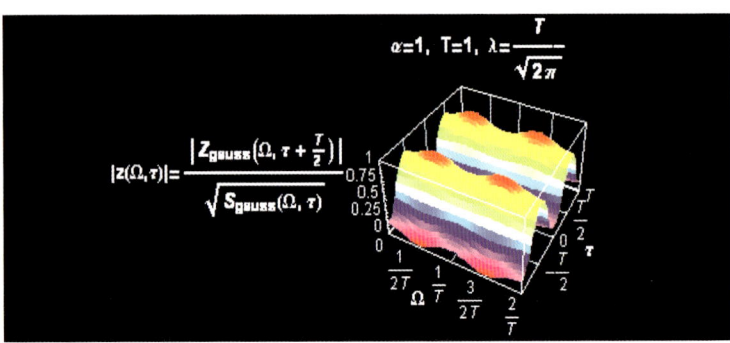

Figure 4.4. The absolute value of the weighted $|z(\Omega, \tau)| = \frac{|Z(\Omega, \tau)|}{\sqrt{S(\Omega, \tau)}}$ for the Gaussian function.

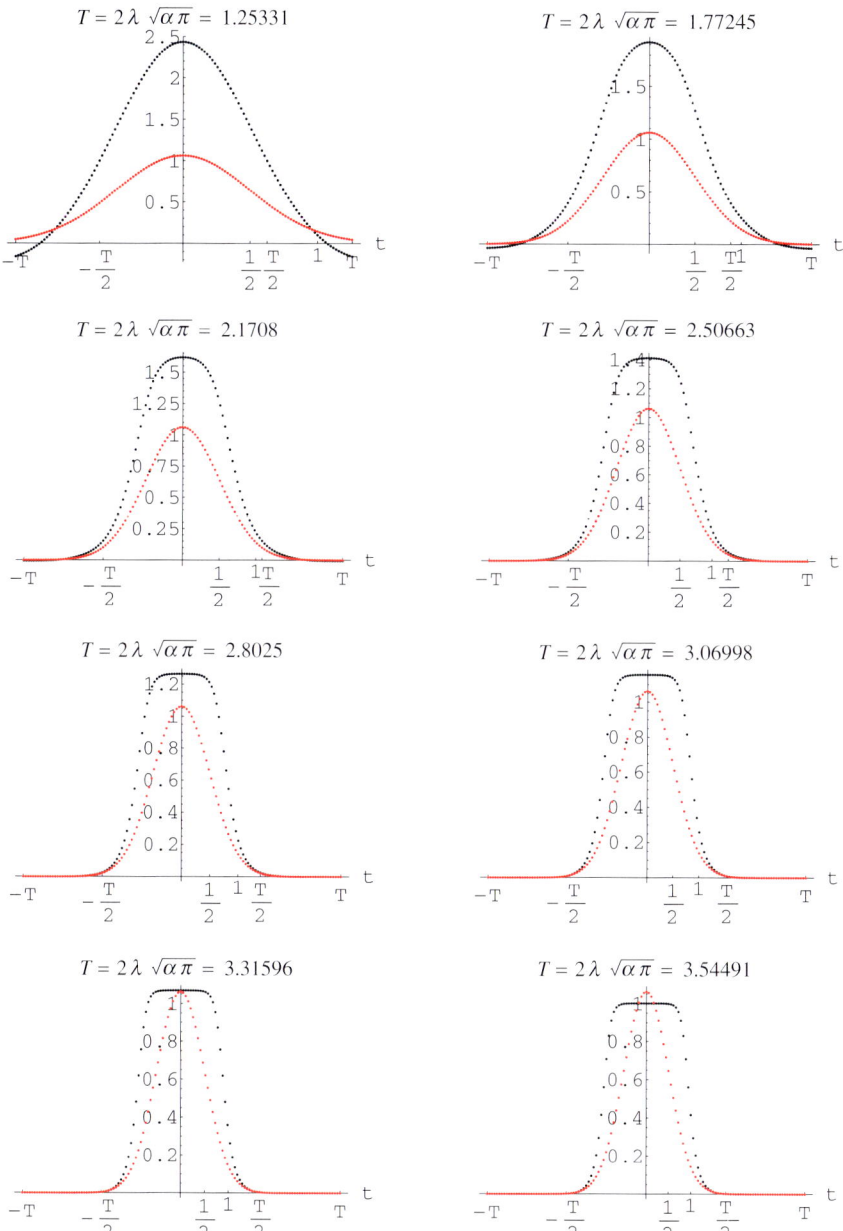

Figure 4.8. The inverse of the weighted $z(\Omega, \tau)$ for larger periods T. The dotted red plot is Gaussian $\lambda = 0.5$. Note the scale of each plot is different: for reference purpose marks of $1, \frac{1}{2}$ on t axis are shown.

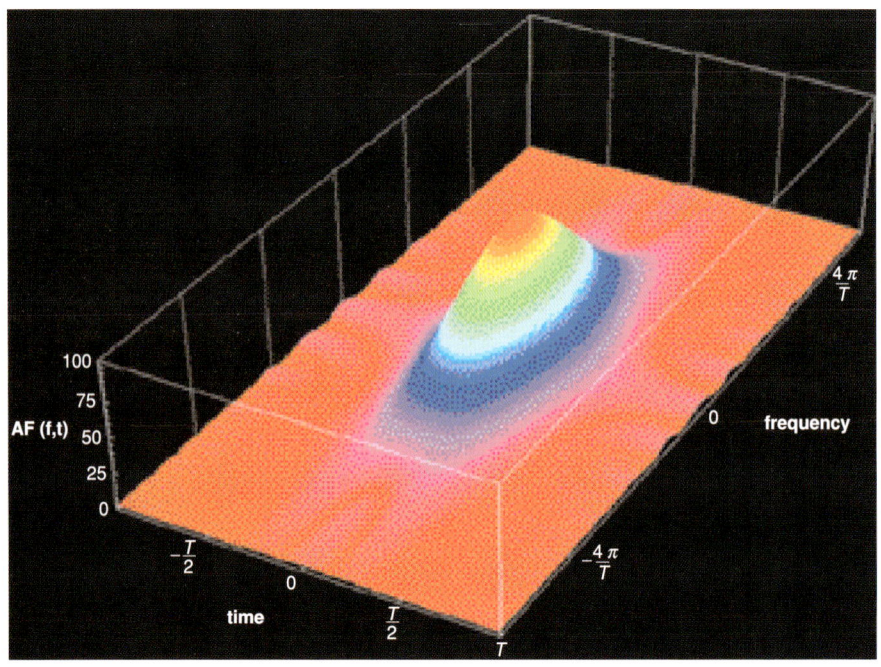

Figure 4.9. The absolute value of the Ambiguity function.

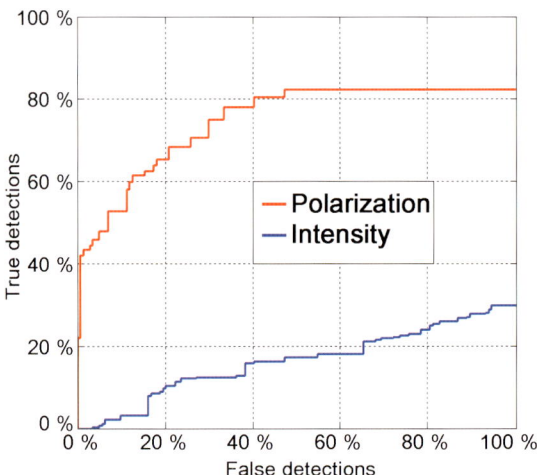

Figure 5.3. Two ROC curves obtained from one data set that show a clear difference in detection performance when using polarization features. In the bottom curve only intensity features are used for detection. In the top curve only polarization features are used.

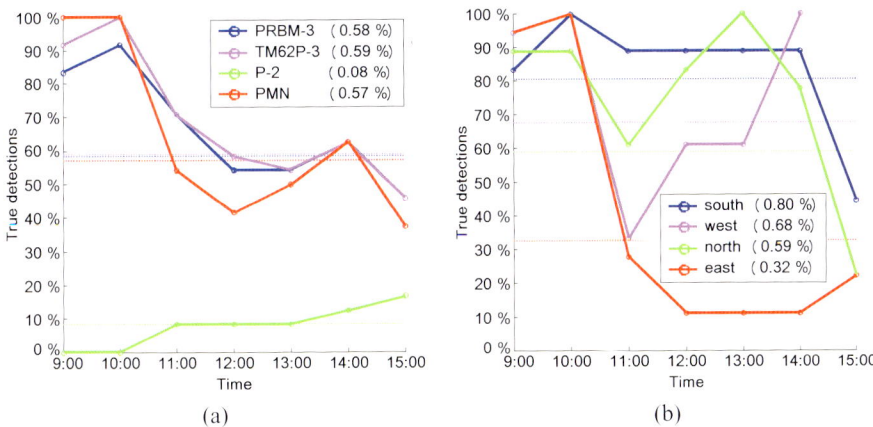

Figure 5.4. (a) Detection rate vs. time of measurement for the four mine types. (b) Detection rate vs. time of measurement for the four viewing directions. The results for all mine types with the exclusion of the P2 mine type are shown. In both figures a false-alarm rate of 3% was chosen. The horizontal dashed lines in the figures are the detection results for the whole measurement period (instead of the results of 1 hour) per mine type (a) or per viewing direction (b).

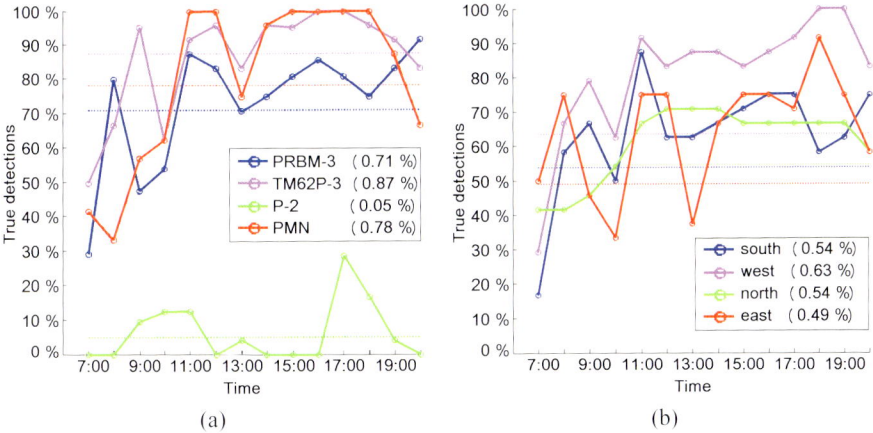

Figure 5.5. (a) Detection rate vs. time of measurement for the four different mine types. (b) Detection rate vs. time of measurement for the four viewing directions. A false-alarm rate of 3% was chosen.

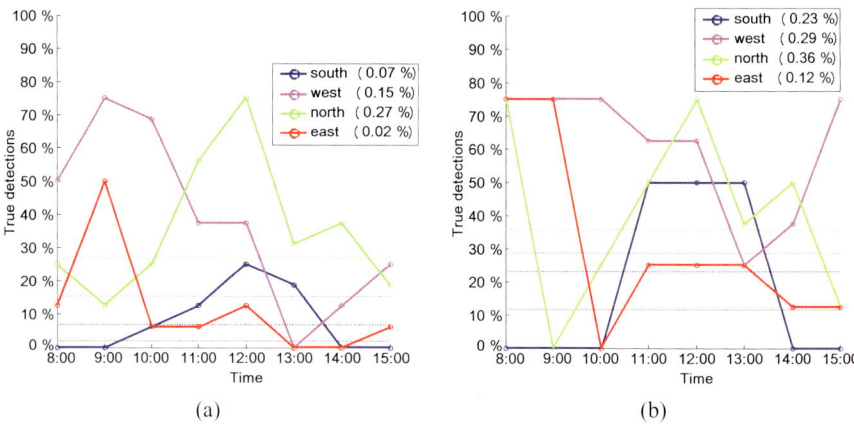

Figure 5.6. Detection rate vs. time of measurement for the four viewing directions. A false-alarm rate of 3% was chosen. (a) Results for mines tilted away from the camera. (b) Results for the nontilted mines.

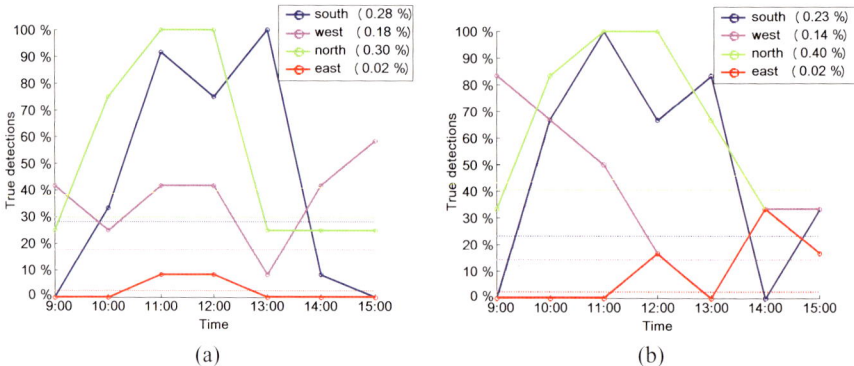

Figure 5.7. Detection rate vs. time of measurement for the four viewing directions. A false-alarm rate of 3% was chosen. (a) Results for mines tilted toward the camera. (b) Results for the nontilted mines.

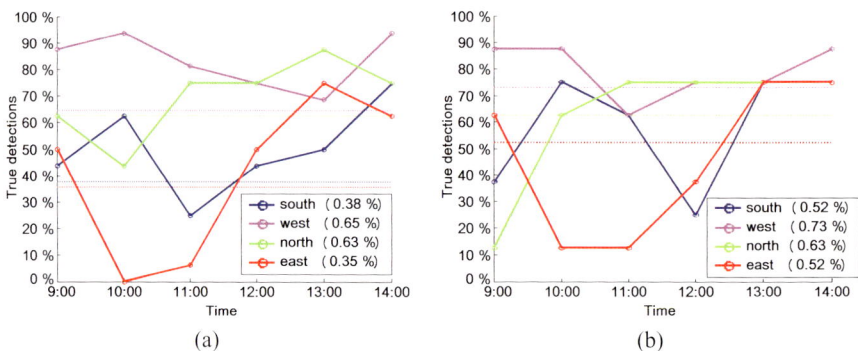

Figure 5.8. Detection rate vs. time of measurement for the four viewing directions. A false-alarm rate of 3% was chosen. (a) Results for mines tilted along the viewing direction. (b) Results for the nontilted mines.

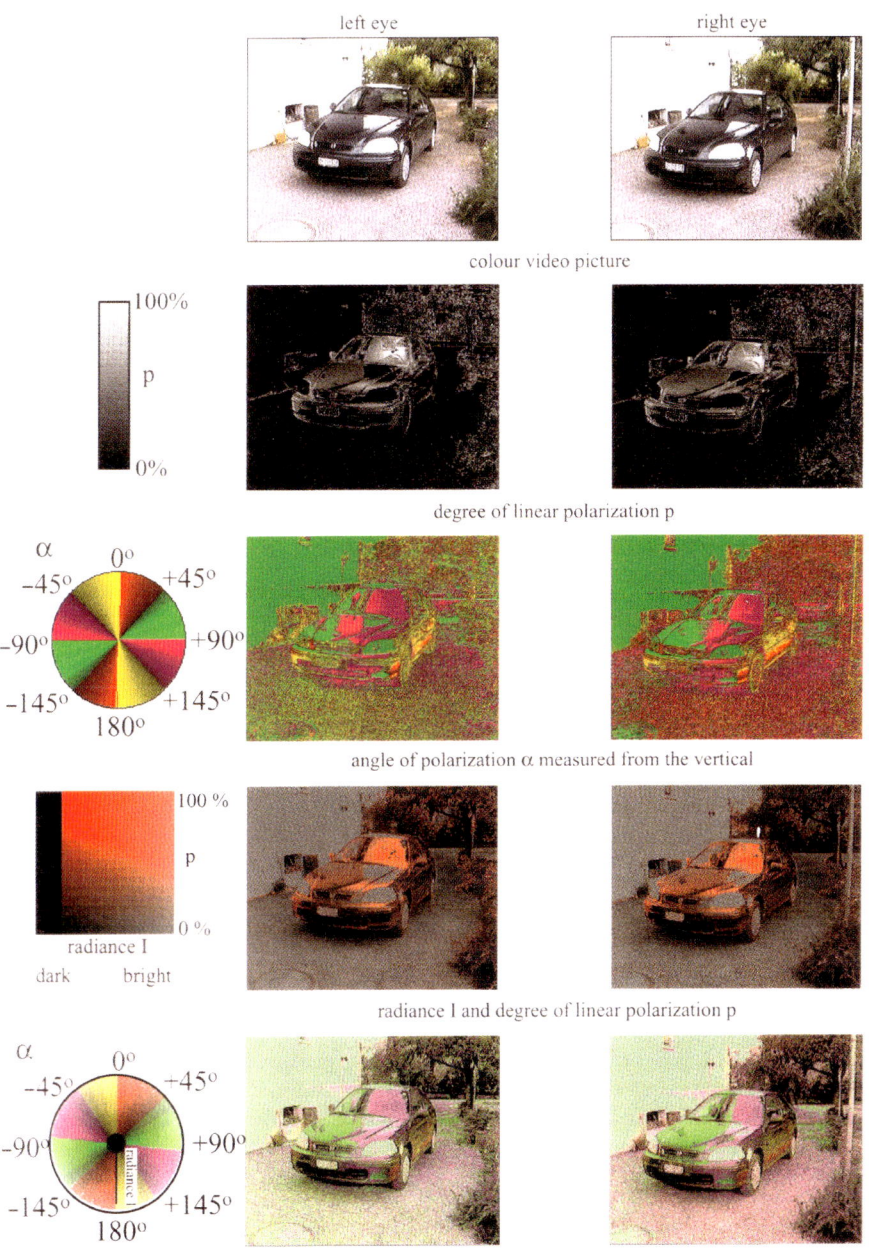

Figure 6.3. Polarization images of a scene with reflected light. [Reprinted with permission of IS&T: The Society for Imaging Science and Technology sole copyright owners of The Journal of Imaging Science and Technology.]

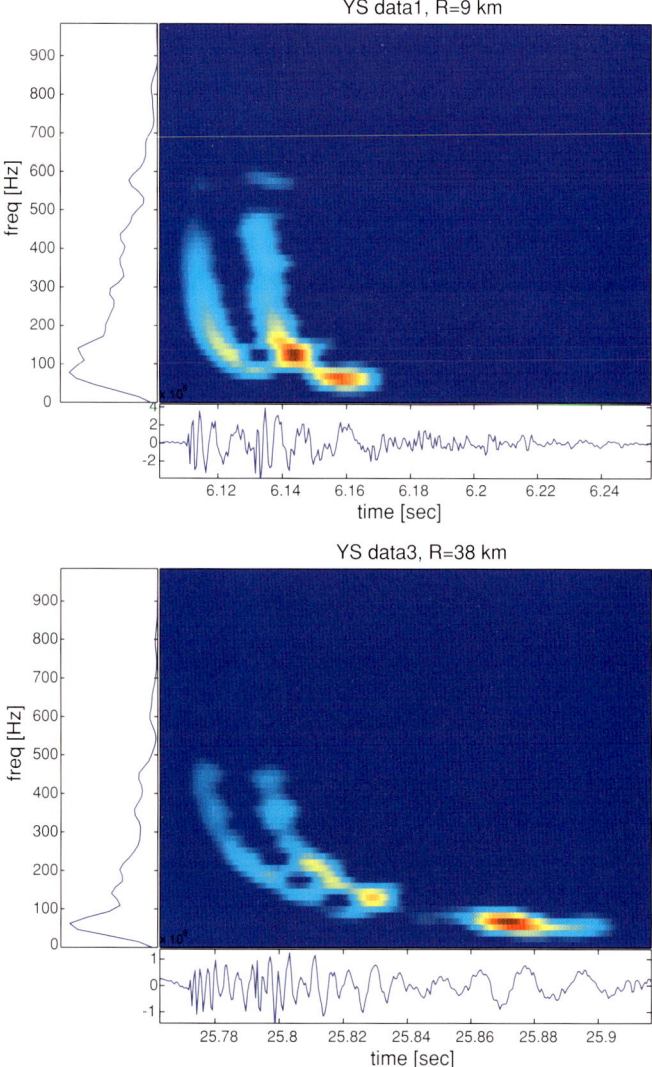

Figure 7.1. Spectrograph (main panel) of an underwater sound pressure wave measured at a distance of 9 km (top) and 38 km (bottom) from the initial source. Note the changes in the wave, which become more pronounced as it propagates, owing to dispersion. [Data courtesy of P. Dahl, Applied Physics Laboratory, University of Washington]

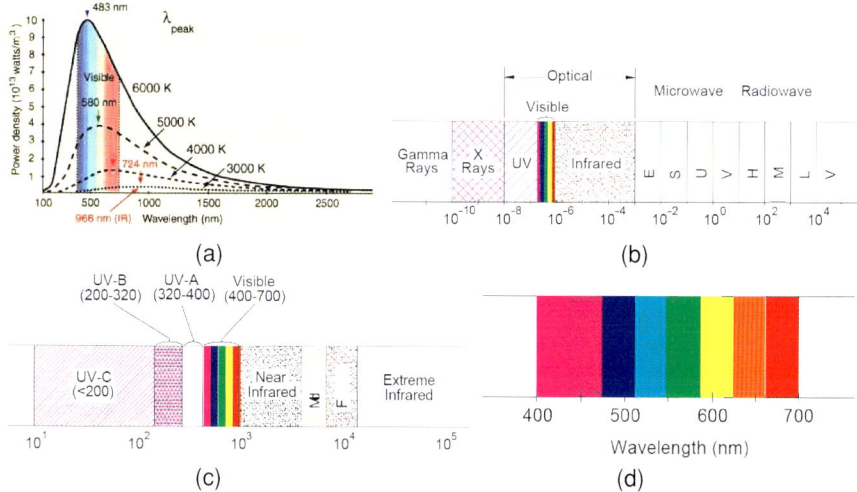

Figure 9.2. Illustrations of the blackbody radiation curve (a), the electromagnetic (b), optical (c), and visible spectrum (d). The wavelengths are in nanometers (nm). The peak of the blackbody radiation curve gives a measure of temperature.

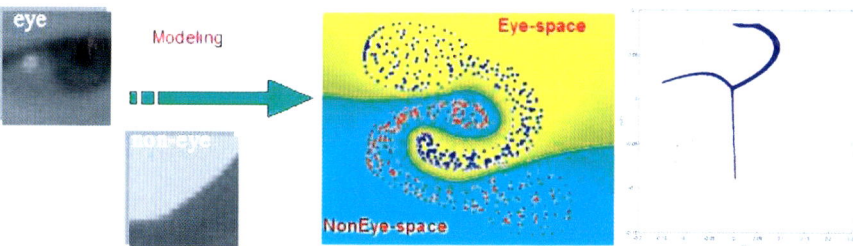

Figure 9.8. Illustration of eye and noneye models and their corresponding areas of membership in a two dimensional feature space. Each eye patch is represented by its projection in the first two eigenvectors. The space is splited into two areas of membership (eye vs. noneye) using SVM technique (middle). In the most right figure, a Gaussian Mixture Model is used to find the subclusters in the eye space, where three clusters were found, each cluster groups similar eye appearances.

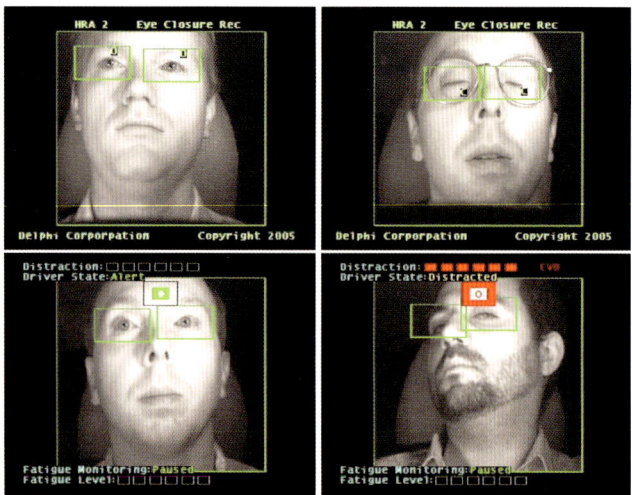

Figure 9.21. Snapshots of two applications of eye tracking: (1) *fatigue assessment*: the eye state is classified as open (top left) or closed (top right); and (2) *distraction assessment*: as the head pose is frontal (bottom left) or non-frontal (bottom right) for a while, the driver is recognized as non-distracted (bottom left) or distracted (bottom right). Both eyes were detected and tracked in real-time. Some human factor algorithms being used to measure distraction from the head pose classification results can be found in [18, 19, 24].

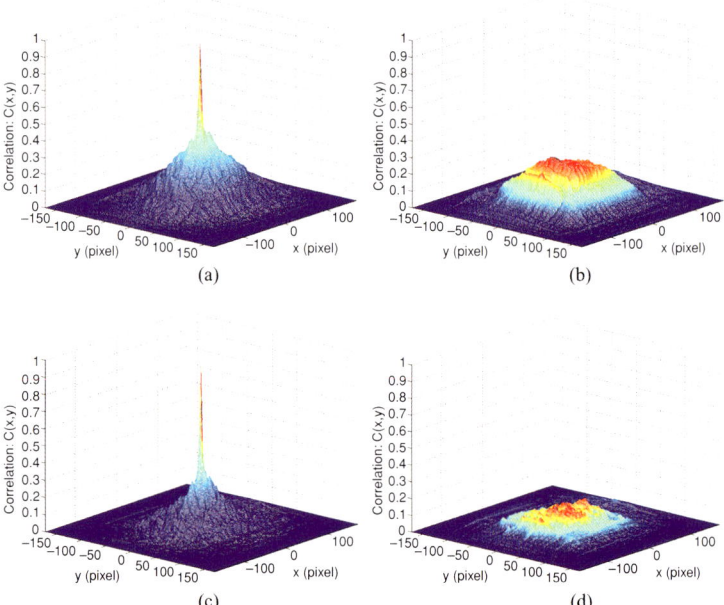

Figure 11.7. (a)–(d) Normalized correlation when input image #8 of the nontraining true targets and input image #13 among the false class objects are used as input scenes for (a)–(b) single exposure on-axis digital holography and (c)–(d) on-axis phase-shift digital holography.

even though the moments of each mode are invariant to dispersion. Accordingly, mode separation techniques are important to apply in order to obtain the dispersion-invariant central moments per mode.

Acknowledgment

The work of Leon Cohen was supported by the Air Force Office of Scientific Research and the NSA HBCU/MI program. The work of Patrick Loughlin was supported by the Office of Naval Research (N00014-02-1-0084 and N00014-06-1-0009).

References

[1] L. Cohen. *Time-Frequency Analysis*. Prentice-Hall, 1995.

[2] L. Cohen. Time-frequency spatial-spatial frequency representations. In: *Nonlinear Signal Analysis and Image Analysis*. Annals of the New York Academy of Sciences, J.R. Buchler and H. Kundrup, eds., pp. 97–115, 1997.

[3] L. Cohen. Characterization of transients. *Proc. SPIE*, 3069:2–15, 1997.

[4] L. Cohen. Pulse propagation in dispersive media. *IEEE Proc. SSAP*, pp. 485–489, 2000.

[5] L. Cohen. The Wigner distribution and pulse propagation. *Proc. SPIE*, 4474:20–24, 2001.

[6] L. Cohen. Why do wave packets sometimes contract? *J. Mod. Optics*, 49:2365–2382, 2003.

[7] K. Graff. *Wave Motion in Elastic Solids*. Oxford University Press, 1975.

[8] J. D. Jackson. *Classical Electrodynamics*. Wiley, 1992.

[9] L. Jacobson and H. Wechsler. Joint spatial/spatial-frequency representations. *Signal Process*, 14:37–68, 1988.

[10] J. Lighthill. *Waves in Fluids*. Cambridge University Press, 1978.

[11] P. Loughlin and L. Cohen. Local properties of dispersive pulses, *J. Mod. Optics*, 49(14/15):2645–2655, 2002.

[12] P. Loughlin and L. Cohen. Moment features invariant to dispersion, Proc. SPIE Automatic Target Recognition XIV, edited by Firooz A. Sadjadi, Vol. 5426, pp. 234–246, 2004.

[13] P. Loughlin and L. Cohen. Phase-space approach to wave propagation with dispersion and damping, 48th Annual Meeting, International Society for Optical Engineering, 2–6 August 2004, Denver, Colorado, vol. 5559, pp. 221–231.

[14] P. H. Morse and K. U. Ingard. *Theoretical Acoustics*. McGraw-Hill, 1968.

[15] I. Tolstoy and C. Clay. *Ocean Acoustics: Theory and Experiment in Underwater Sound*. AIP, New York, 1987.

[16] G. Whitham. *Linear and Nonlinear Waves*. J. Wiley and Sons, New York, 1974.

8

Multisensor Target Recognition in Image Response Space Using Evolutionary Algorithms

Igor Maslov[1] and Izidor Gertner[2]

[1]Department of Computer Science, CUNY/Graduate Center, 365 Fifth Avenue, New York, NY, 10016
[2]Department of Computer Science, CUNY/City College, Convent Avenue and 138th Street, New York, NY, 10031

8.1 Introduction

Development of efficient methods for automatic target recognition on the battlefield is one of the important research areas of electronic imaging. Evolutionary algorithms (EA) have been successfully used for solving a few electronic imaging problems closely related to target recognition such as pattern matching, semantic scene interpretation, and image registration.

A version of the classical Genetic algorithm (GA) with roulette-wheel selection, single point crossover, and uniform mutation is used in [1] to solve incomplete pattern matching problem for two point sets P_1 and P_2 under affine transformation. The approach reduces parameter space by constructing feature ellipses of the point sets P_1 and P_2. A reference point triplet is selected on the feature ellipse of the point set P_1 and the corresponding triplet in the point set P_2 is sought with GA. Trial point triplets in P_2 are binary encoded in a chromosome. The chromosome that minimizes the Hausdorff distance between the point sets provides the proper parameters of the affine transformation between P_1 and P_2.

In [1, 2] a GA-based approach is applied to semantic scene interpretation, i.e., recognizing objects present in the image of a scene, and labeling them according to their appropriate semantic classes. An individual chromosome encodes a feasible mapping between different segments of an image and classes of objects in the semantic net that establishes predicate relations between the classes. The quality of the chromosome, i.e., its fitness is evaluated by calling a set of feature comparison functions, each of which computes the degree of correspondence between the feature sets of an image segment and the feature sets of objects belonging to a particular class. The chromosome that maximizes the sum of all predicates provides the optimal classification scheme.

Registration of two images obtained from different sources includes finding unknown parameters of geometric transformation which properly align both images. A method for automatic registration of medical images with the classical variant of GA is proposed in [2–6]. A chromosome encodes parameters of geometric transformation between the images including rigid body rotation and translation, and limited elastic motion. The quality of the chromosome is evaluated via the quality of the mapping between the images provided by the encoded parameters. In order to facilitate fitness evaluation a small subset of image pixels is randomly selected for evaluation based on statistical sampling. This technique allows to evaluate larger number of candidate solutions during one iteration of the algorithm, as opposed to a more accurate but slower evaluation of all image pixels.

From the standpoint of the method used to evaluate the quality of the candidate solution, current applications of EA in electronic imaging can be broken into two main groups:

1. The first group of methods directly compares distributions of the pixel values in images. If the light conditions are changing between the images, the comparison becomes difficult since no matching pixels can be found. Moreover, comparison of different types of imagery, e.g., infrared and real visual images obtained from different types of sensors (i.e., multisensor image fusion) becomes virtually impossible with these methods.
2. Methods in the second group attempt to find a set of salient characteristics, i.e., features that are common for the compared images. Choosing appropriate features is not a trivial task, which becomes even more complex when images are distorted by some kind of geometric transformation, e.g., affine or perspective.

This chapter evaluates the feasibility of extending the evolutionary approach to the problem of recognizing a target in a scene where the target and the scene have different imagery types, e.g., their images are obtained from different types of sensors. The approach is based on a peculiar image transformation called image local response (ILR) proposed in [7] for automatic selection of image features that are most responsive to a particular geometric transformation. As shown in [7], the ILR transformation helps improve the performance of the hybrid evolutionary algorithm (HEA) in solving electronic imaging problems. Method discussed in this chapter utilizes the fact that ILR is significantly invariant to the type of imagery, which makes it particularly suitable for object and target recognition from multiple sensors. The key idea of the method is to transfer the search for a target from the real visual space into image response space defined by the ILR transformation.

The chapter is organized as follows. Section 8.2 describes the proposed method of target recognition from multiple sensors; it gives a brief overview of the main features of the response-based hybrid model of EA used in this work. Section 8.3 discusses experimental results of target recognition in the case of different imagery types. Section 8.4 concludes the chapter with the summary of the findings.

8.2 Hybrid Model of Evolutionary Algorithm in Target Recognition with Different Imagery Types

Target recognition can be stated as a problem of finding a particular object (i.e., a target) with the known signature (i.e., an image) Img_T in the image Img_S of a scene. Either image can be distorted by, e.g., affine or perspective transformation; the scene might also contain objects other than the target. The problem can be formulated mathematically as a search for an appropriate geometric transformation A providing the correct aligning of the image of the target with the corresponding segment of the scene [8]. This chapter uses a particular model of the HEA to conduct the search. A trial vector of parameters V defining the sought transformation A is real-valued encoded in a chromosome. The chromosome V^* that minimizes a measure of the difference F between the images Img_T and Img_S identifies and aligns the recognized target. The chromosome's fitness, i.e., the measure of the difference F can be defined as the sum of the squared differences of the gray values of the images, i.e.,

$$F = \frac{\sum_{\Omega}(g_T(x',y') - g_S(x,y))^2}{\Omega^2}, \tag{8.1}$$

where $g_T(x',y')$ and $g_S(x,y)$ are the gray values of the images Img_T and Img_S, respectively and Ω is the area of their overlap.

The hybrid model of EA used in this chapter for solving the target recognition problem utilizes hybridization in a broader sense [7–9]. In particular, the following techniques are incorporated in the evolutionary search:

1. Direct local search, i.e. hybridization in the narrower sense of this term.
2. Random search with uniform probability.
3. Preservation and analysis of the history of the search.
4. Utilization of problem-specific knowledge.

Incorporating traditional methods of local search in the evolutionary approach is a well known hybridization technique commonly used to refine the best fraction of the current population [10, 11]. However, direct application of local search in electronic imaging can result in a few computational problems that can be mitigated with a two-phase cyclic search procedure based on the modified Downhill simplex method (DSM) and random search [8].

The classical DSM is one of the popular choices among practitioners because it has low computational cost along with the relatively high robustness and does not require computing the derivatives of the objective function

[12, 13]. However, direct local search with DSM in a multidimensional parameter space can occasionally converge to a suboptimal solution. In order to redefine or improve the current configuration of the simplex, random search with the uniform probability in the locality of the existent simplex is periodically invoked by the algorithm as an interim alternative procedure.

During the search EA samples numerous points in the N-dimensional parameter space. Preserving the history of the search in a way similar to Tabu search introduces additional elements of organization in the selection-driven random evolutionary process by structuralizing the parameter space. In particular, the algorithm used in this chapter performs bookkeeping operations that affect local search and mutation as described below.

The local search procedure keeps track of the neighbors selected during random search by building a binary tree with the selected chromosome $\mathbf{V}_i$ as the initial root of the tree. Once the designated number of the chromosome's neighbors has been evaluated, the tree is rotated to the right, so that the individual $\mathbf{V}_j$ with the lowest value of the fitness function F_j replaces the chromosome $\mathbf{V}_i$, and becomes the new root of the tree. When the local procedure switches to DSM, it chooses the chromosome $\mathbf{V}_j$ as the first vertex of the initial simplex. The other N vertices are picked from the right brunch of the tree; these vertices have the next N lowest values of the fitness function. The selection technique ensures that the best-known simplex in the local neighborhood is formed as the starting point for the refined DSM search.

Sampling parameter space with operations of crossover, mutation, and local search, as well as favoring and promoting the best up-to-date individuals result in the uneven representation of different subregions in the chromosome pool, which leads to what is known as exploitation vs. exploration problem. Bookkeeping is used to keep track of the usage of the search space, and mutation is performed at random only in the subregions that have been least represented up-to-date in the population. This technique helps maintain the fair usage of the search space, especially at the initial stages of the algorithm run.

One of the attractive features of EA is its versatility: the algorithm can be applied to a broad spectrum of problems without involving problem-specific knowledge. The flip side of this versatility is that many ad hoc methods utilizing relevant problem-specific information might outperform EA when solving these problems. There are two principal ways in which problem-specific knowledge can be incorporated in the algorithm in order to improve its performance:

1. To provide the algorithm with the specific data about the problem at hand. This approach typically requires human intervention, and can result in the loss of so much appreciated generality of the algorithm.
2. To provide the algorithm with the general method that can be invoked within the EA framework to autonomously obtain specific data about the problem at hand.

ILR is a general method that can provide evolutionary search with important data related to the particular electronic imaging problem [9]. ILR is defined as the variation of the objective (i.e., fitness) function F which occurs because of a small variation of the parameter vector $\boldsymbol{V}$ in the N-dimensional parameter space. The response is computed over a small pixel area ω as follows:

$$F = \frac{\sum_{\omega}(g(x',y') - g(x,y))^2}{\omega^2}, \qquad (8.2)$$

where $g(x,y)$ and $g(x',y')$ are the respective gray values of the image before and after the variation of the parameter vector has occurred.

ILR can significantly reduce the computational cost of the HEA in electronic imaging applications [7]. There are two important properties of ILR that can be effectively utilized in target recognition:

1. Response emphasizes segments of an image that are most responsive to a particular geometric transformation. These segments identify the salient features of the image that are particularly important in solving various image mapping problems.
2. Response is computed by mapping an image onto itself which makes the result of the mapping largely invariant in relation to imagery type.

The key idea of the proposed method of target recognition from multiple sensors is to utilize the above-formulated properties of ILR, and to transform the originally stated problem of target recognition in real visual space G into a new problem of target recognition in image response space R. The new problem can be formulated now as a search for an appropriate transformation A providing the correct aligning of ILR $r_T(x,y)$ of the target with ILR $r_S(x,y)$ of the scene. The matrices of ILR for both images are computed at the pre-processing stage. As in the original statement of the target recognition problem, a chromosome encodes a trial vector of parameters $\boldsymbol{V}$ defining the sought geometric transformation A. The chromosome $\boldsymbol{V}^*$ that minimizes a measure of the difference F between the responses r_T and r_S of the images Img_T and Img_S identifies and aligns the recognized target. The sum of the squared differences of the response values of the images Img_T and Img_S is used as the measure of the difference F between the images, i.e.,

$$F = \frac{\sum_{\Omega}(r_T(x',y') - r_S(x,y))^2}{\Omega^2}, \qquad (8.3)$$

where $r_T(x',y')$ and $r_S(x,y)$ are the ILR values of the images Img_T and Img_S, respectively, and Ω is the area of their overlap. The use of the imagery-invariant response values $r_T(x',y')$ and $r_S(x,y)$ in formula (8.3) makes the proposed method of target recognition fundamentally different from the traditional methods based on the comparison of the actual pixel values $g_T(x',y')$ and $g_S(x,y)$ as in formula (8.1).

Next section discusses the results of the computational experiments that support the proposed approach to target recognition with the HEA in the response space in the case of different imagery types.

8.3 Computational Experiments with Different Imagery Types

Computational experiments were conducted on a set of two-dimensional synthetic grayscale images shown in Fig. 8.1. The set includes a wireframe M_W and a realistic solid model M_S of a target, and a realistic image of a scene with the target. The wireframe and the solid models exhibit a significant

Figure 8.1. A set of synthetic test images: a wireframe model of the target (top left), a realistic solid model of the target (top right), and a realistic scene with the target (bottom).

Table 8.1. Ranges and step sizes of parameters in computational experimants.

Parameter	DX	DY	θ	SX	SY	SHX	SHY
Min value	0	0	0.0	1.0	1.0	0.0	0.0
Max value	299	299	6.28	4.0	4.0	2.0	2.0
	255	255		3.0	3.0		
Step size	1	1	0.1	0.1	0.1	0.1	0.1
	2	2					

difference in imagery type, which simulates the difference in the images of the same target obtained from the sensors of different types. The images of the target were distorted by affine transformations defined in general case by the vector of parameters

$$\mathbf{V} = \{DX, DY, \theta, SX, SY, SHX, SHY\}, \tag{8.4}$$

where DX and DY are translations along the x and y axes, θ is rotation in the xy plane, SX and SY are scaling factors along the x and y axes, and SHX and SHY are shears along the x and y axes. The target recognition problem was stated as a search for the vector of parameters $\mathbf{V}^*$ minimizing the difference F between the images Img_T and Img_S of the target and the scene, respectively. The search was conducted with the hybrid model of EA described in Section 8.2. The ranges and the step sizes of the sought parameters are presented in Table 8.1. The experiments were conducted in real visual space G, i.e., by comparing the pixel gray values $g(x, y)$ of the original images, and in response space R, i.e., by comparing the response values $r(x, y)$ of the images.

The experiments had the following objectives:

1. To show that different types of imagery cannot be used in real visual space G for target recognition with EA because fitness functions corresponding to the wireframe and solid models have different optimal values in G.
2. To validate the use of image response space R in target recognition with EA by showing that fitness functions corresponding to the solid model have the same optimal values in both real visual G and response R spaces.
3. To show that different types of imagery can be used in image response space R for target recognition with EA because fitness functions corresponding to the wireframe and solid models have similar optimal values in R.
4. To show that the performance of the local DSM search does not degrade when the search is relocated into image response space.
5. To show that target recognition problem for different imagery types can be successfully solved with the HEA in image response space R.

The feasibility of conducting the search in real visual space G for different imagery types was assessed by comparing the gray values of the wireframe and solid models with the gray values of the scene, i.e., fitness F was

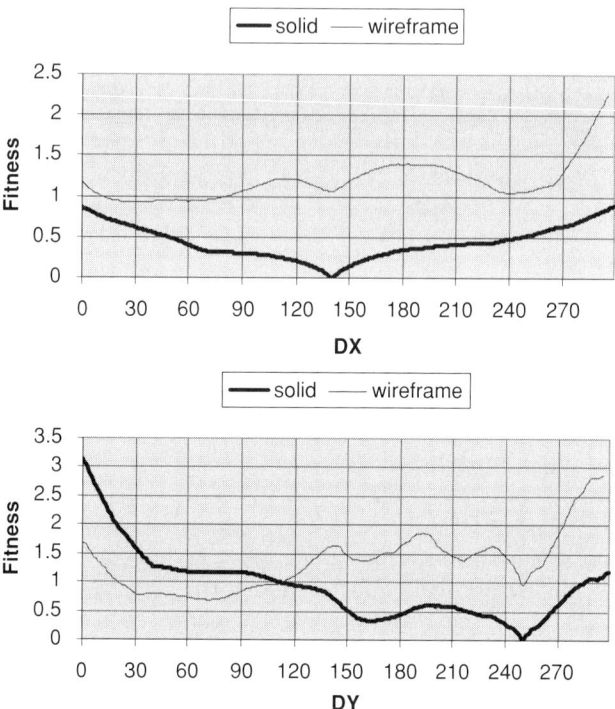

Figure 8.2. Fitness function corresponding to the solid and wireframe models along the optimal cross sections of the scene in real visual space G:$DY = 250$ (top); $DX = 140$ (bottom).

computed according to formula (8.1). The parameters θ, SX, SY, SHX, and SHY of the chromosome V were set to their optimal values, while the translations DX and DY varied between their minimum 0 and maximum 299 values. The computed values of the fitness function along the optimal cross sections $DX = 140$ and $DY = 250$ for both models are shown in Fig. 8.2. The solid model clearly expresses global minima along both optimal cross sections, which corresponds to the correct aligning of the target and the scene. The wireframe model exhibits only local minima at the correct locations along both cross sections; it identifies the erroneous positions of the global minima $DX = 32$ and $DY = 76$. This result clearly indicates that the wireframe model cannot be used for recognizing the realistic solid target in real visual space G because of the significant difference in the imagery type between the models.

In order to validate the use of image response space R in target recognition with EA, ILR was computed for all test images. The images of the response values are shown in Figs. 8.3 and 8.4. The lighter areas with higher intensity (i.e., larger gray values) correspond to larger response values. The responses of the wireframe and solid models exhibit a high degree of similarity;

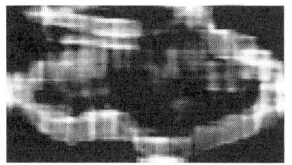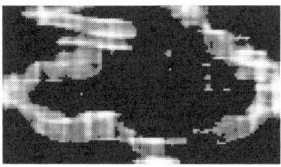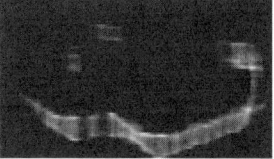

Figure 8.3. Images of ILR computed for test images (from the left to the right): A wireframe model, a wireframe model after applying intensity threshold, a solid model.

both responses highlight the main shape features of the target. The wireframe response has more low-intensity details which can be discarded after applying an intensity threshold.

The values of the fitness function along the optimal cross sections of the scene were computed for the solid model in image response space R using formula (8.3), i.e., by comparing the response values $r(x, y)$ of the solid model and the scene. The computed values plotted in Fig. 8.5 clearly indicate the correct locations of the global optimum in x and y directions, i.e., the optimal values $DX = 140$ and $DY = 250$. The fact that the fitness function corresponding to the solid model does not change its behavior when the search is conducted in image response space validates the feasibility of using response space R instead of real visual space G in target recognition with EA.

Once the validity of relocating the search space from real space G into response space R has been proven, the most important question arises whether the behavior of the fitness function F in response space is similar for both wire-

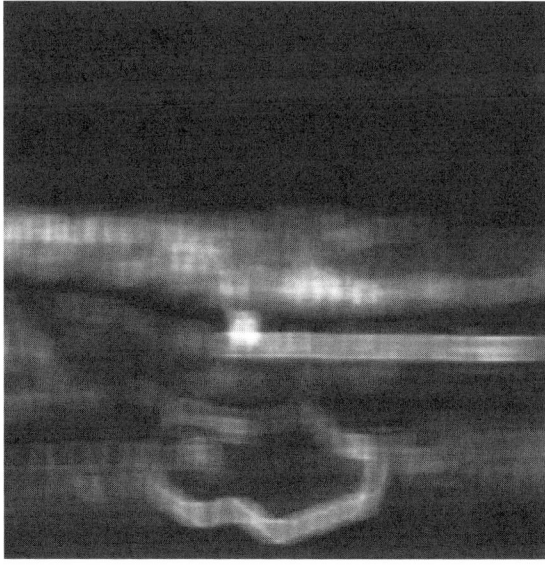

Figure 8.4. Image of ILR computed for the test image of a scene.

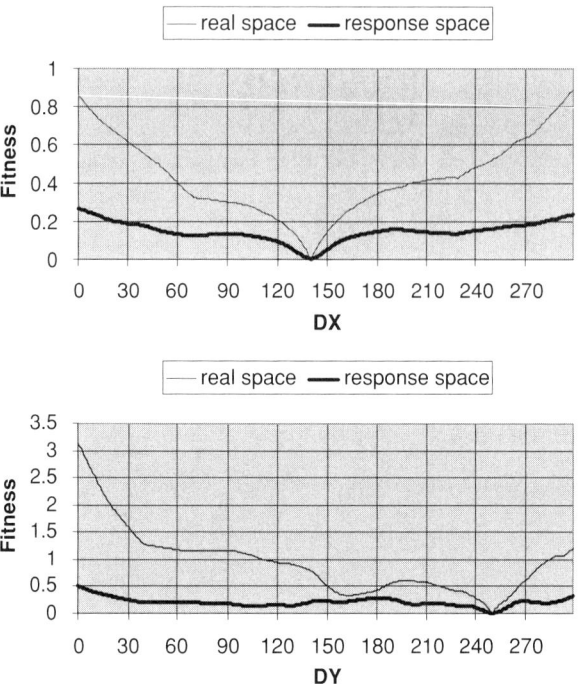

Figure 8.5. Fitness function corresponding to the solid model along the optimal cross sections of the scene in real visual G and response R spaces: $DY = 250$ (top); $DX = 140$ (bottom).

frame and solid models. The computed values of the fitness function along the optimal cross sections for both models are compared in Fig. 8.6. The global minima along the x and y directions computed for the wireframe model correspond to $DX = 144$ and $DY = 254$ which are very close (with the maximum error 3%) to the minimum positions for the solid model. This important result clearly indicates that the wireframe model can be used for recognizing the realistic solid target in image response space R.

One of the main features of HEA is utilization of local search within the global evolutionary procedure. The objective of the next experiment was to show that the performance of the local DSM search did not degrade when the search was conducted in response space R. The performance was assessed as the number of fitness evaluations required to find the optimal solution. A set of 100 test runs of the local DSM search for the solid and wireframe models in real G and response R spaces was performed. For every test run, the initial simplex was placed in the proximity of the optimal solution using the following technique. The value of each parameter of the vector V for each of the vertices was independently drawn at random with the uniform probability from the ($\pm 10\%$) range of the corresponding range centered at the parameter's

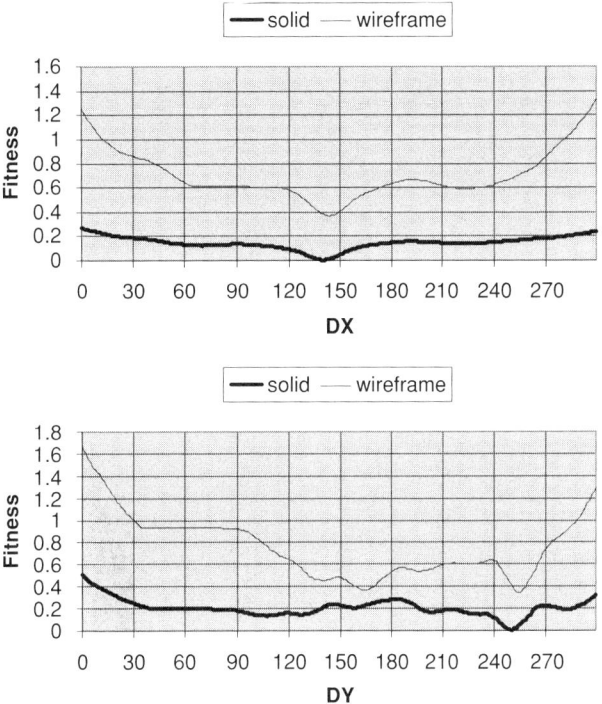

Figure 8.6. Fitness function corresponding to the solid and wireframe models along the optimal cross sections of the scene in response space: $DY = 250$ (top); $DX = 140$ (bottom).

optimal value. For example, the value of the translation DX for a 300×300-pixel image of the scene was drawn from the interval (140.0 ± 30). Figure 8.7 shows a sample image of the transformed solid model.

The comparative results for the number of fitness evaluations for all test runs are shown in Fig. 8.8. The performance of the DSM search for the solid model is similar in both real and response spaces: the total number of fitness evaluations over 100 runs was 8681 in real space and 8472 in response space. The performance of the local DSM search for the wireframe model in response space was even better than the performance of the solid model: The total number of fitness evaluations was 5354. These results assure that the performance of local search at least does not degrade when the search is conducted in response space for imagery of different types.

Finally, the hybrid model of EA discussed in Section 8.2 was used to solve the target recognition problem for imagery of different types. The distorted image of ILR of the wireframe model was used to recognize the target in the image of ILR of the scene containing the solid model. In the first test run of the algorithm the wireframe model was subject to affine transformation

Figure 8.7. Sample transformed image of solid model with 10% parameter range about the optimal position.

defined by the four-dimensional vector

$$V = \{DX, DY, SX, SY\}, \tag{8.5}$$

where $SX = SY$ (i.e., isotropic scaling of the image).

In the second test run the wireframe model was subject to affine transformation defined by the five-dimensional vector

$$V = \{DX, DY, \theta, SX, SY\}, \tag{8.6}$$

where $SY = 2SX$ (i.e., non-isotropic scaling of the image).

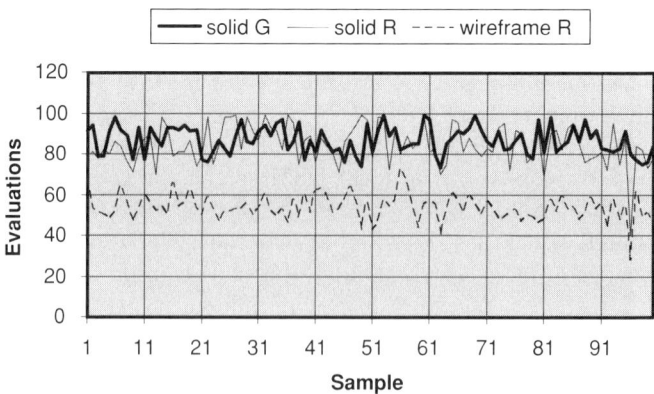

Figure 8.8. Performance of local DSM search for solid and wireframe models in real G and response R spaces.

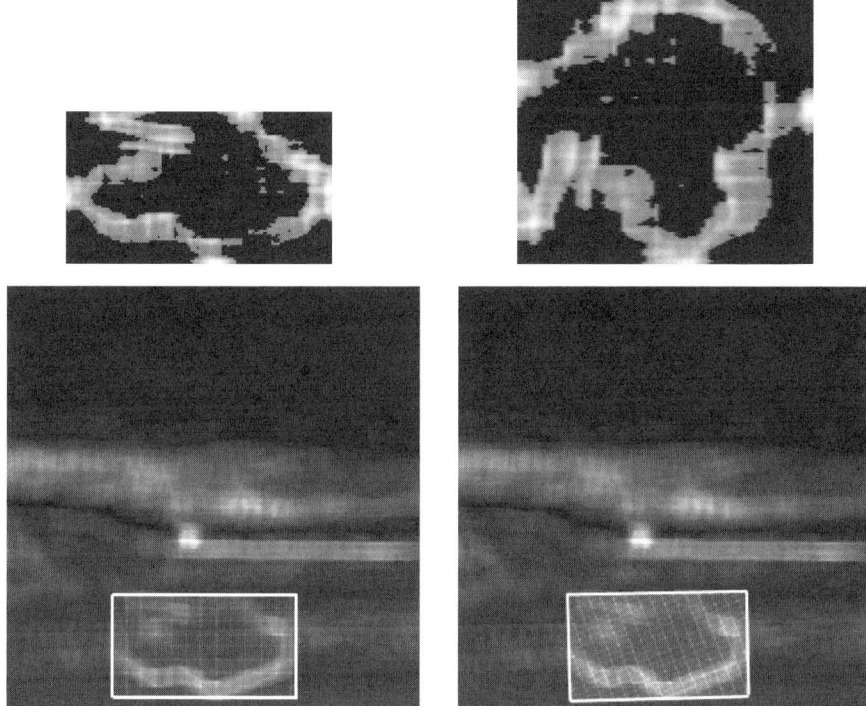

Figure 8.9. Images of the target and results of recognition: four-parameter test (left); five-parameter test (right).

The first test run was performed with the population of 137 chromosomes initially selected at random with the uniform probability. The population size in the second test run was 260. The images of the target used in the search, and the results of the recognition are shown in Fig. 8.9, where the rectangular segment of the scene corresponding to the image of the target is highlighted. Algorithm was able to correctly recognize and align the target in the scene in both test runs. The total number of fitness evaluations in the first test was 119,454 with 12,153 evaluations attributed to the local DSM search. The total number of fitness evaluations in the second test was 117,982 with 13,898 evaluations attributed to the local DSM search.

8.4 Conclusions

EA has been successfully used for solving a few electronic imaging problems closely related to target recognition. However, the existing methods are based either on direct comparison of the pixel values or on extracting the salient features of images. Both approaches can show poor performance when images of the target are obtained from the sensors of different types. Moreover, the images of the target can be geometrically distorted. This chapter evaluates the feasibility of target recognition with the HEA for different imagery types by relocating the search from real visual space into image response space.

Computational experiments conducted on a test set of two-dimensional synthetic grayscale images allow to draw the following conclusions:

1. Different types of imagery cannot be directly used in real visual space for target recognition with HEA because fitness functions corresponding to images of different types have different optimal values in real space.
2. Evolutionary search can be relocated into image response space because fitness functions corresponding to the real image and to its response have similar optimal values in both real and response spaces.
3. Different types of imagery can be used in image response space because fitness functions corresponding to images of different types have similar optimal values in response space.
4. The performance of the local DSM search does not degrade when the search is relocated into image response space.
5. Target recognition problem for different imagery types can be successfully solved with the HEA in image response space.

Further research is needed to extend the proposed approach to more complex problems of target recognition including cluttered and noisy scenes, and recognition of three-dimensional targets.

Acknowledgement

This research was partially sponsored by ONR NO: N000140210122.

References

[1] L. Zhang, W. Xu, and C. Chang. Genetic algorithm for affine point pattern matching. *Pattern Recognit. Lett.*, 24(1–3):9–19, 2003.
[2] C. A. Ankenbrandt, B. P. Buckles, and F. E. Petry. Scene recognition using genetic algorithms with semantic nets. *Pattern Recognit. Lett.*, 11(4):231–304, 1990.
[3] B. P. Buckles and F. E. Petry. *Cloud Identification Using Genetic Algorithms and Massively Parallel Computation*. Final Report, Grant No. NAG 5-2216, Center for Intelligent and Knowledge-Based Systems, Dept. of Computer Science, Tulane University, New Orleans, LA, June 1996.
[4] J. M. Fitzpatrick and J. J. Grefenstette. Genetic algorithms in noisy environments. *Mach. Learn.*, 3(2/3):101–120, 1988.
[5] J. J. Grefenstette and J. M. Fitzpatrick. Genetic search with approximate function evaluations. In *Proceedings of the First International Conference on Genetic Algorithms and Their Applications*. Lawrence Erlbaum, Pittsburgh, PA, pp. 112–120, 1985.

[6] V. R. Mandava, J. M. Fitzpatrick, and D. R. Pickens III. Adaptive search space scaling in digital image registration. *IEEE Trans. Med. Imaging*, 8(3):251–262, 1989.

[7] I. V. Maslov. Reducing the cost of the hybrid evolutionary algorithm with image local response in electronic imaging. In *Lecture Notes in Computer Science, vol. 3103*, Springer-Verlag, Berlin, Heidelberg, pp. 1177–1188, 2004.

[8] I. V. Maslov and I. Gertner. Object recognition with the hybrid evolutionary algorithm and response analysis in security applications. *Opt. Eng.*, 43(10):2292–2302, 2004.

[9] I. V. Maslov and I. Gertner. Reducing the computational cost of local search in the hybrid evolutionary algorithm with application to electronic imaging. *Eng. Optim.*, 37(1):103–119, 2005.

[10] W. E. Hart and R. K. Belew, Optimization with genetic algorithm hybrids that use local search. In *Adaptive Individuals in Evolving Populations: Models and Algorithms: Proccedings of Santa Fe Institute Studies in the Sciences of Complexity*, 26. Addison-Wesley, Reading, MA, pp. 483–496, 1996.

[11] J. A. Joines and M. G. Kay. Utilizing hybrid genetic algorithms. In *Evolutionary Optimization*. Kluwer, Boston, MA, pp. 199–228, 2002.

[12] J. A. Nelder and R. Mead. A simplex method for function minimization. *Comput. J.*, 7(4):308–313, 1965.

[13] M. H. Wright. Direct search methods: Once scorned, now respectable. In *Numerical Analysis, 1995: Proceedings of the 1995 Dundee Biennial Conference in Numerical Analysis*. Addison Wesley Longman, Harlow, UK, pp. 191–208, 1996.

9

Biophysics of the Eye in Computer Vision: Methods and Advanced Technologies

Riad I. Hammoud[1] and Dan Witzner Hansen[2]

[1]Delphi Electronics & Safety, One Corporate Center, P.O. Box 9005, Kokomo, IN 46904-9005, USA, riad.hammoud@delphi.com
[2]IT University, Copenhagen, Rued Langaardsvej 7, 2300 Copenhagen S, Denmark.

9.1 Chapter Overview

The eyes have it! This chapter describes cutting-edge computer vision methods employed in advanced vision sensing technologies for medical, safety, and security applications, where the human eye represents the object of interest for both the imager and the computer. A camera receives light from the real eye to form a sequence of digital images of it. As the eye scans the environment, or focuses on particular objects in the scene, the computer simultaneously localizes the eye position, tracks its movement over time, and infers measures such as the attention level, and the gaze direction in real time and fully automatic. The main focus of this chapter is on computer vision and pattern recognition algorithms for eye appearance variability modeling, automatic eye detection, and robust eye position tracking. This chapter offers good readings and solid methodologies to build the two fundamental low-level building blocks of a vision-based eye tracking technology.

9.1.1 Eye Tracking-Based Applications

The eye tracking research and application community has grown remarkably in the past few years [1], and thus eye tracking systems appear to be ready for utilization in various areas of human computer interaction. In [75] eye tracking is used for examining the effectiveness of marketing material and newspaper layouts. Andersen et al. [4] have used eye tracking in interior design and ergonomic evaluation such as in airplane cockpits. In vision-based automotive applications, human factors use eye tracking systems to assess the driver state such as fatigue and distraction level [18, 19, 24, 45, 64, 65, 77].

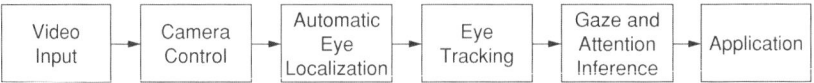

Figure 9.1. Snapshot of main modules in a vision-based eye tracking sensing system.

The use of eye gaze as a pointing device is a straight forward approach when considering real-time interaction. Information about the users intentions (like browsing and reading) can be exploited in applications such as eye typing and as an aid for reading foreign text [49]. More recently gaze determination is used in connection with rendering digital displays for example by only displaying the areas of the gaze fixations in high resolution [6, 53, 55]. Such applications require robust, real-time, and low-cost eye tracking components.

9.1.2 Eye Tracking Modules

A general flowchart of a typical eye tracking system is depicted in Figure 9.1. The camera control module adjusts the intensity distribution to match a predefined model of the eye or face image. This module is very useful for systems that run in all weather conditions and for any subject. The subject eye is then searched in the entire frame. Once it is detected, the tracking module takes over and tracks it in next frames. These two modules, automatic eye detection and eye location tracking, represent the fundamental low-level building blocks in an eye tracking system and will be reviewed in details in this chapter. Afterward, one could analyze the tracked eye region to compute for instance the eye gaze and/or the eye closure state. The gaze direction determines where the subject is looking at. The eye closure state determines the percentage of eye opening or eye closing. Such high-level features are cumulated over time and used in estimating system features like distraction, attention and fatigue level, drowsiness state, etc.

9.1.3 Chapter Organization

In Section 9.2 we will review some background materials about light spectrum and the responses of the pupil eye in some specific near infrared light setup. In Sections 9.3 and 9.4 the important issue of eye appearance changes and eye representation are brought up and discussed in details. Then, some adequate approaches to model the appearance variability in feature spaces are described in Section 9.5. The variability is due to the existing difference in eye shapes among people, contrast, sensitivity, partial eye occlusion, iris orientation, and cluttered background. We will present the state-of-the-art automatic eye detection in Section 9.6. The tracking problem is formulated in Section 9.7. A brief review of tracking methods is given in Section 9.8. In Sections 9.9, 9.9.2 and 9.9.3, a particular focus is placed on particle filter theory and its adaptation to dark-bright pupil observations and iris-contour data. Some advanced eye tracking–based technologies are introduced in Section 9.10. Finally, the chapter is concluded in Section 9.11.

9.2 Background Review: Light, Eye Responses, and Movements

Lighting setup can make the task of computer vision easier or harder. One could say that its position in relation to vision is the same as the position of the axis in relation to the hand-mill. A better lighting equals greater success and less effort in vision sensing algorithm research and development. A poor lighting results into missperception and lack of information about the physical world.

9.2.1 Electromagnetic Spectrum

The light is emitted from an object which is then captured by a light detector such as human eyes and imagers. The light, which is a wave of discrete photons, moves at a specific wavelength λ of the electromagnetic spectrum (EM) range. Photons with a specific wavelength have a specific energy which describes the best the relationship between light and heat or light and electricity. The electromagnetic spectrum is a map of all the types of light that are identified (see Figure 9.2). It separates all the types of light by wavelength

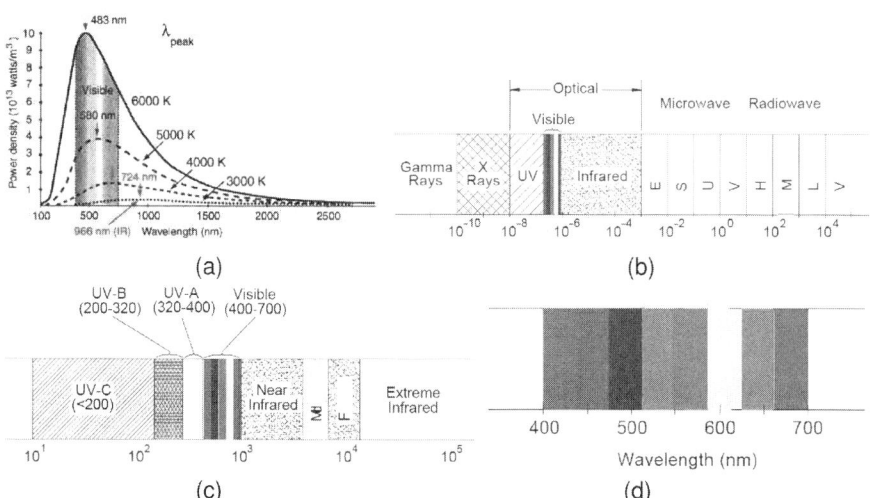

Figure 9.2. Illustrations of the blackbody radiation curve (a), the electromagnetic (b), optical (c), and visible spectrum (d). The wavelengths are in nanometers (nm). The peak of the blackbody radiation curve gives a measure of temperature (see color insert).

because that directly relates to how energetic the wave is. Less energetic waves have longer wavelengths while more energetic waves have shorter wavelengths (see Figure 9.2(a)). The human eye is sensitive to the narrow range of visible light (between 400 and 700 nm, see Figure 9.2), and has its peak sensitivity to bright light at approximately 550 nm, which is close to the peak of the emission curve for the sun.

9.2.2 Pupil Responses

In machine vision, light acquired by the detector is aimed to provide sufficient information (contrast) for detector to separate the primary features of interest, and for the processor to distinguish them. Therefore, the primary goal of machine vision lighting engineers is to maximize feature of interest contrast while minimizing all other features.

In the eye tracking vision-based sensing world, it is found that there are two types of illumination techniques to acquire the human eye pupil: *bright pupil* and *dark pupil*. These phenomena are related to the biophysics of the eye, infrared-eye responses, and illuminator and camera placement. The infrared light impinging on the eye can be reflected, absorbed, and/or transmitted. The bright-pupil technique lines up the optical axes such that the light entered to the retina is bounced back (retro-reflected) within the field of view of the camera. The bright-pupil technique provides a nice boundary of the pupil image; it looks like a "moon-like" image. The other case is using the dark-pupil technique where the majority of the light entered is not bounced back along the axis of the optical measurement system. In order to produce these two physiological eye properties, two banks of infrared sources (LEDs) are mounted on- and off-optical axis of the camera. These two LEDs are turned on and off in order to produce an interlaced image. An illustration of the camera and LEDs setup is shown in Figure 9.3, while Figure 9.4 shows two generated pupil images.

9.2.3 Eye Movements

Eye movements and neuroscientific studies have revealed many important findings of the mammal visual system. Our eyes are constantly directed to a series of locations in the visual field. These movements have been shown to be essential to our visual system and can be subdivided into the following basic types [46]:

1. **Adaption and accommodation** are nonpositional eye movements which are used for pupil dilation and focusing.
2. **Rolling** of the eyes is a rotational motion around an axis passing through the fovea and pupil. It is involuntary, and is influenced by, among other things, the angle of the neck [36].

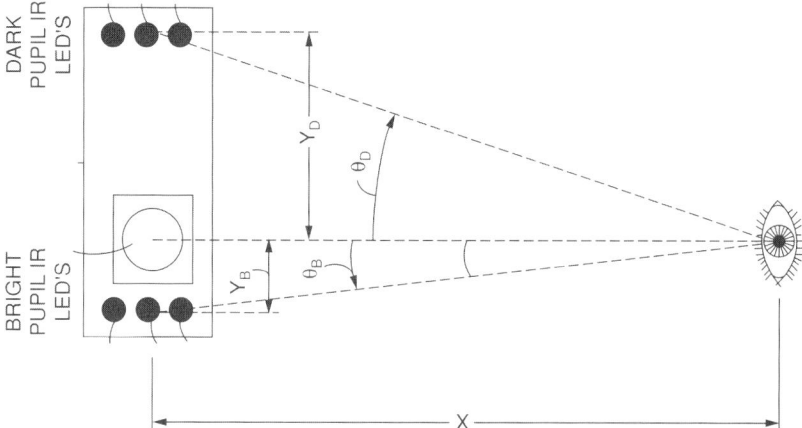

Figure 9.3. Illustration of the LEDs setup as implemented in [27] to generate dark and bright pupil images. The first LED is located at an angle less than 2.5 degrees of the imaging axis, while the second light source is arranged at an angle greater than 4.5 degrees from the imaging axis for illuminating the eye. The first light produces bright pupil, while the second light produces dark pupil. When the light source is placed off-axis (top), the camera does not capture the light returning from the eye.

3. **Saccadic Movements** is the principal method for repositioning the eye (fovea) to a location in the visual field. The movements are rapid and can be both voluntary and reflexive. The duration of saccade movements varies between 10 and 100 ms and takes about 100–300 ms to initiate. The peak velocity of a saccade can be as much as 900 deg/sec [74].

4. **Smooth pursuit** is a much smoother, slower movement than a saccade; it acts to keep a moving object foveated allowing the object to be tracked. It cannot be induced voluntarily, but requires a moving object in the visual field.

5. **Vergence movements** are used to position both eyes when the distance from the observer to the foveated object is changed; the closer the object is, the more the eyes point toward each other. Vergence movement is normally

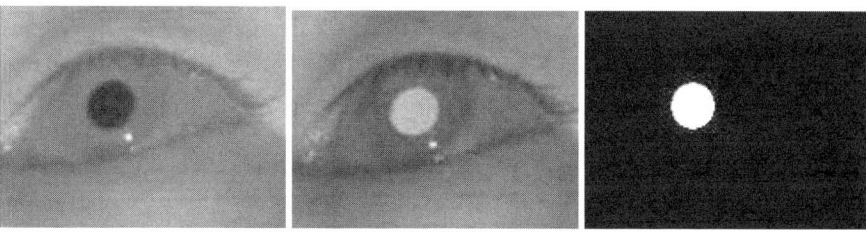

Figure 9.4. Examples of a dark and bright pupil images and their corresponding pixel-to-pixel enhanced subtraction image.

the result of a moving stimulus, but may be voluntary and are thus either saccadic or smooth.

6. **Nystagmus** is a pattern of eye movements that occur as a response to the turning of the head or the viewing of a moving, repetitive pattern. It consists of two alternating components, (1) "smooth pursuit" motion in one direction followed by (2) a saccadic motion in the another direction. For example, in a train where looking out of the window you look at a tree and follow it as it moves past (slow phase), then make a rapid movement back (fast phase) and fixate on the next object which again moves past you.

7. **Fixations** are movements which stabilize the gaze direction on a stationary object.

From the description above it is clear that fixations, smooth pursuit, and saccades are important for establishing visual attention and thus also important in relation to eye tracking.

9.3 Eye Appearance Changes

The problem of visual eye recognition poses a number of challenges to existing machine vision and pattern recognition algorithms due to dramatic changes in the eye appearance among people, poses, and lighting conditions. Such problem exists too in many other applications of computer vision such as Optical Character Recognition.

The eye can be seen as a multistate model; it is subject to transitions from wide open eye to completely closed eye and thus the appearance of the iris and pupil are heavily influenced by occlusions from the eye lids and may often be totally covered. The effects of occlusion and illumination changes are also related to the ethnic origin of the user. Furthermore, the eye shape differs from one subject to another (Asian, European, African, etc.); eyes might appear with or without eyeglasses, sunglasses, eyebrows, and eye makeup. Both Caucasian and Asian people have facial features that may make eye detection and tracking difficult. The eye lids of Asians are generally close together and may thus result in less boundary information. On the other hand, the nasal bone and superciliary arch of Caucasians are usually more pronounced and therefore casts more shadows on the eye. In addition, the distance between the two eyes also varies across ethnic backgrounds.

The industrial push to develop cost-effective and free calibration eye tracking systems contribute furthermore to the variability problem at two levels: (1) the noise level increases when off-the-shelf cheap infrared cameras are used, and (2) the shadow effects, partial eye occlusions, and perspective distortions become more important with nonconstrained 3D head rotations and fast motions.

As with a monocamera system the depth map is mostly inaccurate, any departure of the head from the nominal position may decrease considerably

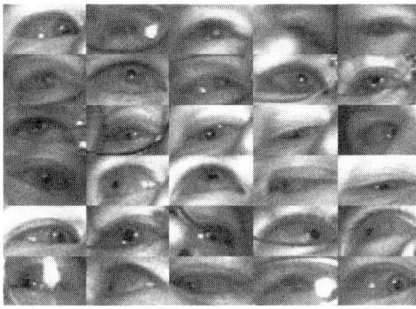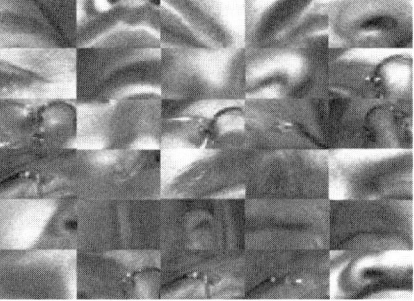

Figure 9.5. Snapshots of eye and noneye patches.

the performance rate of eye detection and eye tracking algorithms. Most algorithms use a predefined eye region size, and eventually with any depth changes, this region would either lose essential eye components such as eyelids, or would gain nondesirable background objects such as eyebrows and edges of glasses (cluttered background). Figure 9.5 shows some examples of cluttered eye regions and noneye patches with close appearance to eyes.

The current focus of eye tracking researchers is shifted toward developing robust algorithms that are capable of handling the listed challenges above. These challenges to be overcome are summarized in Figure 9.6. Interestingly enough the human vision system is naturally adapted to such variations among the visual objects, while it is still challenging to computers to handle easily and completely the iris-shift event illustrated in Figure 9.7.

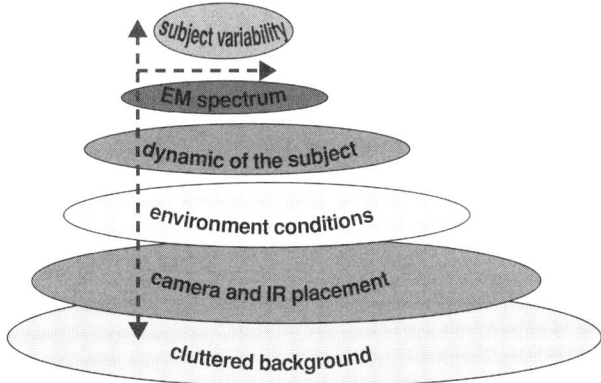

Figure 9.6. Pyramid of factors that contribute to eye appearance variability. A robust detection and tracking algorithm must tolerate to all of these levels (vertical dimension) as well inside all cases within each level (horizontal dimension). As more layers are crossed horizontally and vertically the complexity to develop robust vision algorithm increases in a nonlinear way.

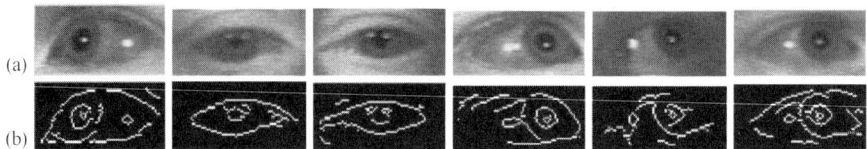

Figure 9.7. Illustration of the eye appearance variability during an iris-shift event in two spaces: (a) gray scale and (b) edge-pixel space.

9.4 Eye Image Representation

The first step toward building an efficient eye tracking system is choosing an adequate representation of the eye. In this section we will review most effective eye representation methods.

9.4.1 Selection Criteria

The eye is coded in a two-dimensional gray-scale image as a set of pixels ordered in multiple rows and columns with intensity values between 0 and 255. A well-formed dark pupil would have pixels of intensity values less than 15. As for the bright pupil the intensity values are expected to be above 240 to distinguish them easily from the sclera or the skin pixels. The representation process of an eye region consists of mapping the pixel format to another space through a defined transformation. The generated space out of a transformation (filter) is known by the "feature space." The consensus in computer vision is that a good image representation is:

1. *Compact*: The dimension of the feature space must be as small as possible, and at least smaller than the original eye region size; the smaller the better for memory allocation and matching procedure;
2. *Unique*: The feature vector describing eye is unique in such a way that the class of eyes and noneyes are well separable and easily distinguishable;
3. *Robust*: When the eye region is subject to an affine transformations (translation, rotation, etc.), its representation in feature space must remain relatively unchanged; and,
4. *Computationally inexpensive*: A filter must be computationally fast to be employed in real-time systems.

Unfortunately, the selection process of an adequate image feature is always a compromise between these four criteria. The higher the complexity of the feature the more computation resources are needed.

9.4.2 Feature Spaces

During the last decade a tremendous effort has been made in the area of image representation. Image representation methods employed by the eye tracking community could be arranged into the following groups: brightness-, texture-, and feature-based representations.

The **brightness-based methods**, known too by appearance-based methods, work with intensity or colors of the pixels and use the image itself as a feature descriptor. In most real-time eye tracking systems the gray-scale vector is employed. This vector is obtained by simple linear transformation of the eye region. It is intuitive, inexpensive, and more importantly it preserves the spatial relationship between the pixels distribution. Other features such as statistical moments could be used, but to be efficient they should be computed at high orders making them less used in real-time eye tracking systems.

In contrast, the **texture-based methods** try to encode the unique texture of the eye. Often a Gabor filter bank is used for this purpose. In [20] it is reported that the best performance in locating facial landmarks may be achieved by concentrating a large number of orientations (8 orientation bands) on very low frequency carriers (spatial frequencies in the order of 5–8 iris widths per cycle). The Gabor filters are orientation and scale-tunable edge detectors, and the statistics of the Gabor filter outputs in a given image region are often used to characterize the underlying texture information. More recently, Haar-like texture features have been employed in [13].

Finally, the **feature-based methods** extract edges, points, and contours and the problem is therefore reduced to matching of these salient features to the model [7, 12, 14, 25, 32, 35, 43, 44, 71, 78]. The model could take the form of a generic template which is matched to the image, for example, through energy minimization. In [73] and [23], generic local descriptors and spatial constraints are used for face detection and infrared eye tracking. A local descriptor is a combination of gray-value derivatives named "local jet." They are computed stably by convolution of the eye image with Gaussian derivatives. Note that derivatives up to Nth order describe the intensity function locally up to that order. The "local jet" of order N at a point $\mathbf{x} = (x_i, y_i)$ for image I and scale σ is defined by $J^N[I](\mathbf{x}, \sigma) = \{L_{j_1 \ldots j_n}(\mathbf{x}, \sigma) \mid (\mathbf{x}, \sigma) \in I \times R^+; n = 0, \ldots, N\}$, where $L_{j_1 \ldots j_n}(\mathbf{x}, \sigma)$ is the convolution of image I with the Gaussian derivatives $G_{j_1 \ldots j_n}(\mathbf{x}, \sigma)$ and $j_k \in \{x, y\}$. To obtain invariance under 2D image rotations and affine illumination changes, differential invariants from the local jet are computed. The local jet is computed at each facial point in the image and this describes the local geometry around a point of interest. The eye neighbor region is then represented by a compact feature vector that is tolerant to translation, 2D eye rotations, global light changes, and small factors of scale changes.

9.5 Modeling of the Eye Variability

It is quite difficult to develop a perfect feature that is capable of representing the eye in a unique way under all image conditions. On the other hand, modeling of the eye appearance changes seems to be a reasonable solution. The modeling aims to capture the intraclass variations in the feature space with a parametric model often called "generic model." It is constructed through supervised machine learning techniques. Such techniques provide a natural and successful framework for bringing automatic eye detection to practical use. Processing brightness, shape, and texture-based eye features, a number of learning approaches have shown success in discriminating eyes vs. noneye [23, 28, 31, 41, 58, 80].

During the learning process, all positive and negative patches (see Figure 9.6) are projected in a selected eye representation space. The learning process aims to find the optimal boundary between the two classes of patches. As a result of such process, two generic eye and noneye models are built. They represent the eye and noneye distributions, in a compact and efficient way. Each generic model is described by a set of parameters. These parameters are employed in the online testing process. An unknown patch is classified into either class according to a specific classification rule such as minimum distance or maximum probability.

King et al. [41] applied a linear principal component analysis approach [70] to find the principal components of the training eye patches. These representative principal components (PCs) are used as templates. An unknown image patch is projected into these PCs, and either Mahalanobis distance or Reconstruction error could be used along with a dissimilarity threshold to classify the patch. The issue with this technique is the selection of appropriate thresholds. To avoid this problem of threshold selection, a noneye model could be used in the comparison. Haro et al. [28] use a probabilistic principal component analysis to model offline the intraclass variability within the eye space and the noneye distribution where the probability is used as a measure of confidence about the classification decision.

More sophisticated learning techniques have been proposed in [23, 58, 73, 80]. The support vector machines (SVM) classifier [9] is used to train dark-pupil eye patches and noneye patches on gray-scale image vectors and invariant local jet [23, 80]. Using a selected eye database, the highest accuracy (above 98%) was obtained when the SVM kernel was chosen as a Gaussian kernel whose standard deviation is 3 [58]. Vogelhuber et al. [73] used a Gaussian Mixture Model (GMM) to learn the different aspect (variability) of the left eye and right eye patches in the local jet feature space. In this latter approach subclusters in the eye space are searched using Expectation Maximization (EM) algorithm. The eye patches are localized manually and represented by invariant descriptor ("local jet" proposed by Koenderink and van Doorn [42]). Invariant descriptors are proved to give more reliable results in recognition than simple gray scale features [61]. The issue with GMM method is that enough training data must be available and that a Gaussian assumption is made on the shape of the distribution, which may not be valid. However, finding compact and homogeneous clusters within a class might be more efficient than fitting a nonlinear multidimensional kernel function. Figure 9.8 illustrates the separation boundary between eye and noneye distributions

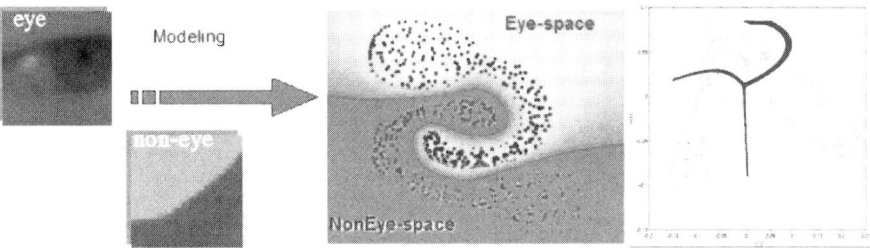

Figure 9.8. Illustration of eye and noneye models and their corresponding areas of membership in a two dimensional feature space. Each eye patch is represented by its projection in the first two eigenvectors. The space is splited into two areas of membership (eye vs. noneye) using SVM technique (middle). In the most right figure, a Gaussian Mixture Model is used to find the subclusters in the eye space, where three clusters were found, each cluster groups similar eye appearances (see color insert).

using SVM. It illustrates also the subclusters found by GMM in the eye space.

While the SVM is a very attractive classifier, manifesting performance comparable to or better than nearly all classifiers, based on experience with a wide range of data sets [62], it does have some drawbacks. For example, the restriction to Mercer kernels can be limiting, and while the number of support vectors (training examples x_n with nonzero Lagrange multiplier α_n that reflects the importance of training example n) is typically a small percentage of the training data, the number of support vectors grows linearly with the available training data [63]. In order to reduce the memory usage and the computation complexity, Hammoud [23] used the Relevance Vector Machine [67] instead of SVM. The RVM is based on a Bayesian formalism, with which we again learn functional mappings of the form $f(x) = \sum_{n=1}^{N} w_n K(x, x_n) + w_0$. However, $K(x, x_n)$, which is in the case of SVM, a kernel that quantifies the similarity between x and x_n in a prescribed sense, is now viewed simply as a basis function, and therefore it need not be a Mercer kernel. More theoretical details about Support Vector Kernels are found in [63].

In contrast to statistical approaches, evolutionary genetic algorithms [31], and parameterized geometric deformable templates have been employed very rarely on modeling of the eye appearance variability. Deformable templates and the closely related elastic snake models are interesting concepts to model the eye appearance changes, but using them is difficult in terms of learning and implementation, not mentioning their computational complexity.

9.6 Automatic Eye Localization

For most eye tracking applications it is highly required to automatically localize both eyes of a test subject without prior calibrations [2, 16, 18, 19, 24, 64,

Figure 9.9. Illustration of some steps of the eye motion–based detection algorithm. First raw: open eye, closed eye, and their corresponding absolute-difference image. Second raw: filtered-binarized image and final set of detected eyes.

65]. Otherwise, a human expert should interfer to identify the eyes through multiple steps such as aligning the subject's head in a window of interest or fit the facial features in a predefined face template. Evidently, employing a semiautomatic eye detection limits the usability of an eye tracking system considerably. The main challenge today is to minimize false positives and false negatives during automatic eye detection.

Two categories of automatic eye detection methods have appeared in the vision community: image-based passive approaches [41, 58] and active IR-based approaches [23, 28, 80]. The first category can be further divided into four subapproaches: appearance-, feature-, model-, and motion-based methods. The second category exploits the spectral properties of pupil under structured infrared illumination.

The common steps among these methods are summarized below: *Extractions of potential eye candidates, Selection of initial set of eyes,* and *Selection of final set of eyes.* Figures 9.9 and 9.10 illustrate the results of eye detection, using motion-based method, and dark-bright pupil technique, respectively.

- *Extractions of potential eye candidates:* The goal of this stage is to reduce the search space of the target eye. The more compact this search space is, the smaller the likelihood of false eye detection. The classic way is to perform a dense search in the entire image. For instance, in [51, 66] a method being adopted in eye detection consists of detecting the face first, then each pixel in the upper block of the face is considered a potential eye candidate

  As opposed to dense methods, sparse methods select intelligently a small set of potential eye candidates. The subtraction method that relies on

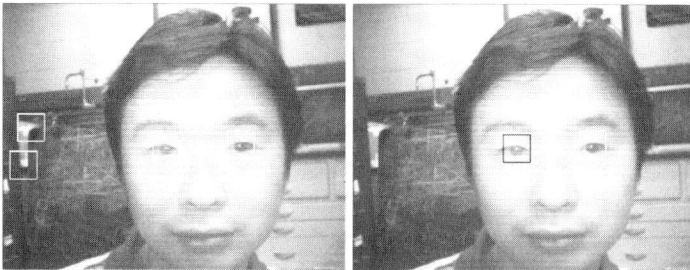

Figure 9.10. Illustration of some steps of the dark-bright pupil-based approach: initial list of potential eye candidates (a), and final list of detected eyes (b). Note that the two false eye candidates are rejected using a SVM classifier.

dark-bright pupil technology [23, 28, 37, 52] has gained a lot of attention recently due to its effective way to extract potential eye candidates. A review of pupil responses to near-structured infrared light is given in Section 9.2. The two images, the dark and bright pupil fields of the same interlaced frame, are subtracted from each other, and only candidates of high contrasts are kept as shown in the right image of Figure 9.4.

Similarly, regions of eye motion are identified by subtracting two successive frames of the subject [23, 39]. The eye motion could be either eye blinking or iris shift. Only regions that satisfy certain size and shape constraints, like circularity and orientation, are kept for further processing.

- *Selection of initial set of eyes:* The initial set of potential eye candidates is filtered and a subset of eyes is generated. This is done through a filtering procedure applied on an entity of single, pair, or triplet of eye candidates. Most methods which do not constraint eye detection in frontal head pose apply a common appearance-based filter. First, a feature vector is extracted from the image to represent a potential eye candidate. Second, the candidate is classified as an eye or a noneye, according to a well-defined classification rule and offline generic eye model. A brief review of eye features and machine learning tools was given in Sections 9.4 and 9.5, respectively.

  For instance, Haro et al. [28] used the gray-scale vector and Probabilistic Principal Component Analysis (PPCA) classifier to represent and classify the eye region candidate as eye or noneye. While in [23] more robust image feature and nonlinear classifier, the local jet, and a derivative of Support Vector Machine, are used.

  At this stage the majority of false eye candidates are eliminated. However, the output of this stage may not be optimal and more than two candidates left. Such a scenario is addressed in next step.

- *Selection of final set of eyes:* In order to select the final set of eyes, a further analysis could be performed. One solution adopted in [28] consists of taking the initial set of eyes, generated in the previous step, track all of them simultaneously, and then eliminate the false candidates over time according

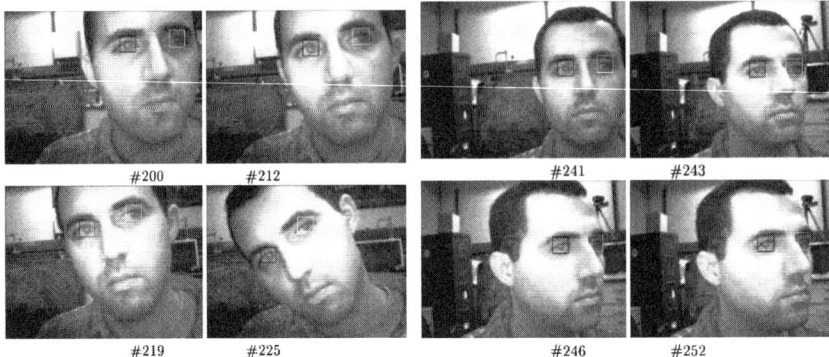

Figure 9.11. Illustration of detection results on both eyes using invariant local features and RVM appearance models, eye motion, and infrared-eye responses. The tracking boxes are overlaid on the bright-pupil images.

to certain criteria like dynamic. Another approach to tackle this problem makes use of information like facial feature geometry, anthropomorphic constraints, and eye blinks. If two points are selected but they seem too far from each other in space, then only the eye that satisfies the ellipse-like constraint is considered in the final set of eyes.

Figure 9.11 shows detection results of both eyes, using invariant local features and RVM appearance models, eye motion, and infrared-eye responses [23].

9.7 Tracking Problem Formulation

The eye tracking process consists of following one or both eyes in consecutive video frames. The choice of tools used to perform a particular tracking task depends on the prior knowledge of the object being tracked and its surroundings. For example, both the appearance and the dynamical properties of the eyes are important modeling problems. This section provides brief background materials on tracking such as state vector, tracking components, and tracking problem formulation.

9.7.1 Components

To formulate the tracking problem properly, the following tracking components shall be introduced first. These terminologies are used in the rest of the chapter and more specific definitions are given in Section 9.9.

- *Target representation:* The target representation is used for discriminating the eye from other objects and is often described and abstracted by a set of parameters, called the *state vector*. In the following x denotes the state.

Often the shape and appearance of the eye is sufficiently discriminative but also the way the eye moves may be useful. Section 9.5 about representation elaborates further on this issue.

- *Observation representation:* The observation representation defines the image evidence of the eye representation; i.e., the image features Y observed in the images. The observation representation is closely related to the target representation, as for example, when the eye is represented by its contour shape, we expect to observe edges of the contour in the image. If the target is characterized by its color appearance, particular color distribution patterns in the images can be used as the observation of the eye.
- *Hypotheses measurement:* The hypotheses measurement evaluates the eye state hypotheses with the image observations, and essentially describe the evolution of the posterior density.
- *Hypotheses generation:* The probability of the state hypotheses generating the image observations is often needed, for example, to determine the hypothesis most likely to produce the image observations. Hypotheses generation is intended to produce new state hypotheses based on the old state estimates and the old observations. The target's dynamics could be embedded in such a predicting process. At a certain time instant, the eye state is a random variable. The posterior density of the eye state, given the history of observations, changes with time.

9.7.2 Bayesian Formulation

Given a sequence of T frames, at time t only data from the previous $t-1$ images are available. Let us represent the states and measurements by $\mathbf{x}_t$ and $\mathbf{y}_t$, respectively, and the previous states and measurements by $\underline{\mathbf{x}}_t = (\mathbf{x}_1, \ldots, \mathbf{x}_t)$ and $\underline{\mathbf{y}}_t = (\mathbf{y}_1, \ldots, \mathbf{y}_t)$. At time t the observation $\mathbf{y}_t$ is assumed independent of the previous state $\mathbf{x}_{t-1}$ and previous observation $\mathbf{y}_{t-1}$ given the current state $\mathbf{x}_t$. Each of the noise sequences is assumed identically distributed (i.i.d.). The probabilistic formulation can be represented by the graphical model in Figure 9.12.

The objective of state space-based tracking is to estimate the state of the eye, given all the measurements up to that moment, or equivalently to construct the probability density function (pdf). The theoretically optimal

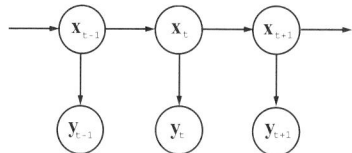

Figure 9.12. The dependency graph of the assumptions made for particle filtering The assumptions used for particle filtering can be represented by a graphical model, similar to the Hidden Markov Model.

solution is provided by the recursive Bayesian filter which solves the problem in two steps.

$$p(\mathbf{x}_{t+1}|\underline{\mathbf{y}}_{t+1}) \propto p(\mathbf{y}_t|\mathbf{x}_t)p(\mathbf{x}_{t+1}|\underline{\mathbf{y}}_t) \qquad (9.1)$$

$$p(\mathbf{x}_{t+1}|\underline{\mathbf{y}}_t) = \int p(\mathbf{x}_{t+1}|\mathbf{x}_t)p(\mathbf{x}_t|\underline{\mathbf{y}}_t)d\mathbf{x}_t. \qquad (9.2)$$

The equation contains the prior distribution, $p(\mathbf{x}_t|\underline{\mathbf{y}}_t)$, the dynamics, $p(\mathbf{x}_{t+1}|\mathbf{x}_t)$ and the likelihood term $p(\mathbf{y}_t|\mathbf{x}_t)$ The prediction step uses the dynamic equation and the already computed pdf of the state at time $t-1$. Then the update step employs the likelihood function of the current measurement to compute the posterior pdf. Therefore, the tracking problem can be viewed as a problem of conditional probability density propagation.

9.8 Models of Tracking Methods

In general there is no closed form solution to the recursive Bayesian filter in Eq. (9.2). However, under linear dynamic models the equation can be optimally solved in closed form, provided in the Kalman filter. Empirically it has been shown that a Gaussian approximation for the observation density in visual tracking may not be sufficient and can lead to tracking failures. When the noise sequences are nonlinear but modeling the posterior density as Gaussian, a locally linear model is found in the Extended Kalman Filter (EKF) [5]. A recent alternative to the EKF is the Unscented Kalman Filter (UKF), which uses a set of discretely sampled points to parameterize the mean and covariance of the posterior density [38]. Hidden Markov Models (HMM) can be applied for tracking when the number of states is finite [60]. To efficiently handle nonrigid transformations without explicitly modeling dynamics Comaniciu [11] proposes a histogram-based tracking, which exploits a spatially smooth cost function and mean-shift.

Eye tracking methods could be arranged into four major categories, namely *appearance-based*, *deformable templates*, *feature-based*, and *IR-based* methods. Deformable template and appearance-based methods rely on building models directly on the appearance of the eye region while the feature-based methods rely on extraction of local features of the region. The latter methods are largely bottom up while template and appearance-based are generally top-down approaches. That is, feature-based methods rely on fitting the image features to the model while appearance and deformable template-based methods strive to fit the model to the image.

The **appearance-based methods** track eyes based on the photometry of the eye region. The simplest method to track the eye is through template-based correlation. Tracking is performed by correlation maximization of the eye model in a search region. The eye model could be too simply an average gray-scale eye template. Grauman et al. [22] use background subtraction and anthropomorphic constraints to initialize a correlation-based tracker. Trackers based on template matching and stereo cameras seem to produce robust results [50, 54]. An essential part of real-time tracking is to find a model which is sufficiently expressive to handle a large variability in appearance and dynamics while still be sufficiently constrained as to only track the given class of objects. The appearance of eye regions share commonalities across race, illumination, and viewing angle. Rather than relying on a single instance of the eye region, the eye model can be constructed from a large set of training examples with varying pose and light conditions. On the basis of the statistics of the training set, a classifier can be constructed for modeling a large set of subjects.

The **deformable template-based methods** [12, 14, 25, 35, 43, 78] rely on a generic template which is matched to the image for example through energy minimization.The deformable template-based methods seem logical and are generally accurate. They are also computationally demanding, require high contrast images, and usually needs to be initialized close to the eye.

In contrast to the above two tracking models, the **feature-based methods** extract particular features such as skin-color, color distribution of the eye region. These features could be tracking the in-between eyes [40] and [76], filter banks [21, 29, 72], iris location using the Hough transform [56, 77], facial symmetries [47]. Eye tracking methods committed to using explicit feature detection (such as edges) rely on thresholds. Defining thresholds can, in general, be difficult since light conditions and image focus change. Therefore, methods relying on explicit feature detection may be vulnerable to these changes.

Finally, **IR-based methods** make use of the unique IR-pupil observations [3, 17, 52, 57, 69]. More details and experimental results are given in Section 9.9.2.

9.9 Particle Filter-Based Eye Tracking Methods

The most general class of filters is represented by particle filters. Particle filters (PF) are based on Monte Carlo integration. Rather than parameterizing the posterior probability (equation 9.2) may sample from it. Particle filters aim at exactly representing the current distribution by a set of N *particles* $S_t \equiv \{s^{(i)} = (\mathbf{X}_t^{(i)}, \pi_t^{(i)})\}$ at time t [15, 34]. The current density of the state is represented by a set of random samples with associated weights and the new density is computed on the basis of these samples and weights. Each particle, $s^{(i)}$, is defined by a state hypothesis $\mathbf{X}$ and a weight $\pi_t^{(i)}$, which reflects the likelihood of that particular state hypothesis.

9.9.1 Particle Filtering

Over the last couple of years several closely related sequential Markov chain algorithms have been proposed under the name of bootstrap filter, condensation, particle filter, Monte Carlo filter, genetic algorithms, and survival of the fittest methods [15]. Sequential Monte Carlo techniques for iterating time series and their use in the specific context of visual tracking are described at length in the literature [10, 33].

- *Filtering Distribution*: The aim of particle filtering is to recursively estimate the posterior distribution $p(\mathbf{X}_t|\mathbf{Y}_t)$ or the marginal distribution $p(\mathbf{x}_t|\mathbf{Y}_t)$, also known as the *filtering distribution*. The associated features of the density such as the expectation can be estimated using:

$$\mathcal{M}_{p(\mathbf{X}_t|\mathbf{Y}_t)}[f_t(\mathbf{X}_t)] \equiv \int f_t(\mathbf{X}_t)p(\mathbf{X}_t|\mathbf{Y}_t)d\mathbf{X}_t \qquad (9.3)$$

for some function of interest $f_t : \mathcal{X}^{(t+1)} \to \mathbf{R}^n$. For instance using the identity function $f_t(\mathbf{X}_t) = \mathbf{X}_t$ yields the conditional mean.
The objective is to apply Bayes' theorem at each time step, obtaining a posterior $p(\mathbf{x}_t|\mathbf{Y}_t)$ based on all the available information, as defined in Eq.(9.2). That is, a sequence of two operations on density functions: a convolution of the filtering distribution at the previous timestep, $p(\mathbf{x}_{t-1}|\mathbf{Y}_{t-1})$ with the dynamics $p(\mathbf{x}'|\mathbf{x})$ followed by a multiplication by the observation density $p(\mathbf{Y}_t|\mathbf{x}_t)$. Rather than approximating the distribution in Eq.(9.2) the idea is to simulate or sample the distribution. The sampling uses a weighted set of samples, the particles $\{\mathbf{s}_t^i\}_{i=1}^N$. A particle $\mathbf{s}_t^i = (\mathbf{x}_t^i, \pi_t^i)$ consists of a state instance $\mathbf{x}^i \in \mathcal{X}$ and a weight $\pi^i \in [0,1]$. The finite set of particles is meant to represent a distribution $p(\mathbf{x}_t|\mathbf{Y}_t)$, in the same manner so as to choose one of the $\mathbf{x}^i$ with probability π_i should be approximately the same as sampling from the distribution $p(\mathbf{x}_t|\mathbf{Y}_t)$. This principle is shown in Figure 9.13.
If the samples at time t, $\mathbf{s}_t^m$, are fair samples from the filtering distribution at time t, the new particles, $\tilde{\mathbf{x}}_t^m$, associated with the weights π_{t+1}^m approximate the filtering distribution. Resampling these particles with replacement proportional to their weights provides a set $\{\mathbf{x}_m^{t+1}\}_{m=1}^M$ of fair samples from the filtering distribution $p(\mathbf{x}_{t+1}|\mathbf{Y}_{t+1})$. The algorithm for the

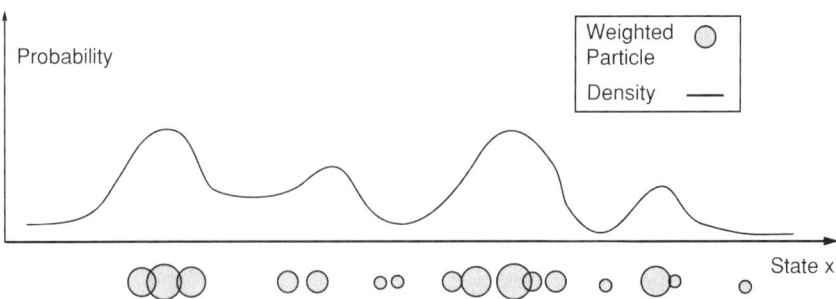

Figure 9.13. Sample representation of a distribution A. weighted set of particles approximate the distribution. Picking one of the samples with probability proportional to the area is similar to drawing randomly from the density function. Adapted from [8].

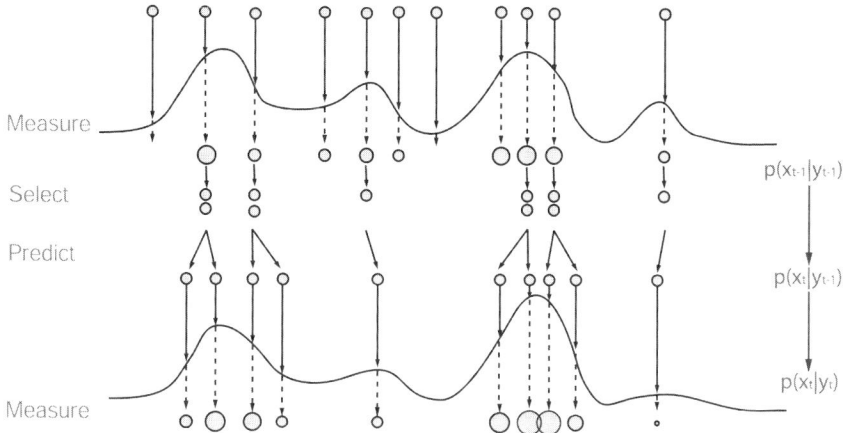

Measure

Select

Predict

Measure

$p(x_{t-1}|y_{t-1})$

$p(x_t|y_{t-1})$

$p(x_t|y_t)$

Figure 9.14. Particle filtering. Starting at time $t - 1$ the particles are unweighted. For each particle the weights are computed using the information at time $t - 1$. These constitute the approximation to $p(x_{t-1}|y_{y-1})$. Subsequently, in the resampling step, the particles are selected according to their weights resulting in a unweighted measure. Finally predicting (resampling) and weighing the samples resulting in a measure which is an approximation of $p(x_t|Y_t)$ [15].

standard particle filter is given in algorithm 9.1 and the iterative process is depicted in Figure 9.14.

• *Components of Particle Filters*: To make a workable particle filter several components have to be modeled.

1. The *state vector* that encodes parameters that are of interest to our modeling problem. In terms of eye tracking, the center of the iris, head orientation, the four eye corner positions, IR glints and reflections, low-dimensional shape parameters are all examples of possible parameters to be estimated. In this chapter we are limited to track the 2D position of the eye only.

Initialization:

– for $i = 1 : N$ sample $x_0^i \sim p(\mathbf{x}_0)$;
– Weigh each sample according to data likelihood $\pi_t^i = p(\mathbf{y}_t|\mathbf{x}_t)$

Iteration:
for $i = 1 : N$

– *Select*: a sample $\tilde{\mathbf{x}}^i$ with probability π_{t-1}^j
– *Predict* by sampling from $p(\mathbf{x}_t|\mathbf{x}_{t-1} = \tilde{\mathbf{x}}^i)$
– *Measure* and weight the new sample in terms of the data likelihood $\pi_t^i = p(\mathbf{y}_t|\mathbf{x}_t)$

Estimate the measures of interest according to equation (9.3) once the N samples are constructed.

2. The *likelihood* is used to evaluate each particle hypothesis as to how well the state instance explains the data.
3. The dynamics that is a model of the object's temporal behavior in state space.

- *The Number of Particles*: The number of particles, N, represents an approximation of the posterior distribution. When $N \rightarrow \infty$, the representation approaches the true distribution. However, because of to the linear time complexity, as few particles as possible are desired. It is difficult to automatically determine the number of samples, but [48] derive the requirement

$$N \geq \frac{\mathcal{D}_{\min}}{\alpha^d} \qquad (9.4)$$

where d is the number of dimensions of the parameter space, $\mathcal{D}_{\min}$ is the smallest number of particles to survive sampling, and the *survival rate*, $\alpha << 1$, is a constant related to the shape of the posterior and prior distributions. As α is very small, the lower bound on N grows exponentially with the number of dimensions. The survival rate will, in general, be lower for noisy distributions with shaper peaks, which, in turn, means that more samples are needed to represent them. In our implementation of a simple PFIR tracker we have used 200 particles (see next section).

9.9.2 Adapted PF Tracker to IR-Pupil Observations

This section describes how to adapt the particle filter theory to tracking of dark-bright pupil. One could see that most predominate eye tracking methods in both research and commercial products are based today on active infrared light [3, 17, 45, 52, 57, 65, 68, 69]. Ebisawa et al. [17] use a novel synchronization scheme in which the difference between images obtained from on-axis and off-axis light emitters are used for tracking. Kalman filtering and Mean shift tracking are more recently applied in similar approaches [79]. The success of these approaches is highly dependent on external light sources and the apparent size of the pupil.

The IR-based PF tracking algorithm is described in details in Figure 9.15. Several simplifying assumptions are made here, to make this example short and comprehensible. Tracking often involves following areas in the image of high probability. This example shows a simple particle filter for following regions of high probability using bright and dark pupil images eye tracking as the main application. The observations are obtained by subtracting the dark and bright eye images and thus the eye regions yield areas of high probability (see Figures 9.4 and 9.16).

The first step is to find the parameters to be estimated. For simplicity, assume that the dimensions of the bounding box is predefined and fixed over time. In other words, there is no scaling or rotations between frames. The only

Initialization:

- Get initial eye location $(x_{\mathrm{init}}, y_{\mathrm{init}})$ and bounding box (d_x, d_y)
- for $i{=}1{:}N$
  set sample $\mathbf{x}^i = (x_{\mathrm{init}}^i + \nu_x, y_{\mathrm{init}}^i + \nu_y)$; where ν_x and ν_y are random Gaussian noise as specified in the dynamical equation (9.9). As the eye location is known the noise covariance is usually set low.
- Weigh each sample according to equation (9.6), by adding all pixel values within the bounding box located at position $\mathbf{x}^i$

Iteration:
for $i = 1 : N$

- get the next images and subtract the dark and bright images
- *Select*: a sample $\tilde{\mathbf{x}}^i$ with probability π_{t-1}^i. This can be done by picking a random number between $[0; \sum_i^N \pi_{t-1}^i]$ and pick the sample in the cumulative histogram which is closest to the random number
- *Predict* each sample by adding random noise to each particle, as given in equation (9.9)
- *Measure* and weight the new sample in terms of the data likelihood $\pi_t^i = p(\mathbf{y}_t|\mathbf{x}_t)$, again by summing all pixel values with the bounding box specified by each sample

   Estimate Calculate the weighted mean of all samples. This will be our state estimate.

Figure 9.15. Adapted particle filter tracking algorithm to dark-bright pupil observations.

parameters to be estimated are the coordinates in the image. The second step is to define a dynamical model. As described in Section, 9.2.3, pupil movements can be very rapid from one image frame to another. As no priori knowledge of the movements is available, the dynamic is modeled as a first-order

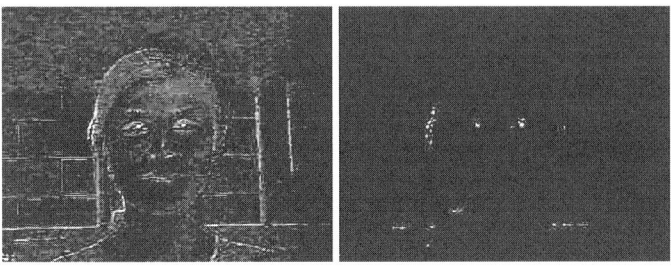

Figure 9.16. Illustration of the subtraction image of dark and bright pupil fields and their corresponding thresholded image (the threshold value is here set heuristically to 60). Another close-up view of the pupil is illustrated in Figure 9.4. Notice how the center of the pupil is highlighted in the thresholded image, in both figures.

auto-regressive process using a Gaussian noise model:

$$\mathbf{x}_{t+1} - \mathbf{x}_t + \mathbf{v}, \qquad \mathbf{v} \sim \mathcal{N}(0, \Sigma), \tag{9.5}$$

where Σ is the covariance matrix of the noise $\mathbf{v}$. That is, each particle is moved in a random direction and its distance is related to the covariance matrix.

Each particle has to be evaluated so as to find its likelihood. We noticed in Figure 9.16 that the pupil reflections generate a bright spot in the difference image. If the intention is to track the eye, we might as well follow bright areas within a bounding box. Assuming that high-intensity values correspond to a high probability that the pixel belongs to the pupil the (pseudo) likelihood can be defined as

$$p(\mathbf{y}|\mathbf{x}) = \int_R \exp(I_d(\mathbf{x})) dX, \tag{9.6}$$

where the integral is defined over the pixel coordinates within the bounding box of the difference image I_d. We are still faced with a couple of problems before we have an actual tracker, namely how is the tracker initialized, how many particles to use and how to represent the object, given the set of particles. Even though the posterior is multimodal, the eye position is represented by the weighted sample mean. The samples in the initial sample distribution are initialized by setting each to the known initialized position plus some random low noise. The number of particle, depends on several factors such as image size and eye dynamics and is often set heuristically. Figure 9.17 shows the tracking results of the left eye using the described algorithm above 9.16. Under various head rotations this algorithm seems to maintain tracking of the eye.

This tracker is easy to implement and is actually capable of tracking the eye region under moderate head, lighting conditions and eye movements, but may lose track when the IR reflections disappear, in outdoor conditions section

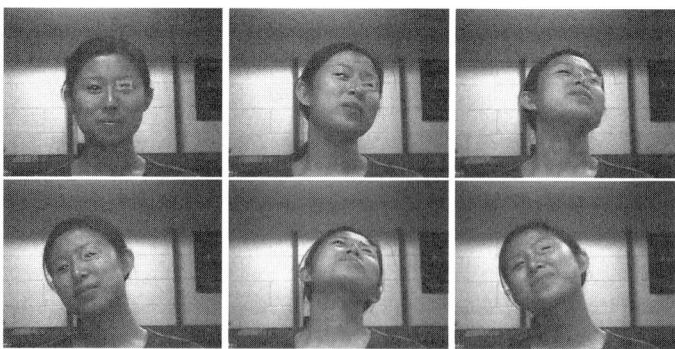

Figure 9.17. Eye position tracking results using the adapted PF tracker (see Figure 9.15) to dark-brigh pupil observations. In this experiment the number of particles used is $N = 200$.

sunlight, or the head movements are too rapid. The reason is that the likelihood measure is not sufficiently discriminative to uniquely represent eyes. An easy extension to the tracker is to add scale changes. This can be done by adding another parameter to the state model. However, one should remember to normalize the likelihood measure when comparing samples of different sizes.

9.9.3 Combined Particle Filter and Iris-Contour Trackers

To make eye trackers more robust, additional eye features must be considered. In this section we will describe an eye tracking method that utilizes the iris as the most prominent feature of the eye. The iris is characterized by being elliptic and by its large contrast to the sclera. This section outlines the components used in a recent contour-based tracker, but with all derivations left out [26].

The method combines particle filtering with the EM (Expectation Maximization) algorithm. Particles filters generally require a large set of particles to accurately determine the pose parameters. By contrast, the method uses a fairly small set of particles to maintain track of the object while using the EM Contour method [59] for precise pose estimation. In this way computation time is lowered while maintaining accuracy.

The iris appears elliptical on the image plane. The state, $\mathbf{x}$, is therefore defined to model an ellipse given by five state variables:

$$\mathbf{x} = (c_x, c_y, \lambda_1, \lambda_2, \theta),$$

where (c_x, c_y) is the center of the iris, λ_1, λ_2 are the major and minor axes, and θ is the angle of the major axis with respect to the vertical. These are the variables being estimated in the method.

The robustness of particle filters lies in maintaining a set of hypotheses. Generally, the larger the number of hypotheses, the better the chances to get accurate tracking results, but the slower the tracking speed. Therefore, there is a trade-off between tracking accuracy and speed. In the approach [26] the particle filter is used to obtain a state estimate on the coarsest image scale. In this way the number of particles can be reduced. The EM Contour method is applied on gradually finer image scales so as to obtain an accurate measurement of the sample mean on finest image scale. An overview of the algorithm is depicted in Figure 9.18.

Many contour-based methods [8], as well as this approach, use the normals to the contour to obtain the observations and assume that they are mutually independent. This means that the likelihood of the entire contour is a multiplication of the likelihoods for each normal.

This contour model is based on the following assumptions:

1. The pdf of the observation depends only on the gray level differences (GLDs).
2. Gray-level differences between pixels along a measurement line are statistically independent.

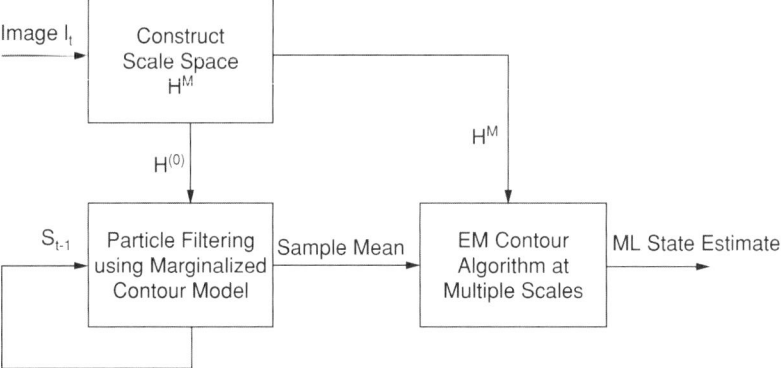

Figure 9.18. Overview of each iteration in the algorithm. Overall tracking is performed on the coarsest image scale through particle filtering; starting from the weighted mean of the particle states (sample mean), maximum likelihood estimation of the object state is performed through the EM contour algorithm over gradually finer image scales.

3. Intensity values of nearby pixels are correlated if both belong to the object being tracked or both belong to the background: thus, a priori statistical dependencies between nearby pixels are assumed.
4. There is no correlation between pixel values if they are on opposite sides of the object boundary.
5. The shape of the contour is subject to random local variability, which means that marginalization over local deformations is required for a Bayesian estimate of the contour parameters.

Taking this into account the log-likelihood contribution from a single measurement line is given by [26]:

$$h(\mathrm{i}|\mu) = -\log(m) + \log \sum_j \frac{f_D(\epsilon_j)}{f_L(\Delta I(j\Delta\nu))} \Delta\nu, \qquad (9.7)$$

where f_D is the Gaussian deformation, f_L is the density of gray-level differences between neighboring pixels, which in turn is well approximated by a generalized Laplacian [30]:

$$f_L(\Delta I) = \frac{1}{Z_L} \exp\left(-\left|\frac{\Delta I}{\lambda}\right|^\beta\right) \qquad (9.8)$$

ΔI is the gray-level difference, λ depends on the distance between the two sampled image locations, β is a parameter approximately equal to 0.5, and Z_L is a normalization constant.

Taking the assumptions together means that no features need to be detected and matched to the model (leading to greater robustness against noise), while at the same time local shape variations are taken explicitly into account.

Contour Tracker: operation for a single image frame
Input: Image scale space H_I^M of I, Motion model Σ and particle set S_{t-1}
Output: Optimized mean state $\bar{\mathbf{x}}$ and new particle set S_t

1. Obtain the sample set S_t by selecting N samples proportionally to their weight from sample set S_{t-1}
2. Predict all samples in S_t according to Eq. (9.9)
3. Update weights in S_t by multiplying all normals with the measurement line likelihood given in equation (9.7) using $H_I^{(0)}$ as reference image.
4. Calculate sample mean

$$\bar{\mathbf{x}}_0 = \frac{\sum_{i=1}^{N} \pi_i \mathbf{x}^{(i)}}{\sum_{i=1}^{N} \pi_i}$$

5. $\tilde{\mathbf{x}}_0 = \mathcal{O}_0(\bar{\mathbf{x}}_0)$ where $\mathcal{O}_i(\mathbf{x}*)$ represents the EM Contour algorithm applied at scale i with initialization given by the argument $\mathbf{x}*$.
6. for each image scale i calculate $\tilde{\mathbf{x}}_i = \mathcal{O}_i(\tilde{\mathbf{x}}_{i-1})$
7. $\bar{\mathbf{x}} = \tilde{\mathbf{x}}_M$

Figure 9.19. The iris tracker as applied to a single frame of an image sequence.

The dynamical model, on the other hand, is slightly different from the one described in the simple IR-based tracker as the covariance matrix Σ_t of the noise $\mathbf{v}_t$ is now made time-dependent matrix so as to compensate for scale changes: when the apparent size of the eye increases, the corresponding eye movements can also be expected to increase. For this reason, the first two diagonal elements of Σ (corresponding to the state variables c_x and c_y) are assumed to be linearly dependent on the previous sample mean. The time-dependent dynamics can therefore be defined to be

$$\mathbf{x}_{t+1} = \mathbf{x}_t + \mathbf{v}_t, \qquad \mathbf{v}_t \sim \mathcal{N}(0, \Sigma_t). \tag{9.9}$$

The iteration step of the contour method is described in Figure 9.19.

Figure 9.20 shows an example of using the iris tracker using images obtained from a standard video camera [26].

While this tracker is capable of tracking the iris without explicit feature detection in both IR and non-IR light, it still faces several issues to be handled to be a final system. One obvious limitation is to use only the iris for gaze determination, as it is impossible to disambiguate head and eye movements from the iris position, thus, in turn, disallow free head movements. Second, the state model is not sufficiently discriminative to uniquely determine whether it is tracking the iris or another circular objects. The extension to facilitate a

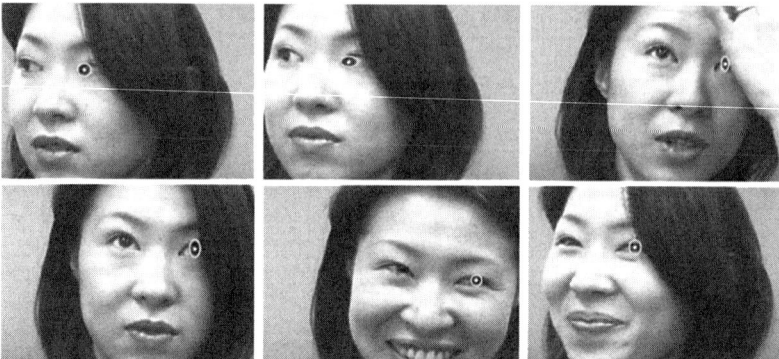

Figure 9.20. Tracking results of the left eye using the PF-iris based tracking algorithm described in Figure 9.19. Note the occlusion of the tracked eye by the subjet's hand and hair.

more committed model has been avoided so as to prove that iris tracking can be done without any explicit feature detection. Even though eye movements are rapid, using a more committed dynamical model may reduce the number of particles as well as improve accuracy, as less particles are lost in least areas of state space.

9.10 Advanced Eye Tracking–Based Technologies

Over the past few years the number of applications of eye tracking has expanded remarkably [1]. Eye tracking-based technologies can broadly be dichotomized from a system analysis point of view as *diagnostic* or *interactive*. In a diagnostic role, the eye tracker provides information about the observer's visual and attentional processes. In an interactive role, the eye tracker serves as an input device. An interactive system responds to the observer's actions and interacts with him. In this instance this is done on the basis of the observer's eye movements. This latter category could be divided further into two types of interactive systems: (a) *selective*, where the point of gaze is used as a pointing device and (b) *gaze-contingent*, where the observer's gaze is used to change the rendering of complex information displays [2].

Once the eye is automatically detected and tracked over time, the eye region could then be analyzed as follows. The eye corners and the pupil position are extracted to determine the orientation of the eyeball. This information is combined with head pose data and depth information to estimate the gaze vector accurately. The head pose estimation could be estimated either using appearance-based techniques which explore the difference between frontal face pattern and profile face pattern, or 3D-based techniques which require detection of facial features and a reference head pose model.

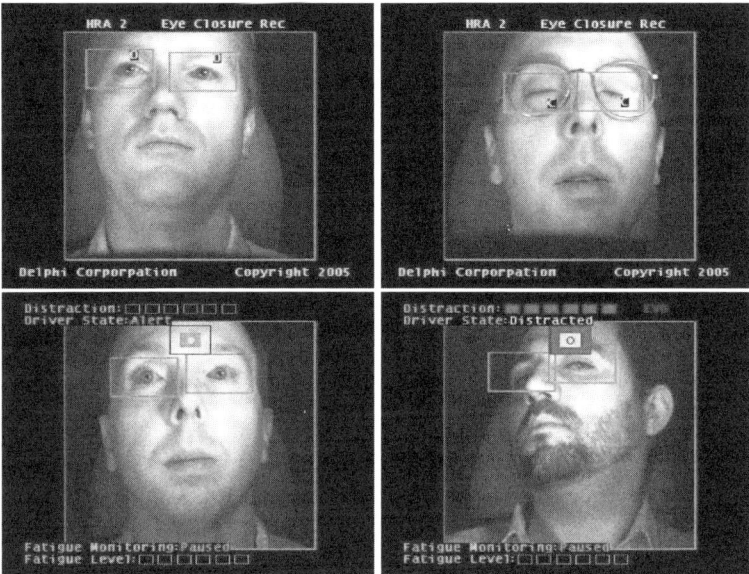

Figure 9.21. Snapshots of two applications of eye tracking: (1) *fatigue assessment*: the eye state is classified as open (top left) or closed (top right); and (2) *distraction assessment*: as the head pose is frontal (bottom left) or non-frontal (bottom right) for a while, the driver is recognized as non-distracted (bottom left) or distracted (bottom right). Both eyes were detected and tracked in real-time. Some human factor algorithms being used to measure distraction from the head pose classification results can be found in [18, 19, 24] (see color insert).

SeeingMachines [64] developed an eye tracking system, called FaceLab, that analyzes head pose and eye behavior, in various complex environments, e.g., on the road, in the air, and in simulators. It provides a robust and flexible stereo-vision solution for research into driver fatigue, distraction, human workload management systems, and smart vehicle applications. Recently, Delphi Corporation announced the readiness of its automotive grade eye-tracking based Driver State Monitoring system [18, 19, 24] which reports in real-time driver fatigue and driver distraction state. Some illustrations of Delphi DSM customer view are depicted in Figure 9.21. It employs an active infrared mono-camera system and a Digital Signal processor. Many other commercial eye tracking products exist today in the market (see for instance [45] and [65]).

Nikolov et al. developed a gaze contingent multimodality display [55]. Their demonstration of focus+context display is very successful in our everyday life. For instance, Multimap [53] offers its users the ability to view online (on their Web site) such displays where a focal region from a street map is overlaid on top of aerial photograph. Information about the user's intentions (like browsing and reading) can be exploited in applications such as eye typing and as an aid for reading foreign text [49]. A complete survey of eye tracking applications is given in [2, 16].

9.11 Conclusion

During the last decay, tremendous effort has been made on developing robust and cheap eye tracking systems for various real-world applications. This chapter presented cutting-edge computer vision methods employed in advanced vision sensing technologies for medical, safety, and security applications, where the human eye represents the object of interest for both the imager and the computer. As the eye scans the environment or focuses on particular objects in the scene, the computer simultaneously localizes the eye position, tracks its movement over time, and reports information about attention level and the gaze direction. The applications of robust eye tracking and gaze estimation methods are growing in importance in many different areas and the current tendency is that eye tracking is sneaking into many domestic appliances such as in TVs and cars and in security systems. Eye tracking seems to have moved out of laboratory use and into our homes and cars.

The main focus was placed on computer vision and pattern recognition algorithms for eye appearance variability modeling, automatic eye detection, and robust eye tracking. This chapter offers methodologies for constructing the fundamental building blocks of a vision-based eye tracking technology: automatic eye detection and eye position tracking. Methods that utilize both active infrared illuminations and ambient light are reviewed in details. Our presentation of these methods was described here in a tutorial-like manner. In particular, detailed examples to illustrate the adaptation of particle filter theory to dark-bright pupil observations and iris-contour models are given. Throughout the chapter a special attention was given to biophysics of the eye in infrared spectrum. The unique responses of pupils to structured infrared lights along with the need to operate during night time are the prominent reasons for choosing infrared light in near production eye tracking system technologies. The main challenge today with such illumination is the glare. The glare occludes the eye when subjects wear glasses or sunglasses. The glare issue is not addressed yet in the literature. However, intensive research effort is under way to tackle this obstacle.

References

[1] *ACM Eye Tracking Research and Applications (ETRA) Symposium*, 2000, 2002, 2004.
[2] *Gaze Tracking Methodology: Theory and Practice*. Springer, London, UK, 2003.
[3] A. Amir, L. Zimet, A. Sangiovanni-Vincentelli, and S. Kao. An embedded system for an eye-detection sensor. *CVIU*, 98(1):104–123, April 2005.
[4] H.H.K. Andersen and G. Hauland. Measuring team situation awareness of reactor operators during normal operation: A technical pilot study. In *Proceedings of the First Human Performance, Situation Awareness and Automation Conference*, pp. 268–273, 2000.

[5] Y. Bar-Shalom and T. Fortmann. *Tracking and Data Association*. Academic Press, 1988.

[6] Patrick Baudisch, Doug DeCarlo, Andrew T. Duchowski, and Wilson S. Geisler. Focusing on the essential: Considering attention in display design. *Communications of the ACM*, 46(3):60–66, 2003.

[7] S. Belongie, J. Malik, and J. Puzicha. Shape matching and object recognition using shape contexts. *IEEE Trans. Pattern Analysis and Machine Intelligence*, pp. 509–522, 2002.

[8] A. Blake and M. Isard. *Active Contours: The Application of Techniques From Graphics, Vision, Control Theory and Statistics to Visual Tracking of Shapes in Motion*. Springer-Verlag, 1998.

[9] C. Burges. A tutorial on support vector machines for pattern revognition. *Data Mining Knowledge Discovery*, 2:121–167, 1998.

[10] K. Choo and D.J. Fleet. People tracking using hybrid Monte Carlo filtering. In *International Conference on Computer Vision*, pp. II: 321–328, 2001.

[11] D. Comaniciu, V. Ramesh, and P. Meer. Kernel-based object tracking. *IEEE Transactions on Pattern Analysis and Machine Intelligence*, 25(5):564–577, 2003.

[12] T. F. Cootes and Taylor. Active shape models—"smart snakes". In *Proceedings. British Machine Vision Conf., BMVC92*, pp. 266–275, 1992.

[13] Ronald Satria Dan Witzner Hansen, Riad Hammoud and Jakob Sorensen. Improved likelihood function in particle-based ir eye tracking. In *IEEE CVPR Workshop on Object Tracking and Classification Beyond the Visible Spectrum*, San Diego, CA, June 2005.

[14] J.Y. Deng and F. Lai. Region-based template deformation and masking for eye-feature extraction and description. *Pattern Recogn*, 30:403–419, 1997.

[15] Arnaud Doucet, Nando de Freitas, and Neil Gordon. *Sequential Monte Carlo Methods in Practice*. Springer-Verlag, ISBN: 0-387-95146-6, 2001.

[16] A. T. Duchowski. A breath-first survey of eye tracking applications. *Behavior Research Methods, Instruments, and Computers (BRMIC)*, 34(4):455–470, 2002.

[17] Y. Ebisawa and S. Satoh. Effectiveness of pupil area detection technique using two light sources and image difference method. In *5th Annual Int. Conf. of the IEEE Eng. in Medicine and Biology Society*, pp. 1268–1269, 1993.

[18] N. Edenborough, R. I. Hammoud, A. Harbach, et al. Drowsy driver monitor from delphi. In *Demon Session, IEEE Computer Vision and Pattern Recognition Conference*, 2004.

[19] N. Edenborough, R. I. Hammoud, and A. Harbach et al. Driver state monitor from delphi. In *Demon session, IEEE Computer Vision And Pattern Recognition Conference*, 2005.

[20] I. R. Fasel and M. S. Bartlett. A comparison of gabor filter methods for automatic detection of facial landmarks. In *Proceedings of International Conference on Automatic Face and Gesture Recognition*, pp. 242–246, 2002.

[21] I.R. Fasel, B. Fortenberry, and J.R. Movellan. A generative framework for real time object detection and classification. *Computer Vision and Image Understanding*, 98(1):182–210, April 2005.

[22] K. Grauman, M. Betke, J. Gips, and G.R. Bradski. Communication via eye blinks: Detection and duration analysis in real time. In *IEEE Computer Vision and Pattern Recognition (CVPR)*, pp. I:1010–1017, 2001.

[23] Riad I. Hammoud. A robust eye position traker based on invariant local features, eye motion and infrared-eye responses. In *SPIE Defense and Security Symposium, Automatic Target Recognition Conference, Proceedings of SPIE Vol. Nb. 5807*, pp. 35–43, Orlando, FL, March 2005.

[24] Riad I. Hammoud, Andrew Wilhelm, Phillip Malawey, and Gerald J. Witt. Efficient real-time algorithms for eye state and head pose tracking in advanced driver support systems. In *IEEE Computer Vision and Pattern Recognition Conference*, 2005.

[25] Dan Witzner Hansen, John Paulin Hansen, Mads Nielsen, Anders Sewerin Johansen, and Mikkel B. Stegmann. Eye typing using markov and active appearance models. In *IEEE Workshop on Applications on Computer Vision*, pp. 132–136, 2003.

[26] D. W. Hansen and A.E.C. Pece. Eye tracking in the wild. *Comp. Vision Image Understand.* 98(1):155–181, April 2005.

[27] Andrew P. Harbach, Gregory K. Scharenbroch, Gerald J. Witt, Timothy J. Newman, Nancy Edenborough, and Hammoud Riad I. Imaging system and method for monitoring an eye. United States, Patent, US 2005/0100191 A1, 2005, (issued).

[28] A. Haro, M. Flickner, and I. Essa. Detecting and tracking eyes by using their physiological properties, dynamics, and appearance. In *IEEE Conf. Comp. Vision and Pattern Recognition*, Hilton Head Island, SC, June 2000.

[29] R. Herpers, M. Michaelis, K. Lichtenauer, and G. Sommer. Edge and keypoint detection in facial regions. In *International Conference on Automatic Face and Gesture-Recognition*, pp. 212–217, 1996.

[30] J. Huang and D. Mumford. Statistics of natural images and models. In *IEEE Computer Vision and Pattern Recognition (CVPR)*, pp. I: 541–547, 1999.

[31] J. Huang and H. Wechsler. Eye location using genetic algorithms. In *2nd Int'l conference in Audio and Video-Based Biometric Person Authentication (AVBPA)*, 1999.

[32] D. Huttenlocher, G. Klanderman, and W. Rucklidge. Comparing images using hausdorff distance. *IEEE Trans. Pattern Anal. Mach. Intell.* 15(9):850–863, 1993.

[33] M. Isard and A. Blake. Condensation—conditional density propagation for visual tracking, 1998.

[34] Michael Isard and Andrew Blake. Contour tracking by stochastic propagation of conditional density. In *European Conference on Computer Vision*, pp. 343–356, 1996.

[35] J.P. Ivins and J. Porrill. A deformable model of the human iris for measuring small 3-dimensional eye movements. *Mach. Vision Appl.* 11(1):42–51, 1998.

[36] R.J.K Jacob. *Eye Tracking in Advanced Interface Design*, Vols. 3–22. Oxford University Press, 1995.

[37] Q. Ji and X. Yang. Real time visual cues extraction for monitoring driver vigilance. In *Workshop on Computer Vision Systems, CVPR*, Vancouver, Canada, 2001.

[38] S. Julier and J. Uhlmann. A new extension of the kalman filter to nonlinear systems, 1997.

[39] S. Kawato and N. Tetsutani. Detection and tracking of eyes for gaze-camera control, 2002.

[40] S. Kawato and N. Tetsutani. Detection and tracking of eyes for gaze-camera control, 2002.

[41] Irwin King and Lei Xu. Localized principal component analysis learning for face feature extraction and recognition. In *Proceedings to the Workshop on 3D Computer Vision*, pp. 124–128, Shatin, Hong Kong, 1997.

[42] J.J. Koenderink and A.J. van Doorn. Representation of local geometry in the visual system. 55:367–375, 1987.

[43] K.M. Lam and H. Yan. Locating and extracting the eye in human face images. *Pattern Recogn.*, 29:771–779, 1996.

[44] L. J. Latecki, R. Lakamper, and U. Eckhardt. Shape descriptors for non-rigid shapes with a single closed contour. In *Proc. IEEE Conf. Comput. Vision and Pattern Recogn.*, pp. 424–429, 2000.

[45] http://www.eyegaze.com. LC Technologies INC., 2004.

[46] Simon P. Liversedge and John M. Findlay. Saccadic eye movements and cognition. *Trends Cogn. Sci.*, 4(1):6–14, January 2000.

[47] G. Loy and A. Zelinsky. Fast radial symmetry for detecting points of interest. *PAMI*, pp. 959–973, August 2003.

[48] John MacCormick and Michael Isard. Partitioned sampling, articulated objects, and interface-quality hand tracking. In *European Conference on Computer Vision*, pp. 3–19, 2000.

[49] Päivi Majaranta and Kari-Jouko Räihä. Twenty years of eye typing: Systems and design issues. In *Symposium on ETRA 2002: Eye Tracking Research Applications Symposium, New Orleans, Louisiana*, pp. 944–950, 2002.

[50] Y. Matsumoto and A. Zelinsky. An algorithm for real-time stereo vision implementation of head pose and gaze direction measurement. In *International Conference on Automatic Face and Gesture Recognition*, pp. 499–504, 2000.

[51] Fan Johnson Messom. Machine vision for an intelligent tutor.

[52] C.H. Morimoto, D. Koons, A. Amir, and M. Flickner. Pupil detection and tracking using multiple light sources. *IVC*, 18(4):331–335, 2000.

[53] http://www.multimap.com. MultiMap, UK aerial photo coverage, 2003.

[54] R. Newman, Y. Matsumoto, S. Rougeaux, and A. Zelinsky. Real-time stereo tracking for head pose and gaze estimation. In *International Conference on Automatic Face and Gesture Recognition*, pp. 122–128, 2000.

[55] Stavri Nikolov, Timothy Newman, Michael Jones, and Iain Gilchrist. Gaze-contingent display using texture mapping and opengl: System and applications. In *ACM Eye Tracking Research and Applications Symposium*, pp. 11–18, 2004.

[56] M. Nixon. Eye spacing measurements for facial recognition. *Applications of Digital Image Processing*, 575(VIII):279–285, 1985.

[57] B. Noureddin, P.D. Lawrence, and C.F. Man. A non-contact device for tracking gaze in a human computer interface. *Comp. Vision Image Understand.* 98(1): 52–82, April 2005.

[58] E. Osuna, R. Freund, and F. Girosi. Training support vector machines: an application to face detection. pp. 130–136, 1997.

[59] A.E.C. Pece and A.D. Worrall. Tracking with the EM contour algorithm. In *European Conference on Computer Vision*, pp. I: 3–17., 2002.

[60] L.R. Rabiner. A tutorial on hidden markov models and selected applications in speech recognition. *Proc. IEEE*, 77(2):257–286, 1989.

[61] C. Schmid and R. Mohr. Local grayvalue invariants for image retrieval. 19(5):530–534, 1997.

[62] B. Scholkopf, C. J. C. Burges, and A. J. Smola. *Advances in Kernel Methods: Support Vector Learning*. MIT Press, 1999.

[63] B. Scholkopf, S. Mika, C. J. C., Burges, P. Knirsch, K.-R. Mueller, G. Raetsch, and A. J. Smola. Input space versus feature space in kernel-based methods. *IEEE Trans. Neural Networks*, 1999.

[64] http://www.seeingmachines.com.au. SEEINGMACHINES, FaceLab, 2003.

[65] http://www.smarteye.se. Smart Eyes A/B, 2004.

[66] K.K. Sung and T. Poggio. Example-based learning for view-based human face detection. 20(1):39–51, 1998.

[67] M. E. Tipping. Sparse bayesian learning and the relevance vector machine. *J. of Mach. Learn. Res.*, 2001.

[68] http://www.tobii.se/. Tobii Technologies, 2004.

[69] A. Tomono, M. Iida, and Y. Kobayashi. A TV camera system which extracts feature points for non-contact eye movement detection. In *SPIE Optics, Illumination, and Image Sensing for Machine Vision*, volume 1194, pp. 2–12, 1989.

[70] M. Turk and A. Pentland. Face recognition using eigenfaces. pp. 586–591, 1991.

[71] R.C. Veltkamp and M. Hagedoorm. State of the art in shape matching. In *Technical Report UU-CS-1999-27*, Utrecht, 1999.

[72] P. Viola and M. Jones. Robust real-time face detection. In *International Conference on Computer Vision*, pp. II: 747, 2001.

[73] V. Vogelhuber and C. Schmid. Face detection based on generic local descriptors and spatial constraints. vol. 1, pp. 1084–1087, 2000.

[74] Colin Ware. *Information Visualization*. Morgan Kaufman Publishers, 2000.

[75] M. Wedel and R. Peiters. Eye fixations on advertisments and memory for brands: A model and findings. *Market. Sci.* 19(4):297–312, 2000.

[76] Jie Yang, Rainer Stiefelhagen, Uwe Meier, and Alex Waibel. Robust detection of facial features by generalized symmetry. In *International Conference on Pattern Recognition*, pp. I:117–120, 1992.

[77] David Young, Hilary Tunley, and Richard Samuels. Specialised hough transform and active contour methods for real-time eye tracking. Technical Report 386, School of Cognitive and Computing Sciences, University of Sussex, 1995.

[78] A. L. Yuille, P. W. Hallinan, and D.S Cohen. Feature extraction from faces using deformable templates. *Int. J. Comput. Vision*, 8(2):99–111, 1992.

[79] Z. Zhu and Q. Ji. Robust real-time eye detection and tracking under variable lighting conditions and various face orientations. *Comput. Vision Image Understand.* 98(1):124–154, April 2005.

[80] Zhiwei Zhu, Qiang Ji, Kikuo Fujimura, and Kuangchih Lee. Combining kalman filtering and mean shift for real tracking under active illumination. In *ICPR 2002*, Québec, Canada, August 11–15 2002.

Two Approaches to 3D Microorganism Recognition Using Single Exposure Online (SEOL) Digital Holography

Seokwon Yeom, Inkyu Moon, and Bahram Javidi

Department of Electrical and Computer Engineering, U-2157, University of Connecticut, Storrs, CT 06269-2157, USA

10.1 Introduction

Pattern recognition in images has broad applications from biometric identification to automatic target recognition (ATR). Optoelectronic imaging and processing has been studied to identify unknown targets in scenes [1–24]. Increased interests in three-dimensional (3D) information processing for pattern recognition are well reflected in the literature [8–24]. For example, digital holography has been utilized for object recognition and image encryption [10–22]. Applications in 3D integral imaging have been extended to ATR as well as depth estimation [23, 24].

There are numerous military and industrial applications of an automated recognition/identification system of microorganism: biological weapon detection in security and defense; diagnosis of diseases and food safety investigation in medical and health care; and ecological monitoring in wastewater treatment or oceanography. However, automated recognition of tiny biological objects in images is very challenging. They are small, simple, and, sometimes, undistinguishable in morphological traits. Moreover, the interclass diversity of microorganisms in size and shape is comparable to the intraclass diversity. The simple geometrical shape can be changed by the motility and growth which are influenced by external factors [25]. A microorganism can appear as an individual object or form a group or clutter with arbitrary complexity.

In this chapter, we introduce 3D visualization and two different approaches to microorganism recognition using single exposure on-line (SEOL) digital holography [17–22]. The frameworks of our system are composed of several stages as shown in Fig. 10.1. For sensing and visualization of 3D objects,

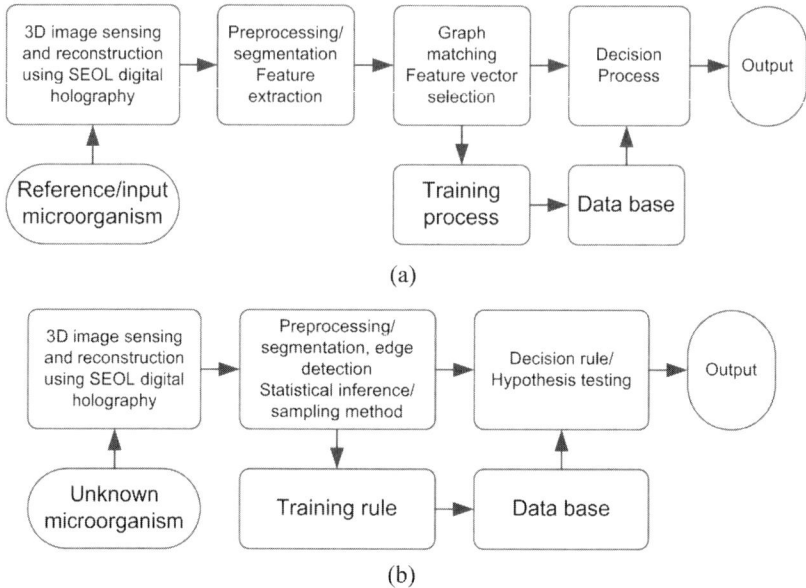

Figure 10.1. Frameworks of the 3D visualization and recognition of microbiological objects: (a) 3D morphology-based recognition, (b) shape-tolerant 3D recognition.

SEOL digital holography captures and reconstructs 3D microbiological objects. Utilizing a microscope-based Mach–Zehnder interferometer, SEOL digital holography records the complex amplitude of microorganisms in the Fresnel diffraction field. The 3D information of the microorganisms can be numerically reconstructed at an image plane of an arbitrary depth [26–30].

In the first approach (See Fig. 10.1(a)), image recognition is performed based on the 3D morphology comprising the complex amplitude of the re-constructed holographic images [21]. For preprocessing, reconstructed complex images are resized and backgrounds are subtracted. We segment foreground objects using the histogram analysis and the maximum transmittance rate of the coherent light. In the next stage, Gabor-based wavelets extract salient features by decomposing them in the spatial frequency domain [31–33].

The following feature matching technique identifies 3D reference shapes in unknown images. The rigid graph matching (RGM) technique measures the similarity of 3D morphologies between a reference microorganism and unknown biological samples. The graph matching technique with Gabor-based wavelets has been studied as a robust template matching technique which is tolerant to shift, rotation, and distortion [34–37].

The second approach utilizes the statistical inference to build a shape-tolerant 3D recognition system as shown in Fig. 10.1(b) [19]. A number of sampling segments are randomly extracted from the reconstructed 3D image of a biological microorganism. These sampling segments are processed using two cost functions: mean-squared distance (MSD) and mean-absolute distance (MAD), and statistical sampling theory for the equality of means and equality of variances between the sampling segments of a reference microorganism and unknown input biological samples. Student's t distribution and Fisher's F distribution are used to analyze the difference of means and the ratio of variances of two populations, respectively [38].

In Section 10.1, we present the sensing and reconstruction of SEOL digital holography. The 3D morphology-based recognition is presented in Section 10.3. In Section 10.4, the shape-tolerant 3D recognition technique is illustrated. In Section 10.5, experimental results are demonstrated. The conclusions follow in Section 10.6.

10.2 Single Exposure Online (SEOL) Digital Holography

In this section, we describe the SEOL digital holography technique. The SEOL digital hologram of a 3D microorganism in the Fresnel diffraction field is recorded by the CCD (charge-coupled device) array as shown in Fig. 10.2. Coherent light from an argon laser (center wavelength of 514.5 nm) is used as a source of illumination. A spatial filter and a collimating lens provide the spatial coherence. A beam splitter divides the plane-parallel wave into object and reference wave. The object wave illuminates the specimen magnified by the microscope objective. The microscope objective produces the focused image at the image plane. The wavefront of the 3D microorganism at the image plane starts to be diffracted. (see Fig. 10.3). The digital hologram of the 3D

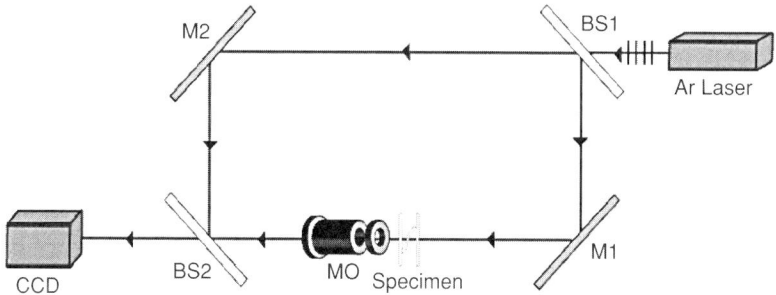

Figure 10.2. Experimental setup for recording the SEOL digital hologram of a 3D microorganism. Ar: argon laser; BS1, BS2: beam splitter; M1, M2: mirror; MO: microscope objective; CCD: charge coupled device array.

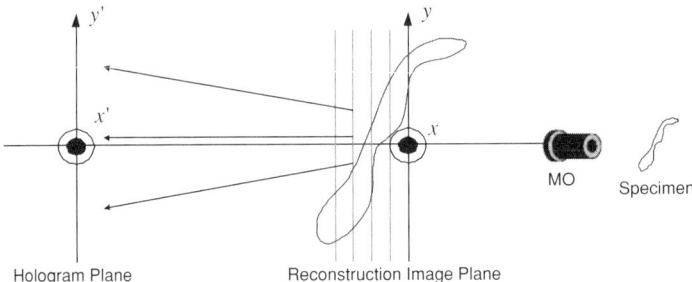

Figure 10.3. Coordinate system for the digital hologram and image reconstruction of 3D microorganisms.

microorganism is the interference intensity pattern generated by the plane-parallel reference wave and the diffracted wavefronts of the specimen.

The Fresnel diffraction field of the 3D microorganism at the CCD plane is represented as follows:

$$\mathbf{O}_H(x, y) = \mathbf{A}_H(x, y) \exp[j\Phi_H(x, y)] \tag{10.1}$$

$$= \int \int \int_{\lambda_d \frac{\delta_F}{2}}^{\lambda_d \frac{\delta_F}{2}} \frac{\exp[j2xz'/\lambda]}{j\lambda z'} \exp\left[j\frac{\pi}{\lambda z'}(x^2 + y^2)\right] O(x', y', z)$$

$$\times \exp\left[j\frac{\pi}{\lambda z'}(x'^2 + y'^2)\right] \exp\left[-j\frac{\pi}{\lambda z'}(xx' + yy')\right] dz' dx' dy',$$

where $\mathbf{O}$ is the 3D object under investigation; d_0 is the distance between the center of the 3D microorganism and CCD plane; and δ_m is the microorganism's depth along the z-axis.

The Fresnel diffraction field of the 3D microorganism can be numerically reconstructed by the inverse Fresnel transformation:

$$\mathbf{O}'(x, y) = \mathbf{A}_o(x, y) \exp[j\Phi_o(x, y)] \tag{10.2}$$

$$= \exp\left[-j\frac{\pi}{\lambda d_o}(\Delta X^2 x^2 + \Delta Y^2 y^2)\right] \sum_{x'=1}^{N_X} \sum_{y'=1}^{N_Y} \mathbf{H}(x', y')$$

$$\times \exp\left[-j\frac{\pi}{\lambda d_0}(\Delta x^2 x'^2 + \Delta y^2 y'^2)\right] \exp\left[j2\pi\left(\frac{xx'}{N_X} + \frac{yy'}{N_Y}\right)\right],$$

where $(\Delta X, \Delta Y)$ are the resolution at the image plane; $(\Delta x, \Delta y)$ are the resolution at the hologram plane; N_X and N_Y are the size of the hologram in the x and y directions; and $\mathbf{H}(x, y)$ is the SEOL digital hologram which is obtained from $\mathbf{O}_H$ [19, 20].

The reconstructed 3D microorganism's image from the SEOL digital hologram contains a conjugate image. This undesired component degrades the quality of the reconstructed 3D image, but this intrinsic and defocused conjugate image also contains information of the 3D microorganism for recognition.

As an additional merit, SEOL digital holography allows us to obtain a dynamic time-varying scene for monitoring and recognizing moving and growing microorganisms.

10.3 3D Morphology-Based Recognition

In this section, we present the segmentation for preprocessing, feature extraction using Gabor-based wavelets, and the graph matching technique for the 3D morphology-based recognition.

10.3.1 Preprocessing, Gabor-Based Wavelets, and Feature Extraction

Since the coherent light is modulated by the semitransparent objects, the intensity on the object surface becomes lower than the background region. Therefore, segmentation can be performed through the histogram analysis of reconstructed images. Detailed processes of the segmentation of the complex amplitude are presented in Refs. [20, 21]. After the segmentation, we apply the Gabor-based wavelets. The Gabor-based wavelets have the form of a Gaussian envelope modulated by the complex sinusoidal function. The impulse response (or kernel) of the 2D Gabor-based wavelet is defined as

$$\mathbf{G}(\mathbf{x}) = \frac{|\mathbf{k}|^2}{\sigma^2} \exp\left(-\frac{|\mathbf{k}|^2|\mathbf{x}|^2}{2\sigma^2}\right) \left[\exp(j\mathbf{k} \cdot \mathbf{x}) - \exp\left(-\frac{\sigma^2}{2}\right)\right], \qquad (10.3)$$

where $\mathbf{x}$ is a position vector; $\mathbf{k}$ is a wave number vector; and σ is proportional to the standard deviation of the Gaussian envelope. By changing the magnitude and direction of the vector $\mathbf{k}$, we can scale and rotate the Gabor kernel to make self-similar forms. The size of the Gaussian envelope is the same in the x and y directions, which is proportional to $\sigma/|\mathbf{k}|$. The second term in the square bracket in equation (10.3), $\exp(-\sigma^2/2)$, subtracts the DC value, thus, it has a zero mean response [32].

We can define a discrete version of the 2D Gabor kernel as $\mathbf{G}_{u\nu}(x, y)$ at $\mathbf{k} = \mathbf{k}_{u\nu}$ and $\mathbf{x} = [x\,y]^t$. The sampling of $\mathbf{k}$ is performed as $\mathbf{k}_{u\nu} = k_{0u}[\cos\phi_\nu \sin\phi_\nu]^t$, $k_{0u} = k_0/\delta^{u-1}$, $\phi_\nu = [(\nu - 1)/V]\pi$, $u = 1, \ldots, U$, and $\nu = 1, \ldots, V$, where k_{0u} is the magnitude of the wave number vector; ϕ_ν is the azimuth angle of the wave number vector; k_0 is the maximum carrier frequency of the Gabor kernels; δ is the spacing factor in the frequency domain; U and V are the total numbers of decompositions along the radial and tangential axes, respectively; and the superscript t stands for the matrix transpose.

Let $\mathbf{Y}_{u\nu}$ be the filtered output of the image $\hat{\mathbf{O}}$ after it is 2D convolved with the Gabor kernel $\mathbf{G}_{u\nu}$. Elements in the output matrix $\mathbf{Y}_{u\nu}$ are called the "Gabor coefficients":

$$\mathbf{Y}_{uv}(x, y) = \sum_{x'=1}^{N_X} \sum_{y'=1}^{N_Y} \mathbf{G}_{uv}(x - x', y - y') \, \hat{\mathbf{O}}(x', y'), \qquad (10.4)$$

where $\hat{\mathbf{O}}$ is the complex amplitude of the segmented image in Refs. [20, 21]. The magnitude of $\hat{\mathbf{O}}(x, y)$ is normalized between 0 and 1.

A rotation-invariant vector is defined at each pixel. The rotation-invariant property can be achieved simply by adding up all the Gabor coefficients along the tangential axes of the frequency domain. We define the component of the rotation-invariant vector for one node (pixel) as the summation of the Gabor coefficients:

$$\mathbf{v}[\mathbf{x}] = \left[\sum_{\nu=1}^{V} \mathbf{Y}_{1\nu}[\mathbf{x}] \cdots \sum_{\nu=1}^{V} \mathbf{Y}_{U\nu}[\mathbf{x}] \right]^{t}. \qquad (10.5)$$

Therefore, the dimension of the node vector $\mathbf{v}$ is U. There is no optimal way to choose the parameters for the Gabor kernels, but several values are widely used heuristically depending on the applications.

10.3.2 Recognition with the Rigid Graph Matching Technique

The RGM technique measures the similarity of 3D geometrical shapes comprising the complex magnitude between a reference microorganism and unknown input samples. The graph is defined as a set of nodes associated in the local area. Let R and S be two identical and rigid graphs placed on the reference image $\mathbf{O}_r$ and unknown sample image $\mathbf{O}_s$, respectively. The location of the reference graph R is pre-determined by a translation vector $\boldsymbol{p}_r$ and a counter clock-wise rotation angle θ_r. Position vectors of K nodes in the graph R are computed as

$$\mathbf{x}_k(\mathbf{p}_r, \theta_r) = \mathbf{A}_{\theta_r}(\mathbf{x}_k^o - \mathbf{x}_c^o) + \mathbf{p}_r, \qquad k = 1, \dots, K, \qquad (10.6)$$

$$\mathbf{A}_\theta = \begin{bmatrix} \cos\theta & -\sin\theta \\ \sin\theta & \cos\theta \end{bmatrix}, \qquad (10.7)$$

where $\mathbf{x}_k^o$ and $\mathbf{x}_c^o$ are, respectively, the position vectors of the node k and the center of the graph without any translation and rotation; and K is the total number of nodes in the graph.

In our database, the reference graph is predetermined in order to represent unique shape features of the microorganism. Assuming that the graph R covers a designated shape of the representing characteristic in the reference microorganism, we search the similar local shape by translating and rotating the graph S on unknown input images. We describe any rigid motion of the graph S by the translation vector $\mathbf{p}_s$ and the clockwise rotation angle θ_s:

$$\mathbf{x}_k(\mathbf{p}_s, \theta_s) = \mathbf{A}_{\theta_s}(\mathbf{x}_k^o - \mathbf{x}_c^o) + \mathbf{p}_s, \qquad k = 1, \dots, K, \qquad (10.8)$$

where $\mathbf{x}_k(\mathbf{p}_s, \theta_s)$ is the position vector of the node k in the graph S. The transformation in equation (10.8) performs a robust detection process of the rotated and shifted reference morphology.

A similarity function between the graphs R and S is defined as the summation of the normalized inner product of two vectors $\mathbf{v}_R[\mathbf{x}_k(\mathbf{p}_r, \theta_r)]$ and $\mathbf{v}_S[\mathbf{x}_k(\mathbf{p}_s, \theta_s)]$:

$$\Gamma_{RS}(\mathbf{p}_s, \theta_s) = \frac{1}{K} \sum_{k=1}^{K} \frac{|\langle \mathbf{v}_R[\mathbf{x}_k(\mathbf{p}_r, \theta_r)], \mathbf{v}_S[\mathbf{x}_k(\mathbf{p}_s, \theta_s)]\rangle|}{\|\mathbf{v}_R[\mathbf{x}_k(\mathbf{p}_r, \theta_r)]\|\|\mathbf{v}_S[\mathbf{x}_k(\mathbf{p}_s, \theta_s)]\|}, \tag{10.9}$$

where $\langle \cdot \rangle$ stands for the inner product; and $\mathbf{v}_R[\mathbf{x}_k(\mathbf{p}_r, \theta_r)]$ and $\mathbf{v}_S[\mathbf{x}_k(\mathbf{p}_s, \theta_s)]$ are the node vectors of the graph R in the reference image and the graph S in the unknown input image, respectively, as in equation (10.5). We adopt a difference cost function to improve discrimination capability between the two graphs R and S. The difference cost is defined as the absolute value of the difference between two vectors:

$$C_{RS}(\mathbf{P}_s, \hat{\theta}_s) = \frac{1}{K} \sum_{k=1}^{K} \|\mathbf{v}_R[\mathbf{x}_k(\mathbf{p_r}, \theta_r)] - \mathbf{v}_S[\mathbf{x}_k(\mathbf{p}_s, \theta_s)]\|. \tag{10.10}$$

The local area which is covered by the graph S is identified with the reference shape if the following two conditions are satisfied:

$$\Gamma_{RS}(\mathbf{p}_s, \hat{\theta}_s) > \alpha_\Gamma \quad \text{and} \quad C_{RS}(\mathbf{p}_s, \hat{\theta}_s) < \alpha_C, \tag{10.11}$$

where α_Γ and α_C are the thresholds for the similarity function and the difference cost, respectively; and $\hat{\theta}_s$ is obtained by searching the best matching angle to maximize the similarity function at the position vector $\mathbf{p}_s$:

$$\hat{\theta}_s = \arg\max_{\theta_s} \Gamma_{RS}(\mathbf{p}_s, \theta_s). \tag{10.12}$$

10.4 Shape-Tolerant 3D Recognition

In the following, we describe the procedure for shape-tolerant 3D microorganism recognition. We apply statistical cost functions and hypothesis testing for the 3D recognition which is independent of the shape of the microorganisms. The shape-tolerant approach may be suitable for recognizing microorganisms that do not have well-defined shapes or profiles. For example, they may be simple, unicellular, or branched in their morphological traits. It could also be applied to biological cells that vary in shape and profile rapidly.

First, we reconstruct the 3D microorganism as a volume image from a SEOL digital hologram corresponding to a reference microorganism. Then, we randomly extract arbitrary sampling segments (patches) in the reconstructed 3D image. We denote the mth reference sampling segment in the pth page as R_{mp} (see Fig. 10.4). We refer to each image plane with different reconstruction

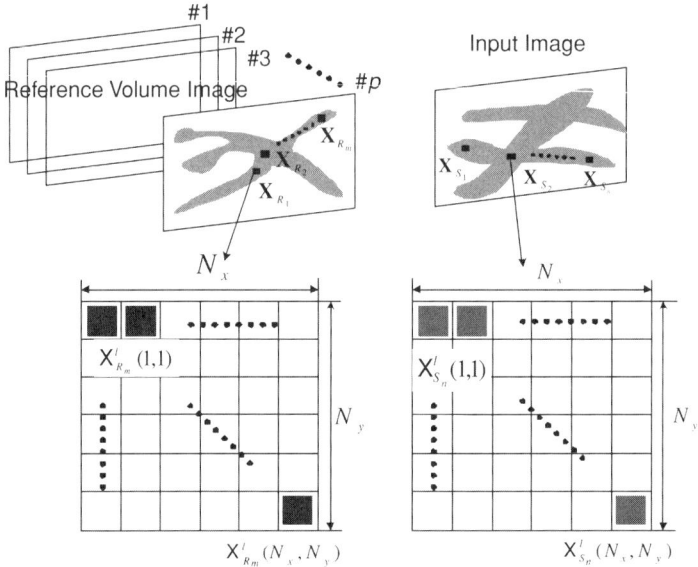

Figure 10.4. The design procedure for the shape-tolerant 3D recognition. The windows of sampling segment are extracted in the reconstructed 3D image from the SEOL digital hologram.

depth as "page." The field distribution $\mathbf{X}_{R_{mp}}$ in the sampling segment R_{mp} is composed of N_x by N_y complex values, where N_x and N_y are the size of the window in the x and y directions.

For testing, we record the SEOL digital hologram of an unknown input microorganism and then computationally reconstruct holographic images. Next, arbitrary sampling segments are extracted in the complex input image. We denote the nth unknown input sampling segment as S_n (see Fig. 10.4). The field distribution $\mathbf{X}_{S_n}$ in the sampling segment S_n is composed of N_x by N_y complex values, where N_x and N_y are the size of the window in the x and y directions.

We utilize two statistical cost functions and statistical inference to recognize the unknown input microorganism. First, we use mean-squared distance (MSD) and mean-absolute distance (MAD) to evaluate the quantitative discrepancy between the reference and the input microorganisms. Two cost functions measure the distances between the reference and unknown input microorganisms in a statistical way. When $p = 1$, the MSD and MAD are

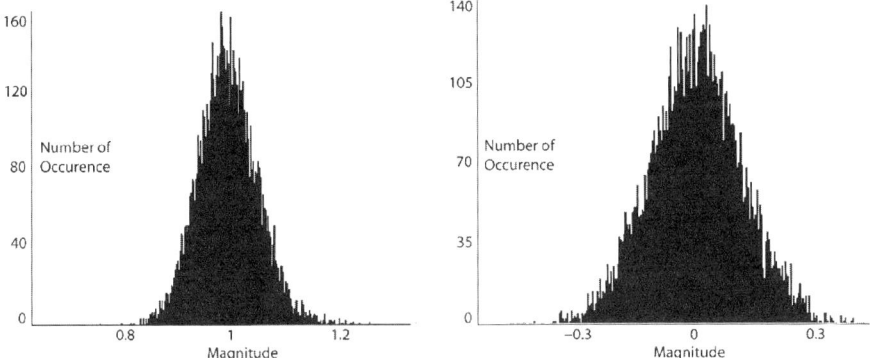

Figure 10.5. The histogram of the (a) real part, (b) imaginary part of the reconstructed image from the SEOL hologram.

defined as

$$\text{MSD} = \frac{\sum\limits_{x=1}^{N_x}\sum\limits_{y=1}^{N_y} |\mathbf{X}_{R_m}(x,y) - E[\mathbf{X}_{S_n}(x,y)]|^2}{\sum\limits_{x=1}^{N_x}\sum\limits_{y=1}^{N_y} |\mathbf{X}_{R_m}(x,y)|^2}, \tag{10.13}$$

$$\text{MAD} = \frac{\sum\limits_{x=1}^{N_x}\sum\limits_{y=1}^{N_y} |\mathbf{X}_{R_m}(x,y) - E[\mathbf{X}_{S_n}(x,y)]|}{\sum\limits_{x=1}^{N_x}\sum\limits_{y=1}^{N_y} |\mathbf{X}_{R_m}(x,y)|}, \tag{10.14}$$

where $E[\cdot]$ stands for the expectation operator which is estimated by the sample mean in the experiments.

Second, we conduct the hypothesis testing for the equality of means (location parameters) and variances (dispersion parameters) between two independent Gaussian populations using a statistical sampling theory [38]. Fig. 10.5 shows the histograms of the real and imaginary parts of the reconstructed complex images after the segmentation and the Sobel edge detection. As shown in this figure, the real and imaginary parts of the reconstructed image may be considered as nearly Gaussian distribution.

We assume that random variables (observations) $X_g^l(1), \ldots, X_g^l(N_g)$, which are elements of the reference sampling segment $\mathbf{X}_R$ or the input sampling segment $\mathbf{X}_S$, are statistically independent with identical Gaussian distribution $N(\mu_{X_g^l}, \sigma_{X_g^l}^2)$, where N_g is the total number of pixels in the sampling segment, i.e., $N_g = N_x \times N_y$; and $\mu_{X_g^l}$ and $\sigma_{X_g^l}^2$ are the mean and the variance of the field distribution, respectively. It is noted that the subscript g can be R or S to represent the reference or input sampling segment and the superscript l can stand for the "real" or the "imaginary" part of the sampling segment.

We also assume that random variables in the reference and the input sampling segments are independent.

For comparing the mean between two sampling segments, we assume that the variances are the same and all three statistical parameters ($\mu_{X_R}, \mu_{X_S}, \sigma^2 = \sigma^2_{X_R} = \sigma^2_{X_S}$) are unknown. The sample mean and sample variance of the reference and input sampling segments are given by

$$\overline{X}^l_g = \frac{1}{N_g} \sum_{n=1}^{N_g} X^l_g(n), \qquad \overline{V}^l_g = \frac{1}{N_g} \sum_{n=1}^{N_g} (X^l_g(n) - \overline{X}^l_g)^2. \qquad (10.15)$$

Since $\frac{N_R \overline{V}^l_R + N_S \overline{V}^l_S}{\sigma^2}$ has Chi-square(χ^2) distribution with $N_R + N_S - 2$ degrees of freedom, the following equation can be claimed:

$$\frac{N_R + N_S - 2}{N_R \overline{V}^l_R + N_S \overline{V}^l_S} \frac{\overline{X}^l_R - \overline{X}^l_S}{(N_R^{-1} + N_S^{-1})^{1/2}} \sim t_{N_R+N_S-2}, \qquad (10.16)$$

where $t_{N_R+N_S-2}$ is the Student's t distribution with $N_R + N_S - 2$ degrees of freedom [38].

To perform the hypothesis testing, we set a null hypothesis H_0 and alterative hypothesis H_1 as follows:

$$H_0 : \mu_{x^l_R} = \mu_{x^l_s}, \quad H_1 : \mu_{X^l_R} \neq \mu_{X^l_s}, \qquad (10.17)$$

where the null hypothesis means that there is no difference between population means. For the null hypothesis H_0, we use the statistic T:

$$T = \frac{1}{\overline{V}_P} \frac{\overline{X}^l_R - \overline{X}^l_S}{(N_R^{-1} + N_S^{-1})^{1/2}}, \qquad (10.18)$$

where $\overline{V}_P$ is the pooled estimator of the variance σ^2.

On the basis of a two-tailed test at a level of significance α_1, we have the following decision rule:

1. Accept H_0 if the statistic T is placed inside the interval $-t_{N_R+N_S-2,1-\alpha_1/2}$ to $t_{N_R+N_S-2,1-\alpha_1/2}$.
2. Reject H_0 otherwise.

We denote the upper $100(\alpha_1/2)\%$ point of $t_{N_R+N_S-2}$ as $t_{N_R+N_S-2,1-\alpha_1/2}$. Therefore, the following probability can be claimed [38]:

$$P\{(\overline{X}^l_R - \overline{X}^l_S) - t_{N_R+N_S-2,1-\alpha_1/2} \cdot \overline{V}_P(N_R^{-1} + N_S^{-1})^{1/2} \qquad (10.19)$$
$$< \mu_{X^l_R} - \mu_{X^l_S} < (\overline{X}^l_R - \overline{X}^l_S) + t_{N_R+N_S-2,1-\alpha_1/2}$$
$$\cdot \overline{V}_P(N_R^{-1} + N_S^{-1})^{1/2}\} = 1 - \alpha_1.$$

This decision rule implies that H_0 is true if the Student's t distribution occurs between the percentile values $-t_{1-\alpha_1/2}$ and $t_{1-\alpha_1/2}$.

For comparing the variance between two samples, we assume that all four parameters ($\mu_{X_R}, \mu_{X_S}, \sigma^2_{X_R}, \sigma^2_{X_S}$) are unknown. The following equation can be claimed from the ratio of the variance of two independent Gaussian populations:

$$\frac{[N_R/(N_R-1)]\bar{V}_R^l/\sigma_R^{l2}}{[N_S/(N_S-1)]\bar{V}_S^l/\sigma_S^{l2}} \sim F_{N_{R-1},N_{S-1}}, \tag{10.20}$$

where F_{N_R-1,N_S-1} is the Fisher's F distribution with N_R-1, N_S-1 degrees of freedom [38]. The null hypothesis H_0 and the alterative hypothesis H_1 are set as

$$H_0 : \sigma_R^{l2} = \sigma_S^{l2}, \qquad H_1 : \sigma_R^{l2} \neq \sigma_S^{l2}, \tag{10.21}$$

where the null hypothesis means that there is no difference between two variances. For the null hypothesis H_0, we use the statistic F:

$$F = \frac{[N_R/(N_R-1)]\bar{V}_R^l}{[N_S/(N_S-1)]\bar{V}_S^l} = \frac{\hat{V}_R^l}{\hat{V}_S^l}. \tag{10.22}$$

On the basis of a two-tailed test at a level of significance α_2, we have the following decision rule:
1. Accept H_0 if the statistic F is placed inside the interval $F_{N_R-1,N_S-1,\alpha_2/2}$ to $F_{N_R-1,N_S-1,1-\alpha_2/2}$.
2. Reject H_0 otherwise.
    We have denoted the upper $100(\alpha_2/2)\%$ point of the F_{N_R-1,N_S-1} distribution as $F_{N_R-1,N_S-1,\alpha_2/2}$. This decision rule implies that H_0 is true if the Fisher's F distribution occurs between the percentile values $F_{\alpha_2/2}$ and $F_{1-\alpha_2/2}$. Therefore, the following probability can be claimed [52]:

$$P\left\{F_{N_R-1,N_S-1,\alpha_2/2}^{-1} \cdot \frac{\hat{V}_R^l}{\hat{V}_S^l} < \frac{\sigma_R^{l2}}{\sigma_S^{l2}} < F_{N_R-1,N_S-1,1-\alpha_2/2}^{-1} \cdot \frac{\hat{V}_R^l}{\hat{V}_S^l}\right\} = 1-\alpha_2. \tag{10.23}$$

10.5 Experimental Results

We show experimental results of the image formation of the following filamentous microorganisms (sphacelaria alga, tribonema aequale alga, and polysiphonia alga) using SEOL digital holography. Recognition process using feature extraction and graph matching are presented to localize the predefined shapes of two different microorganisms. For shape tolerant 3D microorganism recognition, mean-squared distance (MSD), mean-absolute distance (MAD), and the test of hypothesis for the equality of means (location parameters) and equality of variances (dispersion parameters) between two different sampling segments of microorganisms are calculated.

10.5.1 3D Imaging with SEOL Digital Holography

In the following, we present the visualization of 3D microorganisms (sphace-laria alga, tribonema aequale alga and polysiphonia alga) using SEOL digital holography. In the experiments the 3D microorganisms are around 10–100 μm in size. They are recorded using a SEOL digital hologram with a CCD array of 2048 × 2048 pixels and a pixel size of 9 μm × 9 μm, where the specimen is sandwiched between two transparent cover slips. We generate holograms for each alga sample. The results of the reconstructed images from the hologram of the alga samples are shown in Fig. 10.6. Figs. 10.6(a) and

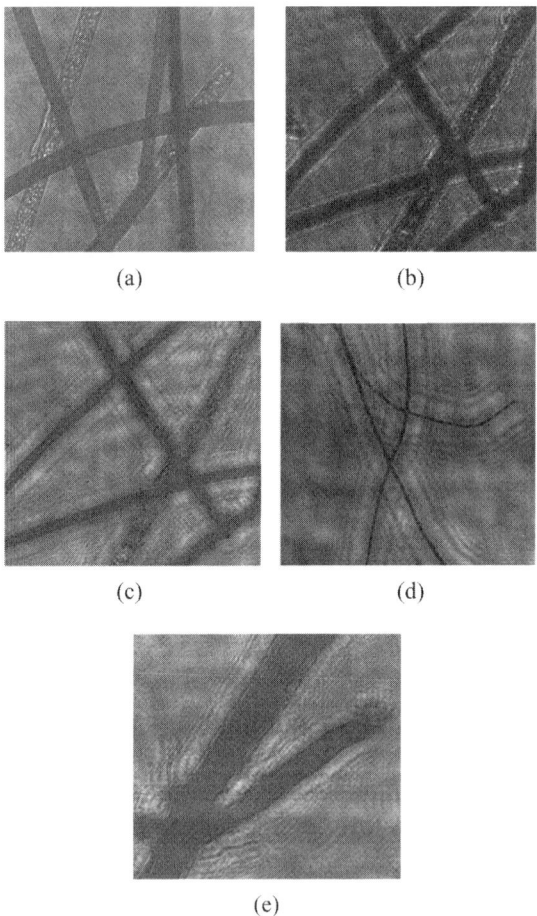

Figure 10.6. Experimental results for microorganism samples by use of a 10× micro-scope objective: (a) sphacelaria's 2D image and (b) sphacelaria's digital hologram by SEOL digital holography, reconstructed image of (c) sphacelaria alga, (d) tribonema aequale alga, (e) polysiphonia alga, (c)–(e) are reconstructed at $d = 180$ mm.

10.6(b) show sphacelaria's 2D image and the digital hologram by means of SEOL digital holography, respectively. Figs. 10.6(c), 10.6(d) and 10.6(e) are sphacelaria's, tribonema aequale's, and polysiphonia's reconstructed images from the blurred digital holograms at a distance $d = 180$ mm, respectively.

10.5.2 3D Morphology-Based Recognition

To test the recognition performance, we generate nine hologram samples each for sphacelaria and tribonema aequale. We denote the nine sphacelaria samples as A1,...,A9 and the nine tribonema aequale samples as B1,...,B9. To test the robustness of the proposed algorithm, we have changed the position of the CCD during the experiments resulting in different depths for the sharpest reconstruction image. The samples A1–A3 are reconstructed at 180 mm, A4–A7 are reconstructed at 200 mm, and A8 and A9 are reconstructed at 300 mm and all samples of tribonema aequale (B1–B9) are reconstructed at 180 mm for the focused images.

Magnitude and phase parts of computationally reconstructed holographic images are cropped and reduced into an image with 256×256 pixels by the reduction ratio 0.25. The probability P_s and the maximum transmittance rate $r_{\max}$ for the segmentation are set at 0.2 and 0.45, respectively. The segmentation applied to the complex holographic images is described in Refs. [20, 21]. We assume that less than 20% of the lower magnitude region is occupied by microorganisms and the magnitude of the microorganisms is less than 45% of the background diffraction field. The parameters for Gabor-based wavelets are set up at $\sigma = \pi, k_0 = \pi/4, \delta = \sqrt{2}, U = 3, V = 6$, and only real parts of Gabor coefficients are used.

To recognize two filamentous objects which have different thicknesses and distributions, we select two different reference graphs and place them on the samples A1 and B1. A rectangular grid is selected as a reference graph for sphacelaria which shows a regular thickness in the reconstructed images. The reference graph is composed of 25×3 nodes and the distance between the nodes is 4 pixels in the x and y directions. Therefore, the total number of nodes in the graph is 75. The reference graph R is located in the sample A1 with $\mathbf{p}_s = [81, 75]^t$ and $\theta_r = 135°$ as shown in Fig. 10.7(a). The threshold α_Γ which is set at 0.6 is only used. The threshold is selected heuristically to produce better results.

Considering the computational load, the graph S is translated by every 3 pixels in the x and y directions for measuring its similarity and difference with the graph R. To search the best matching angles, the graph S is rotated by $7.5°$ from 0 to $180°$ at every translated location. When the positions of rotated nodes are not integers, they are replaced with the nearest neighbor nodes.

Fig. 10.7(b) shows one sample (A9) of test images with the RGM process. The reference shapes are detected 65 times along the filamentous objects. Fig. 10.7(c) shows the number of detections for nine true-class and nine

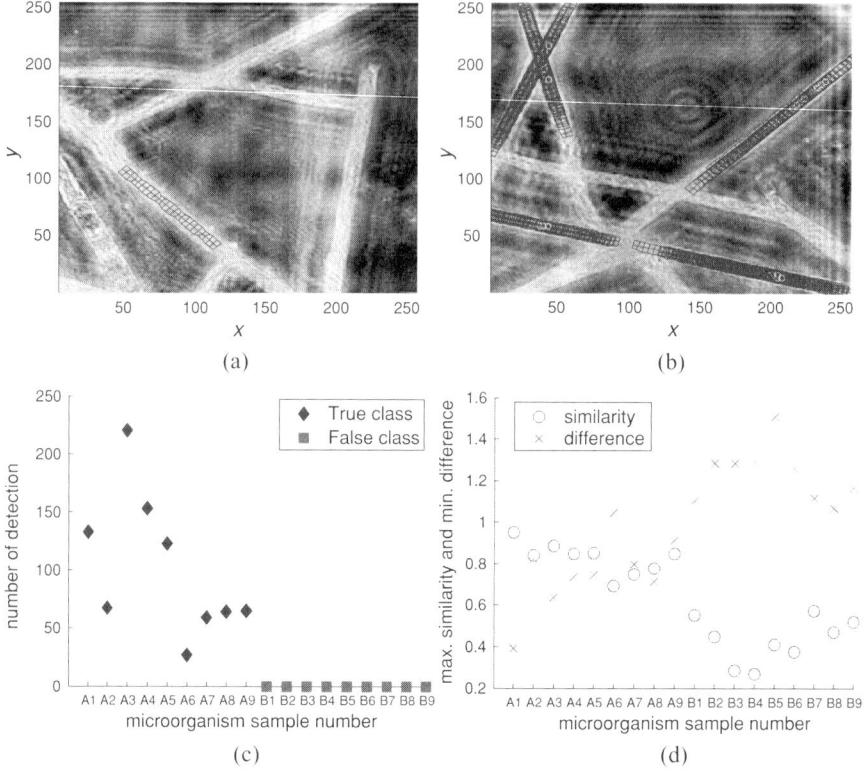

Figure 10.7. Recognition of sphacelaria: (a) reference sample A1 with the graph R, (b) RGM result of one test sample A9, (c) number of detections, (d) maximum similarity and minimum difference cost. (a) and (b) are presented by contrast reversal for better visualization.

false-class samples. The detection number for A1–A8 varies from 27 to 220, showing strong similarity between the reference image (A1) and test images (A2–A9) of the true-class microorganism. There is no detection found in the samples B1–B9, which are the false-class samples. Fig. 10.7(d) shows the maximum similarity and the minimum difference cost for all samples.

To recognize tribonema aequale, a wider rectangular grid is selected to identify its thin filamentous structure. The reference graph is composed of 20×3 nodes and the distance between the nodes is 4 pixels in the x direction and 8 pixels in the y direction; therefore, the total number of nodes in the graph is 60. The reference graph R is located in the sample B1 with $\mathbf{p}_r = [142, 171]^t$ and $\theta_r = 90°$ as shown in Fig. 10.8(a). The thresholds α_Γ and α_C are set at 0.8 and 0.7, respectively.

Figure 10.8(b) shows another sample (B2) of the true-class input image with the graph matching process. The reference shapes are detected 30 times along the thin filamentous object. Fig. 10.8(c) shows the number of detections

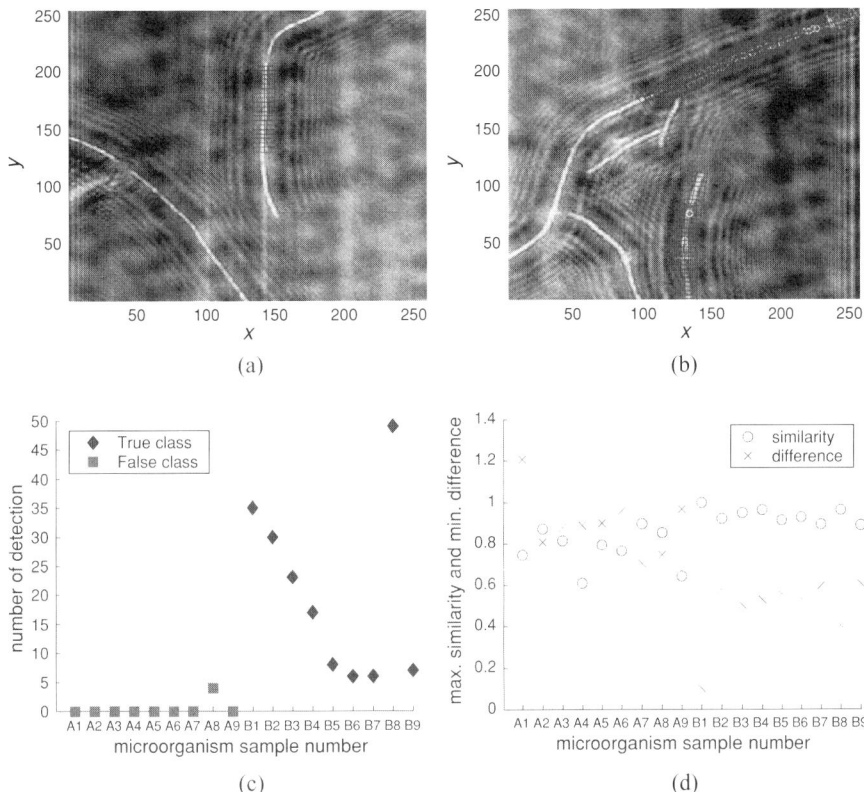

Figure 10.8. Recognition of tribonema aequale alga, (a) reference sample B1 with the graph R, (b) RGM result of one input sample B2, (c) number of detections, (d) maximum similarity and minimum difference cost, (a) and (b) are presented by contrast reversal for better visualization.

for nine true-class and nine false-class microorganisms. The detection number for the true-class samples B1–B9 varies from 6 to 49. Four false detections are found in one of the false-class samples A8. Fig. 10.8(d) shows the maximum similarity and the minimum difference cost for all samples.

10.5.3 Shape-Tolerant 3D Recognition

We evaluate the performance of the shape independent 3D microorganism recognition. First, 100 trial sampling segments are produced by randomly selecting the pixel values after the segmentation and the Sobel edge detection of the sphacelaria alga 3D image as the reference, where we change the size of each trial sampling segment from 2 to 500. Similarly, a number of sampling segments are randomly selected in the sphacelaria alga 3D image as the true-class inputs and in the polysiphonia alga image as the false-class inputs. We produce 100 true-class and 100 false-class input sampling segments to test the performance of the shape independent 3D recognition, respectively, where we change the size of each input sampling segment from 2 to 500. The reference and input images are reconstructed at a distance $d = 180$ mm. Figs. 10.9(a) and 10.9(b) show experimental results of the average MSD and MAD calculated by the complex values between each reference sampling segment and each input sampling segment. As shown in Fig. 10.9, it is noted

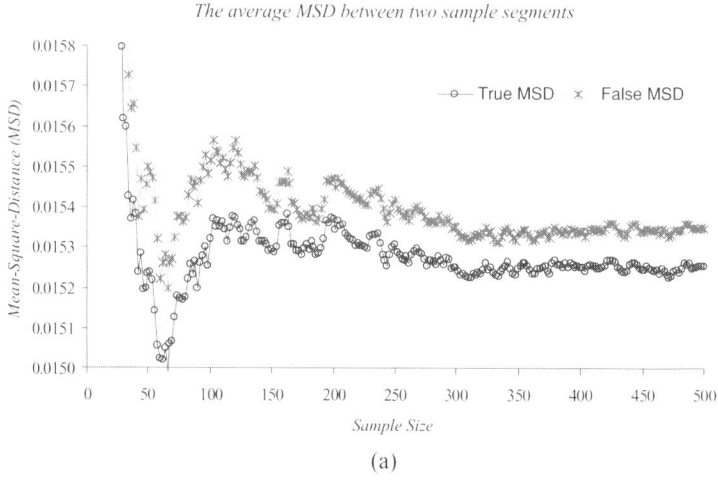

(a)

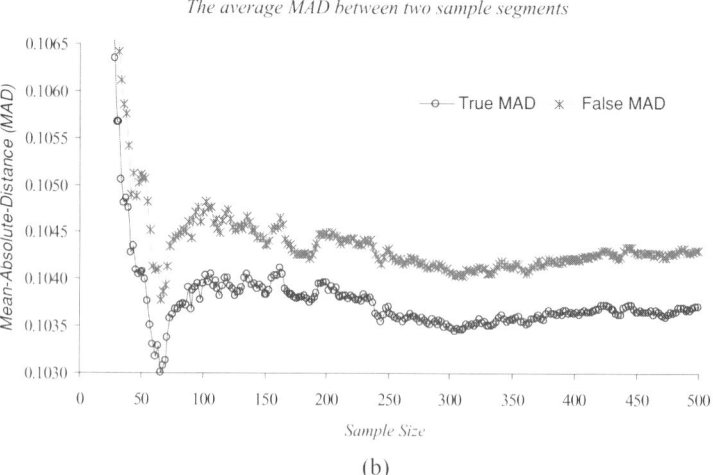

(b)

Figure 10.9. The average (a) MSD, (b) MAD calculated by the complex amplitude between the reference segments and the input segments versus the sample size of sampling segments.

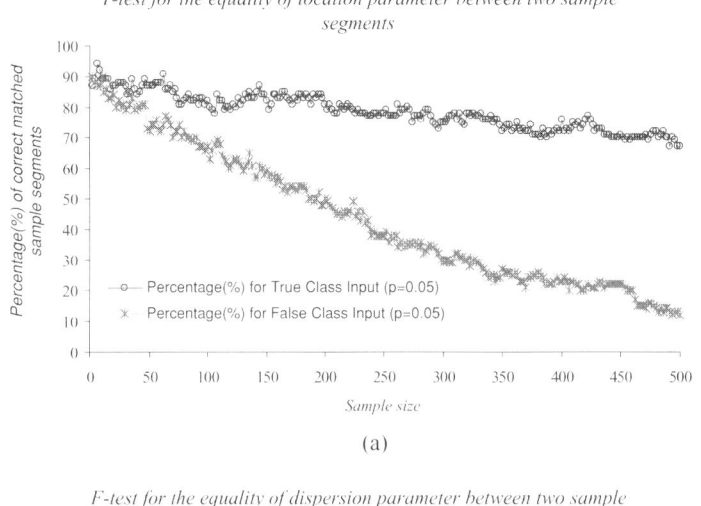

(a)

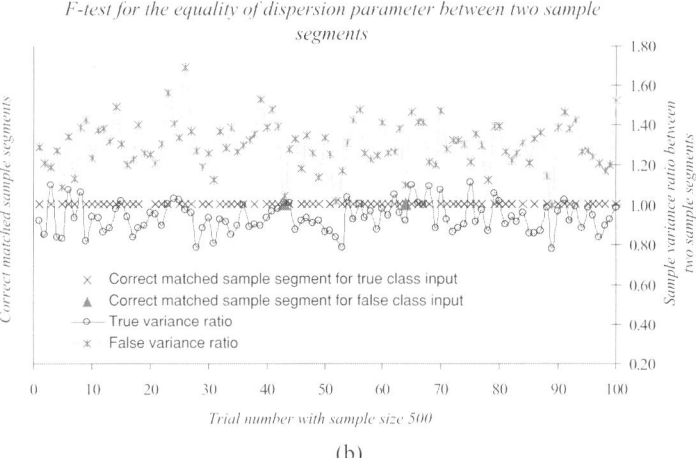

(b)

Figure 10.10. (a) T-test for the equality of the location parameters (means) between two sampling segments versus a sample size, (b) F-test for the equality of the dispersion parameters (variances) between two sampling segments versus a trial number with a sample size 500.

that the average MSD and MAD for the true-class input sampling segments calculated are around 0.01525 and 0.10355, respectively, and for the false-class input sampling segments more than 0.01535 and 0.10425.

Fig. 10.10(a) shows the results of the hypothesis testing for the difference of location parameters between the reference and input sampling segments, where we reject the null hypothesis H_0 if the statistic T defined in equation (10.18) is outside the range $-t_{0.975}$ and $t_{0.975}$ on the basis of a two-tailed test at 0.05 confidence level. It is noted that the percentages of the correct matched segments by the decision rule $(-t_{0.975} \leq T \leq t_{0.975})$ for the true-class input segments are around 80%, while for the false-class input sampling segments, the percentages rapidly decrease as a sample size increases. We also conduct the hypothesis testing for the ratio of dispersions of the reference and input microorganism, where we reject the null hypothesis H_0 if the statistic F defined in equation (10.22) is outside the range $F_{0.025}$ and $F_{0.975}$ on the basis of a two-tailed test at 0.05 confidence level. Fig. 10.10(b) shows the ratio of the dispersions of sampling segments between the reference and the number of the correct matched segments with a sample size 500 by the following decision rules $(F_{0.025} \leq F \leq F_{0.975})$. It is noted that the number of the correct matched segments for the true-class inputs is 87, while for the false-class inputs the number is 2.

10.6 Conclusion

In this chapter, we have overviewed the approaches for 3D sensing, imaging, and recognition of microorganisms. 3D sensing and reconstruction by means of the SEOL digital holography is robust to dynamic movement of microscopic objects and noisy environments as compared with the multiple-exposure phase-shifting digital holography.

In the first approach to the recognition of the biological microorganisms, 3D morphology-based recognition is presented. The recognition process of the first approach examines the simple morphological traits comprising the complex amplitude of microorganisms. Feature extraction is performed by the segmentation and Gabor-based wavelets. They are followed by the graph matching technique to localize the specific 3D shape of different classes of microorganisms.

In the second approach to the recognition, the shape-tolerant 3D recognition using the statistical costs and inference is presented. A number of sampling segments are randomly extracted and processed with the cost functions and statistical inference theory. By investigating the Gaussian property of the reconstructed images of microorganisms, we are able to distinguish the sample segments of the reference from the input of the different class of microorganism.

The automatic recognition and classification of microorganisms is very challenging. By presenting two fundamental approaches using a 3D imaging technique, we hope to put forward a first step toward fully automated biological microorganism surveillance and classification.

References

[1] A. Mahalanobis, R. R. Muise, S. R. Stanfill, and A. V. Nevel. Design and application of quadratic correlation filters for target detection. *IEEE Trans. AES*, 40:837–850, 2004.

[2] F. A. Sadjadi. Infrared target detection with probability density functions of wavelet transform subbands. *Appl. Opt*, 43:315–323, 2004.

[3] B. Javidi and P. Refregier, eds. *Optical Pattern Recognition*. SPIE, Bellingham, WA, 1994.

[4] H. Kwon and N. M. Nasrabadi, "Kernel RX-algorithm: a nonlinear anomaly detector for hyperspectral imagery," *IEEE Trans. Geosci. Remote Sens.*, vol. 43, pp. 388–397, 2005.

[5] F. Sadjadi, ed. *Milestones in Performance Evaluations of Signal and Image Processing Systems*. SPIE Press, Belligham, WA, 1993.

[6] P. Refregier, V. Laude, and B. Javidi, "Nonlinear joint transform correlation: an optimum solution for adaptive image discrimination and input noise robustness," *J. Opt. Lett.*, vol. 19, pp. 405–407, 1994.

[7] F. Sadjadi, "Improved target classification using optimum polarimetric SAR signatures," *IEEE Trans. AES.*, vol. 38, pp. 38–49, 2002.

[8] B. Javidi and F. Okano, eds. *Three-Dimensional Television, Video, and Display Technologies*. Springer, New York, 2002.

[9] B. Javidi, ed. *Image Recognition and Classification: Algorithms, Systems, and Applications*. Marcel Dekker, New York, 2002.

[10] B. Javidi and E. Tajahuerce, "Three dimensional object recognition using digital holography," *Opt. Lett.*, vol. 25, pp. 610–612, 2000.

[11] O. Matoba, T. J. Naughton, Y. Frauel, N. Bertaux, and B. Javidi, "Real-time three-dimensional object reconstruction by use of a phase-encoded digital hologram," *Appl. Opt.*, vol. 41, pp. 6187–6192, 2002.

[12] Y. Frauel and B. Javidi, "Neural network for three-dimensional object recognition based on digital holography," *Opt. Lett.*, vol. 26, pp. 1478–1480, 2001.

[13] E. Tajahuerce, O. Matoba, and B. Javidi, "Shift-invariant three-dimensional object recognition by means of digital holography," *Appl. Opt.*, vol. 40, pp. 3877–3886, 2001.

[14] B. Javidi and D. Kim, "Three-dimensional-object recognition by use of single-exposure on-axis digital holography," *Opt. Lett.*, vol. 30, pp. 236–238, 2005.

[15] D. Kim and B. Javidi, "Distortion-tolerant 3-D object recognition by using single exposure on-axis digital holography," *Opt. Exp.*, vol. 12, pp. 5539–5548, 2005. Available: http://www.opticsexpress.org/abstract.cfm?URI=OPEX-12-22-5539.

[16] S. Yeom and B. Javidi, "Three-dimensional object feature extraction and classification with computational holographic imaging," *Appl. Opt.*, vol. 43, pp. 442–451, 2004.

[17] B. Javidi, I. Moon, S. Yeom, and E. Carapezza, "Three-dimensional imaging and recognition of microorganism using single-exposure on-line (SEOL) digital holography." *Opt. Exp.*, vol. 13, pp. 4492-4506, 2005. Available: http://www.opticsexpress.org/abstract.cfm?URI=OPEX-13-12-4492.

[18] B. Javidi, I. Moon, S. Yeom, and E. Carapezza. Three-dimensional imaging and recognition of microorganism using single-exposure on-line (SEOL) digital holography, In B. Javidi (ed.), *Optics and Photonics for Homeland Security*. Springer, Berlin, 2005.

[19] I. Moon and B. Javidi, "shape-tolerant three-dimensional recognition of microorganisms using digital holography," *Opt. Exp.*, vol. 13, pp. 9612–9622, 2005. Available: http://www.opticsexpress.org/abstract.cfm?URI=OPEX-13-23-9612.

[20] S. Yeom and B. Javidi, "Three-dimensional recognition of microorganisms," *J. Bio. Opt.*, vol. 11, pp. 024017-1 8, 2006.

[21] S. Yeom, I. Moon, and B. Javidi, "Real-time 3D sensing, visualization and recognition of dynamic biological micro-organisms," *Pro. IEEE*, vol. 94, pp. 550–566, 2006.

[22] B. Javidi, S. Yeom, I. Moon, and M. Daneshpanah, "Real-time automated 3D sensing, detection, and recognition of dynamic biological micro-organic events," *Opt. Exp.*, vol. 13, pp. 3806–3829, 2006. Available: http://www.opticsinfobase.org/abstract.cfm?URI=oe-14-9-3806.

[23] S. Kishk and B. Javidi, "Improved resolution 3D object sensing and recognition using time multiplexed computational integral imaging," *Opt. Exp.*, vol. 11, pp. 3528–3541, 2003. Available: http://www.opticsexpress.org/abstract.cfm?URI=OPEX-11-26-3528.

[24] S. Yeom, B. Javidi, and E. Watson, "Photon counting passive 3D image sensing for automatic target recognition," *Opt. Exp.*, vol. 13, pp. 9310–9330, 2005. Available: http://www.opticsexpress.org/abstract.cfm?URI=OPEX-13-23-9310.

[25] J. W. Lengeler, G. Drews, and H. G. Schlegel. *Biology of the Prokaryotes*. Blackwell, New York, 1999.

[26] J. W. Goodman. *Introduction to Fourier Optics*, 2nd ed. McGraw-Hill, Boston, 1996.

[27] P. Ferraro, S. Grilli, D. Alfieri, S. D. Nicola, A. Finizio, G. Pierattini, B. Javidi, G. Coppola, and V. Striano, "Extended focused image in microscopy by digital holography," *Opt. Exp.*, vol. 13, pp. 6738–6749, 2005. Available: http://www.opticsexpress.org/abstract.cfm?URI=OPEX-13-18-6738.

[28] T. Zhang and I. Yamaguchi, "Three-dimensional microscopy with phase-shifting digital holography," *Opt. Lett.*, vol. 23, pp. 1221, 1998.

[29] T. Kreis, ed. *Handbook of Holographic Interferometry*. Wiley, VCH, Germany, 2005.

[30] J. W. Goodman and R. W. Lawrence, "Digital image formation from electronically detected holograms," *Appl. Phy. Lett.*, vol. 11, pp. 77–79, 1967.

[31] J. G. Daugman, "Uncertainty relation for resolution in space, spatial frequency, and orientation optimized by two-dimensional visual cortical filters," *J. Opt. Soc. Am.*, vol. 2, pp. 1160–1169, 1985.

[32] T. S. Lee, "Image representation using 2D Gabor wavelets," *IEEE Trans. PAMI.*, vol. 18, pp. 959–971, 1996.

[33] J. G. Daugman, "How iris recognition works," *IEEE Trans. Circuits Syst. Video Technol.*, vol. 14, pp. 21–30, 2004.

[34] M. Lades, J. C. Vorbruggen, J. Buhmann, J. Lange, C. v.d. Malsburg, R. P. Wurtz, and W. Konen, "Distortion invariant object recognition in the dynamic link architecture," *IEEE Trans. Comput.* vol. 42, pp. 300–311, 1993.

[35] R. P. Wurtz, "Object recognition robust under translations, deformations, and changes in background," *IEEE Trans. PAMI*, vol. 19, pp. 769–775, 1997.

[36] B. Duc, S. Fischer, and J. Bigun, "Face authentification with Gabor information on deformable graphs," *IEEE Trans. Image Process.*, vol. 8, pp. 504–516, 1999.

[37] S. Yeom, B. Javidi, Y. J. Roh, and H. S. Cho, "Three-dimensional object recognition using x-ray imaging," *Opt. Eng.*, vol. 43, Feb. 2005.

[38] N. Mukhopadhyay. *Probability and Statistical Inference*. Marcel Dekker, New York, 2000.

Distortion-Tolerant 3D Object Recognition by Using Single Exposure On-Axis Digital Holography

Daesuk Kim and Bahram Javidi

Electrical and Computer Engineering Department, University of Connecticut, CT 06269-2157, USA

11.1 Introduction

Optical systems have been extensively studied for object recognition [1–8]. With rapid advances in CCD array sensors, computers, and software, digital holography can be performed efficiently as an optical system for 3D object recognition. Digital holography has also been a subject of great interest in various fields [9–16]. In digital holography, the off-axis scheme has been widely used since it is simple and requires only a single exposure in separating the desired real image term from the undesired DC and conjugate terms [9]. However, 3D object recognition by use of the off-axis scheme has inherent limitations in terms of robustness to the variation of the 3D object position due to the superposition of the real image with undesired terms. The tolerance problem to the variation of the 3D object positioning can be resolved by employing an on-axis scheme. Recent studies on 3D object recognition by use of phase-shifting digital holography have provided a feasible approach for implementing the on-axis 3D object recognition system [6–8].

The phase-shifting digital holographic method requires multiple hologram recordings [10], however. Therefore, the 3D object recognition system based on the on-axis phase-shifting approach has inherent constraints in the sense that it needs a vibration-free environment that is convenient only in a laboratory. Also, it is sensitive to recording a moving object. It was proposed that such limitations of 3D object recognition can be overcome by use of single exposure on-axis digital holography [17]. In this study, we present a distortion-tolerant 3D object recognition method based on single exposure on-axis digital holography [18]. The main subject in this study is to show that the single exposure on-axis based 3D object recognition can provide a distortion-tolerant 3D object recognition capability compared with that obtainable by use of on-axis phase-shifting digital holography, while maintaining practical advantages of the single exposure on-axis scheme, such as robustness to environmental disturbances and tolerance to moving targets.

This chapter is organized as follows. In Section 11.2, we introduce a novel scheme on single exposure on-axis digital holography. In Section 11.3, a detailed composite filter design process is described. In Section 11.4, we carry out the performance tests of the distortion-tolerant 3D object recognition by using single exposure on-axis digital holography. The conclusions are presented in Section 11.5.

11.2 Single Exposure On-Axis Digital Holography

The basic experimental setup for the single exposure on-axis digital holographic system is based on a Mach–Zehnder interferometer as shown in Fig. 11.1. B/S means a beam splitter and M represents a plane mirror. Polarized light from an Ar laser source tuned to 514.5 nm is expanded and divided into object and reference beams. The object beam illuminates the object and the reference beam propagates on-axis with the light diffracted from the object. In order to match the overall intensity level between the reference wave and the reflected wave from the 3D object, a tunable density filter is used in the reference beam path. The interference between the object beam and the reference beam is recorded by a CCD camera. In contrast to on-axis phase shifting digital holography, no phase-shifting components are required. The conventional on-axis phase-shifting digital holography can be implemented

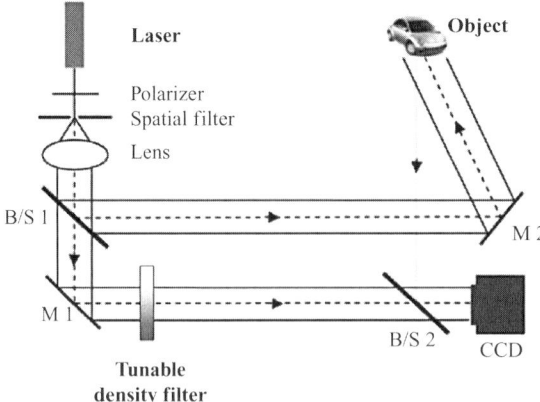

Figure 11.1. Experimental setup of the distortion-tolerant 3D object recognition system based on a single exposure on-axis scheme.

by inserting a quarter wave plate in the reference beam path. For on-axis phase-shifting digital holography, the hologram recorded on the CCD can be represented as follows:

$$H(x,y,\theta) = |O(x,y)|^2 + |R(x,y)|^2 + \exp(-i\theta)O(x,y)R^*(x,y)$$
$$+ \exp(i\theta)O^*(x,y)R(x,y). \tag{11.1}$$

Here, $[O(x,y)]$ and $[R(x,y)]$ represent the Fresnel diffraction of the 3D object and the reference wave functions, respectively. θ is an induced phase shift. In on-axis phase-shifting digital holography based on multiple holograms, the desired object wave function $[O(x,y)]$ can be subtracted by use of two holograms having $\lambda/4$ phase difference and two DC terms $|O(x,y)|^2$ and $|R(x,y)|^2$ as follows:

$$u_i^M(x,y) = O(x,y)R^*(x,y)$$
$$= \frac{1}{2}\left\{[H(x,y,0) - |O(x,y)|^2 - |R(x,y)|^2] + i\left[H\left(x,y,\frac{\pi}{2}\right)\right.\right.$$
$$\left.\left. - |O(x,y)|^2 - |R(x,y)|^2\right]\right\}. \tag{11.2}$$

Here, $u_i^M(x,y)$ represents the synthesized hologram obtained by on-axis phase-shifting digital holography which requires multiple recordings.

Likewise, the following expression represents the synthesized hologram obtained by use of single exposure on-axis digital holography:

$$u_i^S(x,y) = O(x,y)R^*(x,y) + O^*(x,y)R(x,y) \tag{11.3}$$
$$= H(x,y,0) - |R(x,y)|^2 - \frac{1}{N^2}\sum_{k=0}^{N-1}\sum_{l=0}^{N-1}$$
$$\times \{H(k\Delta x, l\Delta y, 0) - |R(k\Delta x, l\Delta y)|^2\}.$$

Here, N is the windowed hologram size used for image reconstruction. Δx and Δy represent the directional spatial resolution of the CCD. The reference wave can be assumed to be a known function because it is not changed while the object intensity distribution varies according to the input scene. The DC term $|O(x,y)|^2$ generated by the object can be reduced effectively by employing an averaging over the whole hologram as described in the third term of equation (11.3) [14]. Thus, the index $u_i^S(x,y)$ can be regarded as a synthesized result obtained by use of only a single hologram. The single exposure on-axis approach suffers from the superposition of the conjugate image which degrades the reconstructed image quality. However, for 3D object recognition, the inherent conjugate term in single exposure on-axis digital holography may not substantially degrade the recognition capability since the conjugate image term also contains information about the 3D object [17].

In order to apply the distortion-tolerant 3D object recognition, we first need to numerically reconstruct 3D object images. 3D multi perspective section images on any parallel plane can be reconstructed by using the following

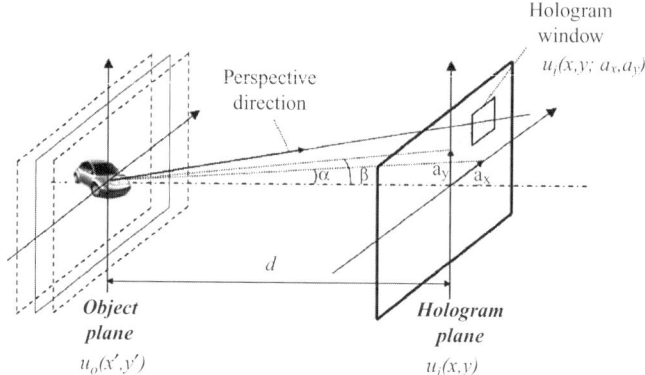

Figure 11.2. Three-dimensional multiple image sectioning and multi perspectives by digital holography.

inverse Fresnel transformation:

$$
u_0(x', y') = \frac{\exp(ikd)}{id\lambda} \exp\left[\frac{ik}{2d}(x'^2 + y'^2)\right] \times \mathbf{F}\left\{u_i(x, y) \exp\left[\frac{ik}{2d}(x^2 + y^2)\right]\right\}.
$$
(11.4)

Here, $\mathbf{F}$ denotes the 2D Fourier transformation. $u_i(x, y)$ represents the synthesized hologram which can be either $u_i^M(x, y)$ in equation (11.2) or $u_i^S(x, y)$ in equation (11.3). $u_0(x', y')$ is the reconstructed complex optical field at a parallel plane which is at a distance d from the CCD. By changing d, we can reconstruct multiple section images from the single hologram without using any optical focusing. Different regions of the digital hologram can be used for reconstructing different perspectives of the 3D object with the angle of view (α, β) as depicted in Fig. 11.2. $u_i(x, y; a_x, a_y)$ denotes the amplitude distribution in the window of the hologram used for the 3D object reconstruction. Here, a_x and a_y are central pixel coordinates of the hologram window.

11.3 Composite Filter Design

The reconstructed complex wave $u_0(x', y')$ in equation (11.4) can be directly used to construct a variety of correlation filters [1–4] for 3D object recognition [17]. However, direct use of the reconstructed image for constructing a matched filter may prevent us from recognizing correctly distorted true class targets [8]. In order to improve the distortion-tolerant recognition capability, removing phase terms in the reconstructed complex field decreases high sensitivity. Also, we can employ some preprocessing such as averaging and median filtering to remove speckle. The main goal of the proposed distortion-tolerant 3D object recognition by use of single exposure on-axis digital holography is to recognize a distorted reference 3D object as a true input while correctly rejecting false class objects. For this purpose, a filter must be designed to provide a high performance distortion-tolerant 3D object recognition capability.

We now describe the detailed process of creating the composite filter by use of single exposure on-axis digital holography. We measure M different holograms as we change the perspectives on a 3D reference object. After capturing each single hologram, we can make a synthesized hologram $u_i^S(x, y)$ which can be used for reconstructing the complex wave function $u_o(x', y')$. Then, in order to decrease the high sensitivity of the reconstructed wave function, we remove phase terms in the reconstructed complex field and apply some preprocessing techniques of averaging and median filtering. By use of such preprocessed M training images represented by $s_1(x, y), s_2(x, y), \ldots, s_M(x, y)$, we can generate p−dimensional column vectors $\mathbf{s}_1, \mathbf{s}_2, \ldots, \mathbf{s}_M$ by taking the rows of the $m \times n$ matrix and linking their transpose to create a $p = m \times n$ vector. We compute the Fourier transformation of each training image and obtain a new set of column vectors $\mathbf{S}_1, \mathbf{S}_2, \ldots, \mathbf{S}_M$. Then, we nonlinearly modify the amplitude of each column vector by use of the power law nonlinearity by which those column vectors become $\mathbf{S}_1^k, \mathbf{S}_2^k, \ldots, \mathbf{S}_M^k$ [18]. This power-law operation is defined for any complex vector $\mathbf{v}$ as follows:

$$\mathbf{v}^k = \begin{bmatrix} \mathbf{v}[1]^k \exp(j\phi_\mathbf{v}[1]) \\ \mathbf{v}[2]^k \exp(j\phi_\mathbf{v}[2]) \\ \vdots \\ \mathbf{v}[p]^k \exp(j\phi_\mathbf{v}[p]) \end{bmatrix}. \tag{11.5}$$

Now, we define a new matrix $\mathbf{S}^k = [\mathbf{S}_1^k, \mathbf{S}_2^k, \ldots, \mathbf{S}_M^k]$ with vector $\mathbf{S}_i$ as its ith column. Using such notation, the k-law-based nonlinear composite filter $\mathbf{h}^k$ can be defined in the spatial domain as follows:

$$\mathbf{h}^k = \mathbf{F}^{-1}\{S^k[(S^k)^+ S^k]^{-1} c^*\}. \tag{11.6}$$

Here $(\mathbf{S}^k)^+$ is the complex conjugate transpose of $\mathbf{S}^k$. The vector $\mathbf{c}$ contains the desired cross-correlation output origin values for each Fourier transformed training data $\mathbf{S}_i$ and the notation $\mathbf{c}^*$ denotes the complex conjugate of $\mathbf{c}$ [18].

11.4 Experimental Results

We compare the distortion tolerant 3D object recognition capability of the single exposure scheme with that of the multiple exposure on-axis phase-shifting holographic approach. Experiments have been conducted with a 3D object as the reference object which is $25 \times 25 \times 35$ mm and several false 3D objects with similar size. The reference car is located at a distance $d = 880$ mm from the CCD camera. We use an argon laser of 514.5 nm. The CCD has a pixel size of $12 \times 12\,\mu$m and 2048×2048 pixels. However, throughout the experiment, we use only 1024×1024 pixels for each reconstruction process since multi perspectives are required from a single hologram in making a composite filter.

11.4.1 Rotation Tolerance

First, we acquire nine synthesized holograms of the reference true class object labeled from 1 to 9 (synthesized holograms: #1–#9) for the on-axis phase-shifting approach as well as single exposure on-axis scheme. For every synthesized hologram capturing step, the reference object is rotated by $0.2°$ around the axis orthogonal to Fig. 11.1. Figures. 11.3(a) and 11.3(b) show the reconstructed image of the reference 3D object from one of the synthesized holograms by use of single exposure on-axis digital holography and on-axis phase-shifting digital holography, respectively. They are reconstructed by using the central window area of 1024×1024 pixels from the total synthesized hologram of 2048×2048 pixels. As can be seen, even the reconstructed image with only amplitude data contains high frequency speckle patterns. In order to smooth the high frequency amplitude distribution, we need to employ preprocessing techniques. Specifically, we use the window of 6×6 pixels for averaging which is followed by a median filtering with the window size of 5×5 pixels. After the averaging process, the window of 1024×1024 pixels is reduced to 171×171 pixels since we reconstruct a new image with averaged values calculated for each averaging window. After the median filtering process, the image size of 171×171 pixels remains unchanged.

Among the nine synthesized holograms of the true class reference object, we use three (synthesized holograms: #3, #6, and #9) to construct a nonlinear composite filter. For each of those three synthesized holograms, we select one centered window of 1024×1024 pixels and two different windows of the same size to reconstruct three training images with different perspectives from each synthesized hologram. Figures 11.4(a)–11.4(c) depict three preprocessed training images reconstructed from the synthesized hologram #6.

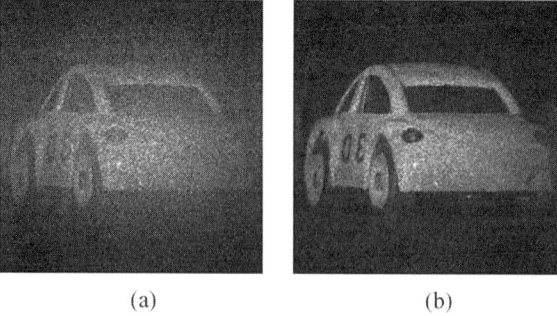

(a)   (b)

Figure 11.3. Reconstructed images of the reference true target prior to pre-processing by use of (a)single exposure on-axis digital holography and (b) multiple exposure on-axis phase shifting digital holography.

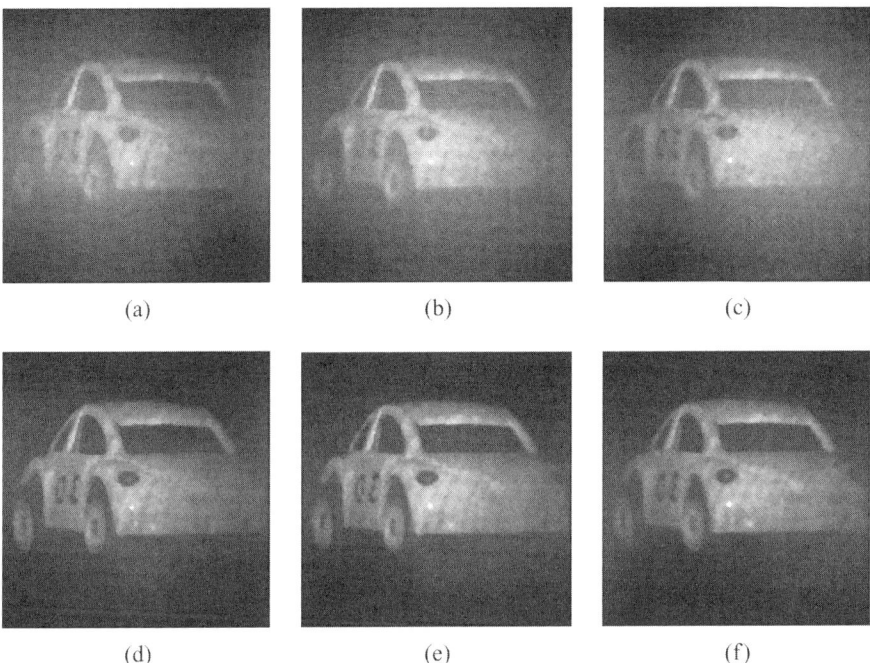

Figure 11.4. (a)–(f) Three reconstructed training images from the synthesized holo-gram #6 by use of (a)–(c) single exposure on-axis digital holography and (d)–(f) multiple exposures on-axis phase-shift digital holography.

Figures 11.4(a) and 11.4(b) have a viewing angle difference of 0.4°. The view-ing angle difference between Figs. 11.4(a) and 11.4(c) is 0.8°. Figure 11.4(d)–11.4(f) represent the three views obtained by use of phase-shifting on-axis digital holography. In the same way, we can obtain nine training images from the three synthesized holograms (synthesized holograms: #3, #6, and #9). Then, we construct a nonlinear composite filter by following the procedures described in Section 3. We use a nonlinear factor $k = 0.1$ in equation (11.6) to improve the discrimination capability of the nonlinear filter. Figures 11.5(a) and 11.5(b) depict the final nonlinear composite filters that can be used for the single exposure approach and the multiple exposure approach, respectively.

Once we construct the composite filter, we perform the distortion-tolerant 3D object recognition process for various input objects. We acquire more synthesized holograms for several false class objects labeled from 10 to 15 (synthesized hologram: #10–#15). Now, we have 15 synthesized holograms consisting of true and false class objects. We can reconstruct 15 images by using the central window area of 1024×1024 pixels from the total synthesized hologram of 2048×2048 pixels and perform the preprocessing.

Finally, those 15 preprocessed images (input images: #1–#15) are put in the input scene for distortion-tolerant 3D object recognition. Figure 11.6

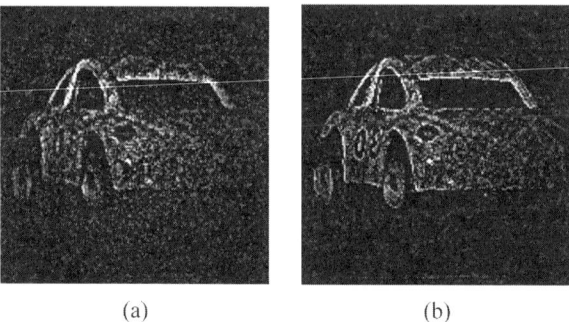

(a)   (b)

Figure 11.5. Constructed nonlinear composite filters for distortion-tolerant 3D object recognition by use of (a) single exposure on-axis digital holography and (b) on-axis phase-shift digital holography.

shows some of the false class objects (images: #10, #13, and #15). Six input images (images: #1, #2, #4, #5, #7, and #8) among the nine true class inputs (images: #1–#9) are used as our nontraining distorted true class objects.

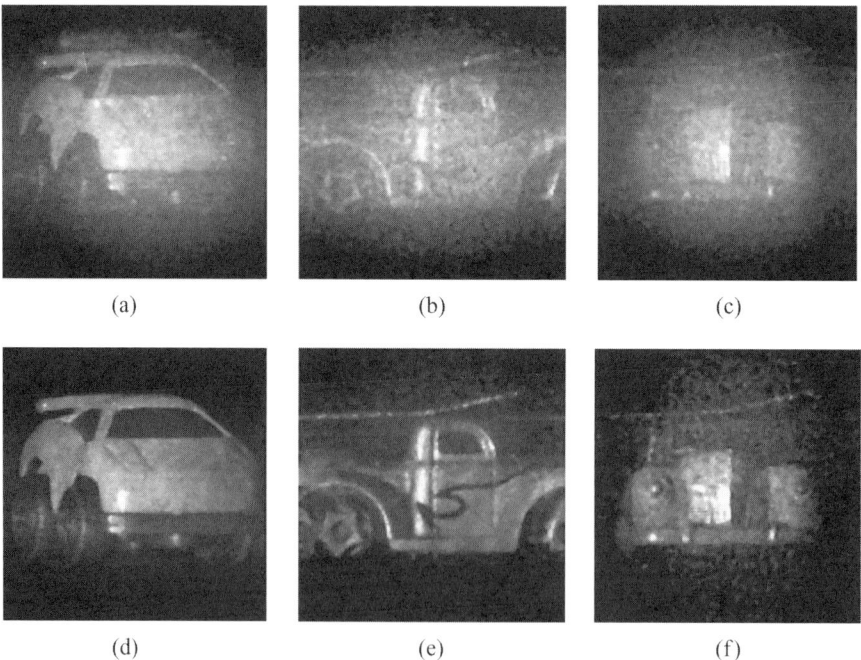

(a)   (b)   (c)

(d)   (e)   (f)

Figure 11.6. Reconstructed images of some flase class oobjects (images: #10, #13, and # 15) by use of (a)–(c) single exposure on-axis digital holography and (d)–(f) on-axis phase-shift digital holography.

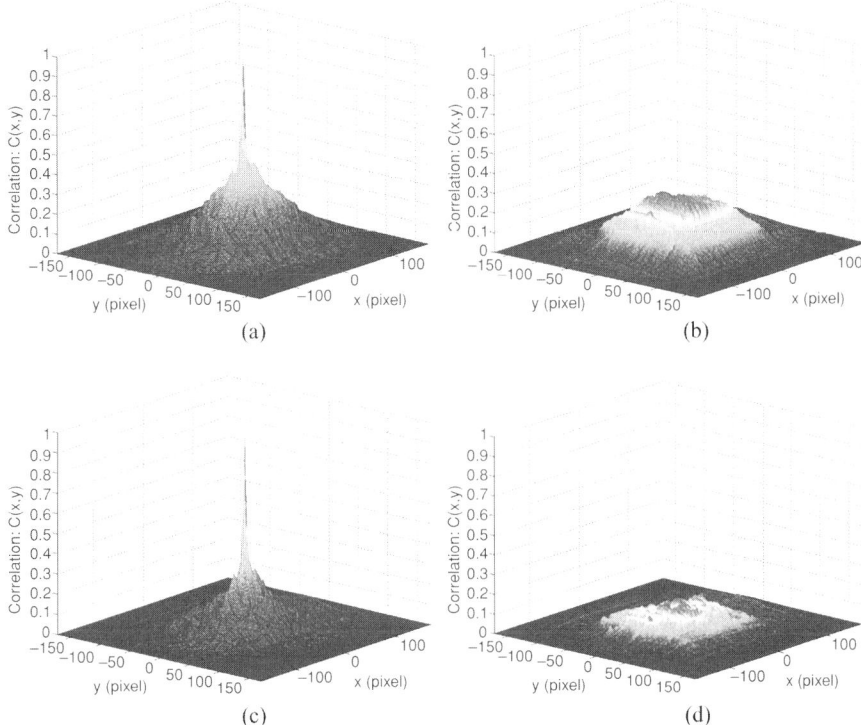

Figure 11.7. (a)–(d) Normalized correlation when input image #8 of the nontraining true targets and input image #13 among the false class objects are used as input scenes for (a)–(b) single exposure on-axis digital holography and (c)–(d) on-axis phase-shift digital holography (see color insert).

Figure 11.7(a),(b) and 11.7(c),(d) represent the normalized correlation obtained by use of single exposure on-axis digital holography and phase-shifting on-axis digital holography, respectively. All plots are normalized by the correlation peak value obtained when the training image #3 is used as input scene. As can be seen in Figs. 11.4(a)–11.4(c) and Figs. 11.6(a)–11.6(c), for the single exposure on-axis digital holography, the reconstructed image quality degrades due to the presence of its conjugate image. The 3D object recognition by use of single exposure on-axis digital holography, which employs both the phase and amplitude information, is capable of 3D object recognition. However, the distortion-tolerant capability of the single exposure on-axis scheme somewhat degrades after the averaging and median filtering since the 3D object information contained in the conjugate image term is disappeared due to the preprocessing. The degradation of the recognition capability could be from the existence of the conjugate term in the single exposure approach.

Figure 11.8 illustrates the distortion tolerant 3D recognition capability of the single exposure on-axis digital holographic approach. The degradation

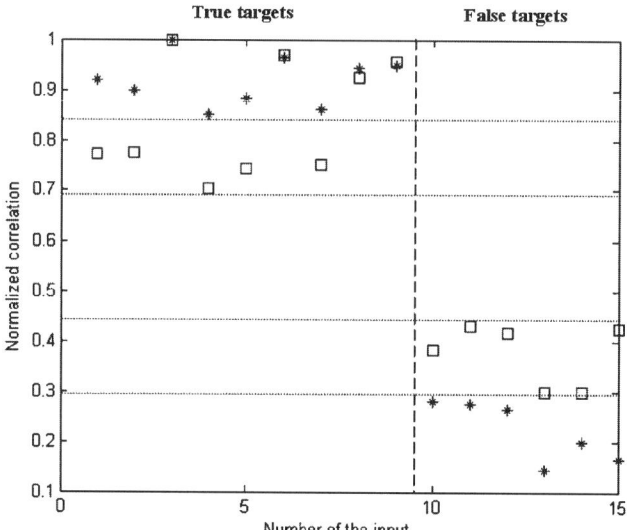

Figure 11.8. Normalized correlation distribution for various input images of true targets (#1–#9) and false class objects (#10–#15) for single exposure on-axis digital holography (□) and on-axis phase-shifting digital holography(*).

in recognition capability is not substantially less than that of on-axis phase-shifting digital holography.

11.4.2 Longitudinal Shift Tolerance

We have examined the longitudinal shift tolerance of the single exposure scheme in order to compare with that of phase-shifting digital holography. For the practical use of digital holography in distortion-tolerant recognition, it must also be less sensitive to longitudinal shifts along the z axis for recognizing a true class input target. It provides benefits in terms of the reduced number of reconstructions that must be performed for recognizing the input 3D object.

For this longitudinal shift tolerance, we construct another nonlinear composite filter which includes a number of out-of-focus reference reconstruction images. For constructing this filter, we use only one hologram (synthesized hologram: #3) among the three used for constructing the composite filter in rotational distortion tolerance. We use the same three reconstruction windows for generating training images. However, for each window, we also use three reconstructed images with a defocus of −20 and 20 mm as well as the reconstructed image of the reference object in the focus plane. Likewise, nine views are used to construct the new nonlinear composite filter for longitudinal shift tolerance.

Figure 11.9 represent, the normalized correlation peak values versus longitudinal shift for both the single exposure on-axis scheme and the

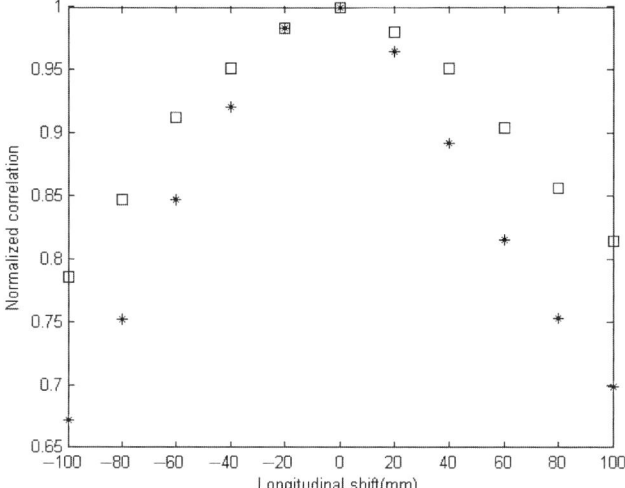

Figure 11.9. Normalized correlation peak values versus longitudinal shift along the z axis for single exposure on-axis digital holography ($\square$) and on-axis phase-shifting digital holography ($*$).

phase-shifting digital holographic approach when a distorted reference target is in the input scene. It shows that the single exposure on-axis scheme is less sensitive to longitudinal shift than the on-axis phase-shifting approach. The reason that the sensitivity is less for the single exposure scheme compared with that of the phase-shifting approach may be due to the inherent image degradation of the single exposure scheme as stated earlier. The single exposure on-axis approach seems to be somewhat more robust to the variation of longitudinal shift.

11.5 Conclusion

We have presented a distortion-tolerant 3D object recognition system using single exposure on-axis digital holography. The proposed method requires only a single hologram recoding for the distortion-tolerant 3D object recognition while the on-axis phase-shifting-based approach requires multiple recordings. The main benefit of the proposed single exposure method is that it can provide more robustness to environmental noise factors such as vibration since it uses only a single hologram, which can be measured and analyzed in real time. Experimental results show that the distortion-tolerant recognition capability of the single exposure approach is somewhat worse than that of the phase-shifting digital holographic method. Nevertheless, the proposed single exposure method makes the recording system simpler and more tolerant to object parameters such as moving targets while maintaining a distortion-tolerant 3D object recognition capability.

References

[1] J. W. Goodman. *Introduction to Fourier Optics*. McGraw-Hill, New-York, 1968.
[2] P. Réfrégier and F. Goudail. Decision theory approach to nonlinear joint-transform correlation. *J. Opt. Soc. Am.* A, 15:61–67, 1998.
[3] A. Mahalanobis, Correlation pattern recognition: an optimum approach. *Image Recognition and Classification*. Marcel Dekker, New-York, 2002.
[4] F. Sadjadi, eds. *Automatic Target Recognition*. Proc. SPIE 5426, Orlando, FL, April 2004.
[5] J. Rosen. Three dimensional electro-optical correlation. *J. Opt. Soc. Am.* A, 15:430–436, 1998.
[6] B. Javidi and E. Tajahuerce. Three-dimensional object recognition by use of digital holography. *Opt. Lett.*, 25:610–612, 2000.
[7] E. Tajahuerce, O. Matoba, and B. Javidi. Shift-invariant three-dimensional object recognition by means of digital holography. *Appl. Opt.*, 40:3877–3886, 2001.
[8] Y. Frauel, E. Tajahuerce, M. Castro, and B. Javidi. Distortion-tolerant three-dimensional object recognition with digital holography. *Appl. Opt.*, 40:3887–3893, 2001.
[9] U. Schnars and W. Jupter. Direct recording of holograms by a CCD target and numerical reconstructions. *Appl. Opt.*, 33:179–181, 1994.
[10] I. Yamaguchi and T. Zhang. Phase-shifting digital holography. *Opt. Lett.*, 22:1268–1270, 1997.
[11] J. W. Goodman and R. W. Lawrence. Digital image formation from electronically detected holograms. *Appl. Phys. Lett.*, 11:77–79, 1967.
[12] J. Caulfield. *Handbook of Optical Holography*. Academic, London, 1979.
[13] G. Pedrini and H. J. Tiziani. Short-coherence digital microscopy by use of a lensless holographic imaging system. *Appl. Opt.*, 41:4489–4496, 2002.
[14] U. Schnars and W. Juptner. Digital recording and numerical reconstruction of holograms. *Meas. Sci. Technol.*, 13:R85–R101, 2002.
[15] Y. Takaki, H. Kawai, and H. Ohzu. Hybrid holographic microscopy free of conjugate and zero-order images. *Appl. Opt.*, 38:4990–4996, 1999.
[16] P. Ferraro, G. Coppola, S. De Nicola, A. Finizio, and G. Pierattini. Digital holographic microscope with automatic focus tracking by detecting sample displacement in real time. *Opt. Lett.*, 28:1257–1259, 2003.
[17] B. Javidi, D. Kim. 3-D object recognition by use of single exposure on-axis digital holography. *Opt. Lett.*, 30:236–238, 2005.
[18] B. Javidi and D. Painchaud. Distortion-invariant pattern recognition with Fourier-plane nonlinear filters. *Appl. Opt.*, 35:318–331, 1996.

Design of Distortion-Invariant Optical ID Tags for Remote Identification and Verification of Objects

Elisabet Pérez-Cabré[1], María Sagrario Millán[1], and Bahram Javidi[2]

[1]Department of Optics & Optometry, Universitat Politècnica de Catalunya, Violinista Vellsolà 37, 08222 Terrassa, Spain
[2]Department of Electrical & Computer Engineering, University of Connecticut, 371 Fairfield Road, Unit 2157, Storrs, CT 06269-2157, USA

12.1 Introduction

Optical identification (ID) tags [1] have a promising future in a number of applications such as the surveillance of vehicles in transportation, control of restricted areas for homeland security, item tracking on conveyor belts or other industrial environment, etc. More specifically, passive optical ID tag [1] was introduced as an optical code containing a signature (that is, a characteristic image or other relevant information of the object), which permits its real-time remote detection and identification. Since their introduction in the literature [1], some contributions have been proposed to increase their usefulness and robustness. To increase security and avoid counterfeiting, the signature was introduced in the optical code as an encrypted function [2–5] following the double-phase encryption technique [6]. Moreover, the design of the optical ID tag was done in such a way that tolerance to variations in scale and rotation was achieved [2–5]. To do that, the encrypted information was multiplexed and distributed in the optical code following an appropriate topology. Further studies were carried out to analyze the influence of different sources of noise. In some proposals [5, 7], the designed ID tag consists of two optical codes where the complex-valued encrypted signature was separately introduced in two real-valued functions according to its magnitude and phase distributions. This solution was introduced to overcome some difficulties in the readout of complex values in outdoors environments. Recently, the fully phase encryption technique [8] has been proposed to increase noise robustness of the authentication system.

This chapter aims to revise the main features introduced in the design of optical ID tags for distortion-invariant detection and verification as well as to introduce new results to show the feasibility of the proposal. The chapter is organized as follows: First, a description of the whole authentication system is provided (Section 12.2). Afterwards, a procedure of the encryption technique is detailed (Section 12.3). We focus our attention on the design of distortion-invariant ID tags in Section 12.4. Section 12.5 introduces the signature verification process which is based on correlation. Then, authentication results and the analysis of noise robustness are provided in Section 12.6. Finally, Section 12.7 summarizes the conclusions.

12.2 Authentication System

Real-time remote object surveillance can be achieved by using an optical code located on a visible part of the item under control. The optical code manufactured with retroreflective materials can be inspected with a receiver to verify the authenticity of the object (Fig. 12.1). Usually, the information stored in the tag will consist of an identification number, an image, or other relevant information, named a signature, that permits the recognition of the object. The signature is obtained after the decryption of the image contained in the ID tag. Encryption and decryption can be done either electronically or optically as it is to be described in Section 12.3. The verification system can be a correlator [9]. The signature to be verified is imaged onto the input plane, and compared with a stored reference function. An intensity peak in the output plane will be used to decide whether the object is authenticated or not. Authentication should be achieved when the ID tag is captured under its original position or even if it appears scaled or rotated.

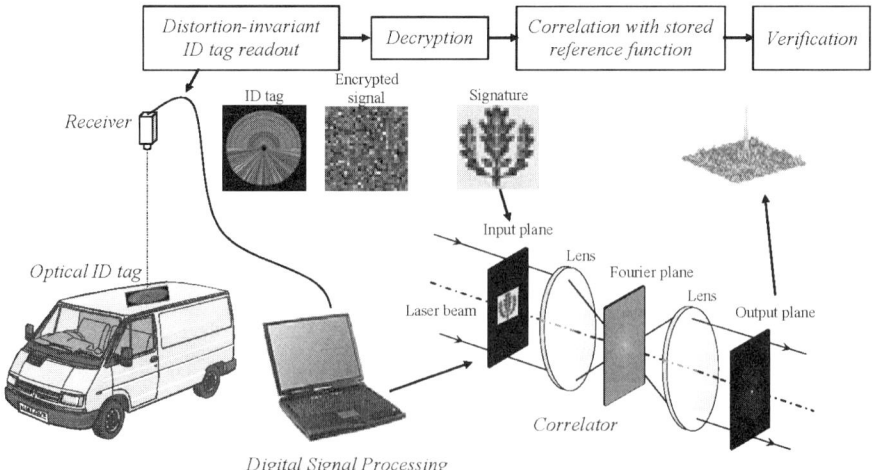

Figure 12.1. Block diagram of the complete authentication system.

A complete diagram of the proposed remote authentication system is depicted in Fig. 12.1. First, an optical code is built and placed on the object under surveillance. Then, a distortion-invariant ID tag readout is carried out by a receiver. And finally, the signature is decrypted and verified by correlation.

12.3 Signature Encryption and Decryption

Prior to introduce the signature in the ID tag, its information is encrypted to increase security and avoid counterfeiting. Commonly, images to be encrypted are intensity representations. Encryption allows one to convert the primary image into stationary white noise, so that the ciphered function does not reveal the appearance of the signature at human sight. Double-phase encryption [6] is used in most of the results shown in this chapter. This technique provides robustness against different types of ID tag degradation such as noise, occlusion, scratches, etc. [10, 11]. Therefore, we choose this technique among other encryption methods, such as the standard private key system. Double-phase encryption permits us to appropriately cipher grayscale images for optical tags without conversion to binary signatures which is the case for XOR encryption with a stream of pseudo random key [10].

Let us describe the algorithm of the double-phase encryption technique [6] used in this work. We denote by $f(x, y)$ the signature to be encrypted, which is normalized ($0 \leq f(x, y) \leq 1$) and sampled to have a total number of pixels N. The coordinates in the spatial and in the frequency domain are (x, y) and (μ, ν), respectively. The double-phase encryption technique uses two random phase codes to convert the input information $f(x, y)$ into stationary noise. One phase code is used in the input plane, and the second phase code is used in the frequency domain (Fourier plane).

Let $p(x, y)$ and $b(\mu, \nu)$ be two independent white sequences, uniformly distributed in the interval [0, 1]. Two operations are performed to obtain the encrypted information. First, the signature $f(x, y)$ is multiplied by the input phase mask $\exp[i2\pi p(x, y)]$. Then, this product is convolved by the impulse response $h(x, y)$ whose phase-only transfer function $H(\mu, \nu) = \exp[i2\pi b(\mu, \nu)]$ is denoted as the Fourier plane phase mask. Thus, the double-phase encrypted information, $\psi(x, y)$, is given by

$$\psi(x, y) = \{f(x, y) \exp[i2\pi\, p(x, y)]\} * h(x, y), \qquad (12.1)$$

where $*$ denotes the convolution operation. The encrypted signature $\psi(x, y)$ is a complex-valued function that needs to be represented by its magnitude and phase distributions. This encryption technique may be implemented electronically or optically. Figure 12.2(a) shows a diagram of the optical setup capable of ciphering the information.

In order to decrypt the encoded signature, $\psi(x, y)$, its Fourier transform must be multiplied by the complex conjugated phase mask $\exp[-i2\pi b(\mu, \nu)]$

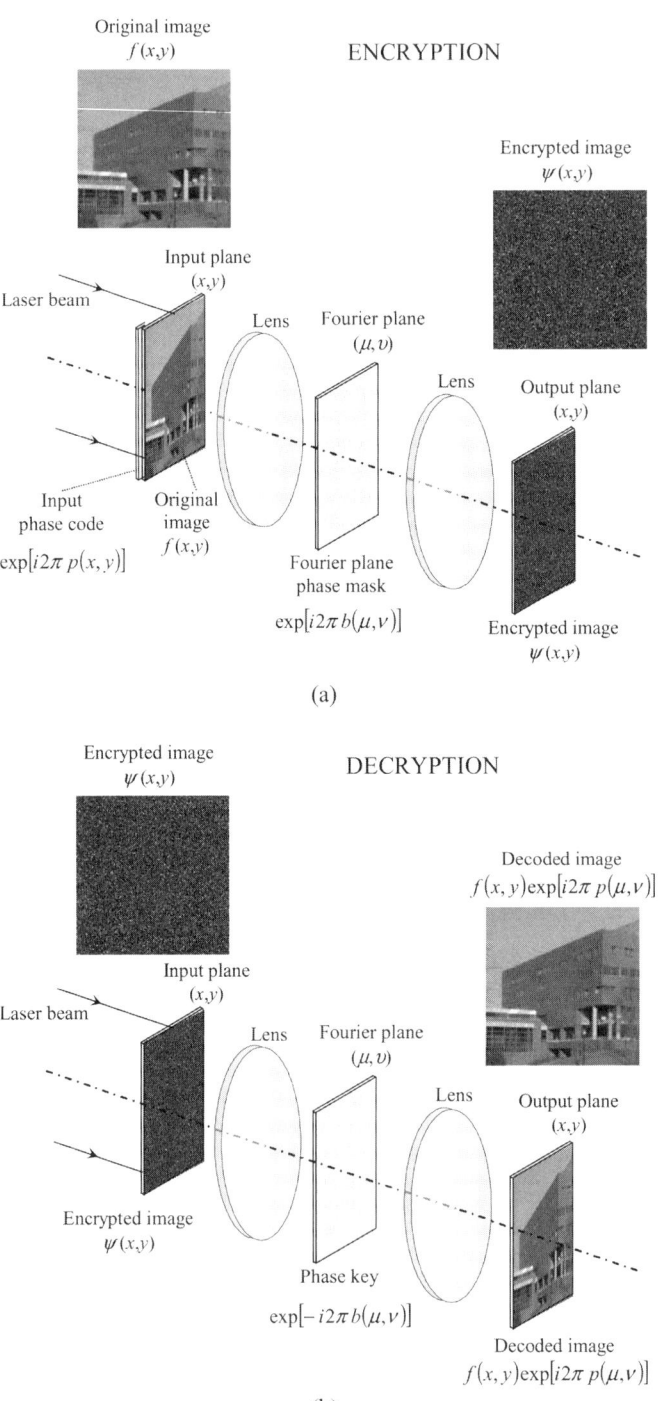

Figure 12.2. Optical setup for double phase: (a) encryption and (b) decryption.

and then inverse Fourier transformed, which produces [6]

$$IFT \{FT [\psi (\mu, \nu)] \exp [-i2\pi b (\mu, \nu)]\} = f (x, y) \exp [i2\pi p (x, y)] . \quad (12.2)$$

Finally, multiplication by the conjugated phase mask $\exp [-i2\pi p (x, y)]$ will recover $f (x, y)$. Alternatively, because $f (x, y)$ is real and positive in general, the signature can be recovered by computing the magnitude of $f (x, y) \exp [i2\pi p (x, y)]$ or by using an intensity-sensitive device such as a video camera or a CCD camera.

From equation (12.2) it can be seen that the encoded signature can only be decrypted when the corresponding phase code $\exp [i2\pi b (\mu, \nu)]$ is known by the processor and used for the decryption. We refer to this phase code as the key.

The same optical setup used for the encryption procedure, can also be used for decrypting the signature. Figure 12.2(b) shows the corresponding steps carried out in the decoding process.

A different encryption technique, the fully phase encryption method [8], has some advantages in terms of noise robustness when additive Gaussian noise corrupts the ID tag [7]. If the fully phase encryption is applied, the images or signatures are represented as phase-only functions [12] in the double-phase encryption technique [6]. That is, the phase encoded signature $\exp [i\pi f (x, y)]$ is introduced in equation (12.1) instead of $f (x, y)$.

12.4 Distortion-Invariant ID Tags

We aim to achieve a detection and verification of the encrypted signature included in the ID tag, even when the receiver captures it from an unexpected location and orientation. To do so, the proposed ID tag should be invariant to distortions such as scale variations and rotations. A number of contributions that deal with scale and rotation-invariant systems have already been made in the literature for a wide variety of purposes [13–23]. In general, sophisticated methods are needed to achieve enough tolerance to different distortions simultaneously. Information of several distorted views of a given target can be included in the design of a filter to obtain a distortion-tolerant system. The level of complexity of the recognition system usually increases notoriously with the number of considered distortions. In this work, distortion invariance is achieved by both multiplexing the information included in the ID tag and taking advantage of the ID tag topology. This procedure permits certain reduction of the system complexity [2–5, 7].

The complex-valued encrypted signature $\psi (x, y) = |\psi (x, y)| \exp \{i\phi_\psi (x, y)\}$ can be represented either by a single optical code [2–4], or by two optical codes [5,7]. In the last case, the real-valued magnitude function and phase distribution of $\psi (x, y)$ are separately reproduced in the final ID tag. This second representation makes the authentication process more robust, particularly when remote identification and verification of objects have to be made outdoors. In such a case, a large number of environmental conditions may affect the information contained in the ID tag as they introduce noise or even invalidate the captured signal. For instance, shading effects or nonuniform illumination affect mainly the signal magnitude, and air turbulences or rain affect the signal phase. Moreover, the phase content of the signature can be easily neutralized and the ID tag sabotaged if an adhesive transparent tape is stuck on it. For all these reasons, it appears to be convenient to further encode the phase content of the signal in intensity variations. Thus, we consider encoding both the magnitude and phase of function $\psi(x, y)$ in grayscale values [5,7].

Let us consider the encrypted signature $\psi(x, y)$ in array notation $\psi(t) = |\psi(t)| \exp \{j\phi_\psi(t)\}$ where $t = 1, 2, \ldots, N$, and N is the total number of pixels of the encrypted signature (Fig. 12.3). We build two vectors: the magnitude vector $|\psi(t)|$ and the phase vector $\phi_\psi(t)$. The information included in the ID

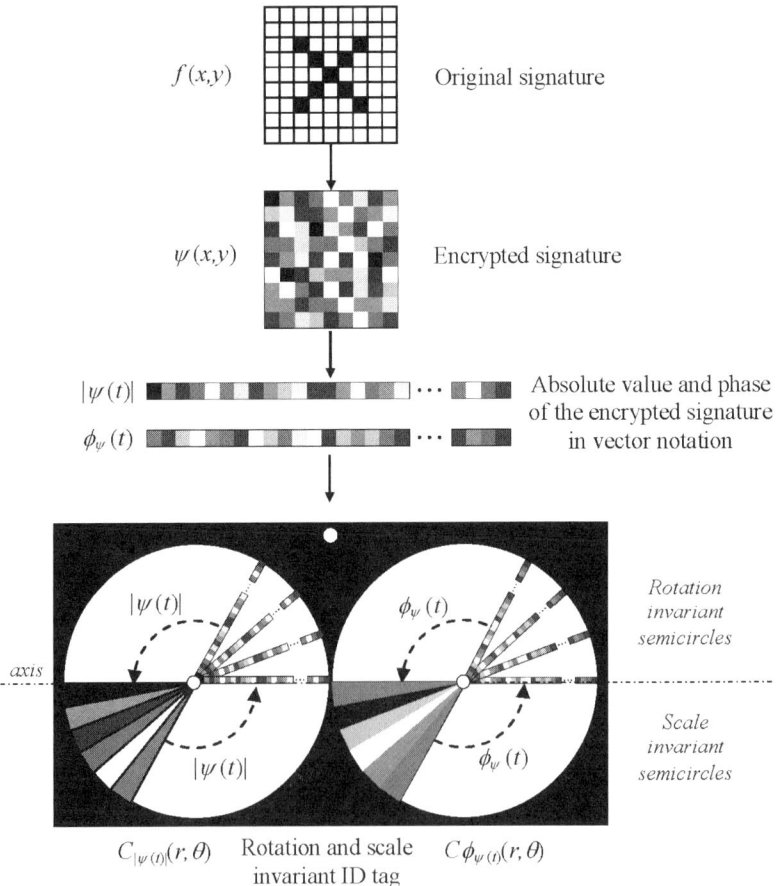

Figure 12.3. Design of a rotation- and scale-invariant ID tag.

tags is distributed in two circles. Different possibilities can be considered to rearrange the information contained in these two circles of the ID tags. In our case, one circle contains the magnitude vector $|\psi(t)|$ and the other circle the phase vector $\phi_\psi(t)$ (Fig. 12.3). The upper semiplane of both circles is the area for rotation-invariant identification whereas the bottom semiplane is the area for scale-invariant identification, just following a distribution similar to that considered in Refs. [2–4]. In an alternative rearrangement, one circle allows the rotation-invariant readout of both the magnitude and phase of the encrypted signature and the other circle the scale-invariant readout of the same signal [5, 7]. The choice of a particular distribution of the signal information depends on practical considerations of a given problem.

Following with the arrangement of the signal shown in Figure 12.3, the circle on the left of the ID tag consists of the magnitude of the encrypted signature $\psi(t)$ written in a radial direction and repeated angularly in the top semicircle so that rotation invariance could be achieved. The bottom semicircle of the optical code on the left contains the same signal, i.e., the magnitude of the encrypted signature, but written circularly and repeated in concentric rings. Therefore, in this area of the circle, the information of a given pixel of the encrypted signature will correspond to a sector of a semicircle in the optical code. Thus, the readout of the ciphered information in the bottom part of the circle will be tolerant to variations in scale.

The second circle of the ID tag, the optical code shown on the right of Fig. 12.3, contains the phase of the encrypted signature $\psi(t)$ with a distribution of the information similar to that described for the circle on the left. Figure 12.3 shows a possible arrangement of both circles. Their centers are white dots that, along with a third white dot in the upper part, build a reference triangular-shaped pattern. The triangle basis or longest side establishes an axis (the horizontal axis in Fig. 12.3) and the triangle vertex defines the semiplane (upper semiplane in Fig. 12.3) where rotation invariance is achieved. The bottom semiplane of both circles is the area that provides scale invariance.

A mathematical description of the proposed ID tag, in polar coordinates, is done according to the circle $C_{|\psi(t)|}(r, \theta)$ corresponding to the absolute value of the encrypted signature, and to the circle $C_{\phi_\psi(t)}(r, \theta)$ corresponding to its phase distribution.

The optical code located on the left of the ID tag corresponds to the magnitude of the encrypted signature and can be written in polar coordinates as

$$
C_{|\psi(t)|}(r, \theta) = \begin{cases} |\psi(r)|, & \text{for } r = (1, \ldots, N)\left(\dfrac{R_L - R_0}{N}\right), & \forall \theta \in (0, \pi], \\ |\psi(\theta)|, & \text{for } \theta = (N+1, \ldots, 2N)\left(\dfrac{\pi}{N}\right), & \forall r \in (R_0, R_L] \\ V_{\max}, & \text{for } r \leq R_0 \quad \text{(white central dot)}, \end{cases}
$$

$$(12.3)$$

where the functions $|\psi(r)|$ and $|\psi(\theta)|$ are discretized in 2^n values, and $V_{\max} = 2^n$ is the maximum value of the grayscale (white). In our simulations, we will consider $n = 8$, that is, a 8-bit grayscale. In equation (12.3), R_L is the radius of the ID tag circle, and R_0 is the radius of the white central dot.

The optical code on the right of the ID tag corresponds to the information of the phase of the encrypted signature and is mathematically described in polar coordinates as

$$
C_{\phi_\psi(t)}(r,\theta) = \begin{cases} \phi_\psi(r), & \text{for } r = (1,\ldots,N)\left(\dfrac{R_L - R_0}{N}\right), & \forall \theta \in (0,\pi] \\[2ex] \phi_\psi(\theta), & \text{for } \theta = (N+1,\ldots,2N)\left(\dfrac{\pi}{N}\right), & \forall r \in (R_0, R_L] \\[2ex] V_{\max}, & \text{for } r \leq R_0 & \text{(white central dot)}. \end{cases}
$$

(12.4)

In equations (12.3) and (12.4), N indicates either the radial partition (upper semicircle) or the angular partition (lower semicircle), which is limited by the receiver resolution at the smallest pixels that surround the central dot. Since the receiver resolution is not generally known a priori, the image of the triangular-shaped pattern consisting of three white dots can be used as a reference to know if the receiver has enough resolution to read the encrypted information. For instance, the ID tags can be designed to ensure an appropriate readout for those receivers that measure a distance between the circle centers (or the triangle basis) greater than a certain value. The triangle pattern could give information about scale and rotation and, therefore, one could think that there is no need to codify the encrypted signature $\psi(x, y)$ in the distortion-invariant ID tags defined by equations (12.3) and (12.4). But we must take into account that if the encrypted signature $\psi(x, y)$, written in a matrix array similar to the one shown in Fig. 12.3, is affected by rotation and/or scale variation, then it needs to be sampled again and rearranged into the matrix form before decryption. This operation entails interpolations that can produce errors such as aliasing. For this reason, we consider that the distortion-invariant ID tags, provided they are correctly built, allow more accurate readouts of the encrypted information under rotation and/or scale variations.

The encrypted information is recovered following the procedure depicted in Fig. 12.4. In each circle, the border between the rotation-invariant and the scale-invariant regions is determined by the axis defined by the two circle centers. Once this border is determined, the third white dot marks the semiplane where the rotation invariance is achieved (upper semiplane in Fig. 12.4). The other semiplane corresponds to the scale-invariant region (bottom semiplane in Fig. 12.4). The signature in vector notation $\psi(t)$ can be decoded by reading out the information of the optical code from either the rotation-invariant or the scale-invariant areas. The magnitude and phase of the encrypted signature are extracted from both circles of the ID tag.

From the semicircles corresponding to the rotation-invariant areas, the magnitude and phase of the encrypted signature could be read out by using a linear array detector placed in any radial direction of the imaged semicircles. The semicircle on the left provides the information of the magnitude

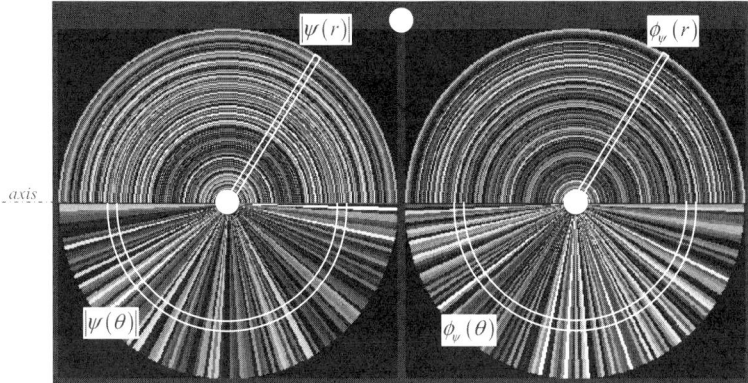

Figure 12.4. Reading out process from the rotation- and scale-invariant ID tag.

and the semicircle on the right, the phase distribution. Not only is a single pixel value read along a unique radius, but a set of different radial codes are read to increase noise robustness. One possibility consists of computing the code that is the mean value of this set [2-4]. Other results can be obtained if the median value is used instead [5, 7]. Both possibilities have been used in previous works, and we are going to compare their performance in this chapter.

From the semicircles corresponding to the scale-invariant areas, the magnitude and phase of the encrypted signature in vector notation $\psi(t)$ are recovered by reading out the pixels of the ID tag in semicircular rings. To minimize errors in the reading process, not only is one pixel taken into account for each position of a single semicircular ring, but also a set of pixels located in neighbor concentric rings in the radial direction. Again, the results obtained taking either the mean value or the median value are compared and analyzed in this work.

The retrieved information of the pixels should be written back into matrix notation prior to decoding the signature by using the decryption technique [6]. Following this procedure, the encrypted signature $\psi(x, y)$ will be recovered when the ID tag is captured in its original orientation and size or in rotated or scaled formats.

For encrypted signatures with a large number of pixels (Fig. 12.5), the information of the scale-invariant region can be distributed using different concentric semicircles to assure a minimum number of pixels for each sector to properly recover the information (Fig. 12.5d). Consequently, the tolerance to scale variation is affected in accordance to the number of concentric semicircles used in the ID tag. In such a case, the procedure to recover the encrypted signature is basically the same, but the existence of concentric semicircles and their size must be taken into account in the readout.

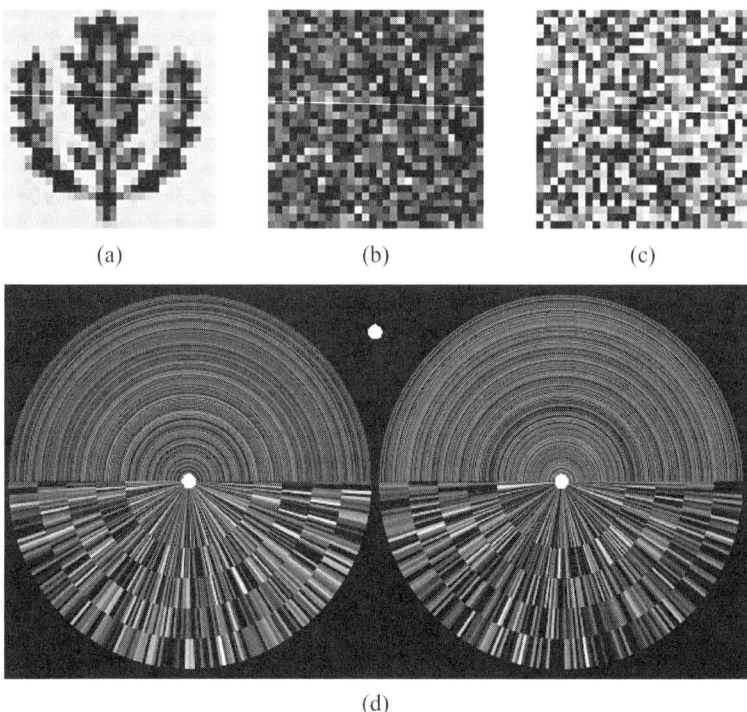

(a)   (b)   (c)

(d)

Figure 12.5. (a) Signature $f(x,y)$ that identifies the object under control. (b) Magnitude and (c) Phase distributions of the encrypted signature $\psi(x,y)$ by using the double-phase encryption technique. (d) Distortion-invariant ID tag built from the encrypted signature. The information of the optical code in the scale-invariant tag is distributed in concentric circles.

12.5 Signature Verification

To verify the authentication of an object, a final step of verification of the captured and decrypted signature is necessary. A correlation-based processor [9, 24] compares the decoded information with a previously stored reference signal. A comparison of these two functions is based on a nonlinear correlator [25].

Let $f(x,y)$ denote the decoded signal and $r(x,y)$ the reference signature. Let $|F(\mu,\nu)|$ and $|R(\mu,\nu)|$ be the modulus of their Fourier transforms, respectively, and let $\phi_F(\mu,\nu)$ and $\phi_R(\mu,\nu)$ denote their phase distributions in the frequency domain. The nonlinear correlation between the input and the reference signals is obtained by using the equation

$$c(x,y) = IFT\left\{|F(\mu,\nu)R(\mu,\nu)|^k \exp\left[i\left(\phi_F(\mu,\nu) - \phi_R(\mu,\nu)\right)\right]\right\}, \quad (12.5)$$

where the parameter k defines the strength of the applied nonlinearity [25]. We use kth-law nonlinearity for computational efficiency. The nonlinearity determines the performance features of the processor, such as its discrimination

capability, noise robustness, peak sharpness, etc. and it can be chosen according to the performance required for a given recognition task [25–27]. Optimum
nonlinear transformations allow one to enhance the detection process by optimizing a given performance metric [28].

Correlation-based detection is feasible when an output peak above a noise
floor is obtained. A threshold operation, applied to the correlation output,
determines the identity of the object. The processor performance can be evaluated using different metrics. The metrics that are taken into account in this
work are well-known parameters described in the literature [29–31]. We consider, as a measure of the system discrimination capability, the cc/ac metric,
which is the ratio between the maximum peak value of the correlation output,
cc, and the maximum autocorrelation value, ac, for the reference signature.
Similarity between the decoded information and the reference signature will
be great if the cc/ac ratio approaches the value of unity.

Another possibility is to directly evaluate the quality of the recovered
image, $f(x, y)$, from the decryption procedure in comparison with a reference
signature $r(x, y)$ by using the mean-squared error (mse) [8]:

$$mse\left[f\left(x, y\right)\right] = E\left\{\frac{1}{N}\sum_{x=0}^{N_x}\sum_{y=0}^{N_y}\left[\left|f\left(x, y\right) - r\left(x, y\right)\right|^2\right]\right\}, \qquad (12.6)$$

where $N = N_x \cdot N_y$ is the total number of pixels of the original image and
$E\{\cdot\}$ is the expected value.

12.6 Authentication Results

To explore the feasibility of the proposal, the designed distortion-invariant ID
tag was tested in the presence of a unique distortion (either rotation or scale
variations). Afterwards, results were obtained for both distortions modifying
the ID tag simultaneously. In the readout process of the information included
in the ID tag, two mathematical operations are separately analyzed and compared: the mean and median values, as they were introduced in Section 12.4.

12.6.1 Rotation Invariance

To test the tolerance of the verification system to rotations of the ID tag, we
digitally rotate the ID tag shown in Fig. 12.5(d) from 0 to 360° in steps of
20°. For all the rotated ID tags, encrypted signatures in vector notation $\psi(t)$
are recovered from the rotation-invariant area (upper semicircle) of the ID
tag following the procedure described in Section 12.4. Then, a series of decrypted signatures $f(x, y)$ are obtained by using the double-phase decryption
technique [6]. Some of these decrypted signatures are depicted in Fig. 12.6.
Results shown in Fig. 12.6(a) correspond to the case where the mean value of
a set of pixels is considered in the readout process. The decrypted signatures
depicted in Fig. 12.6(b) are the results obtained when the median value is
computed during the readout.

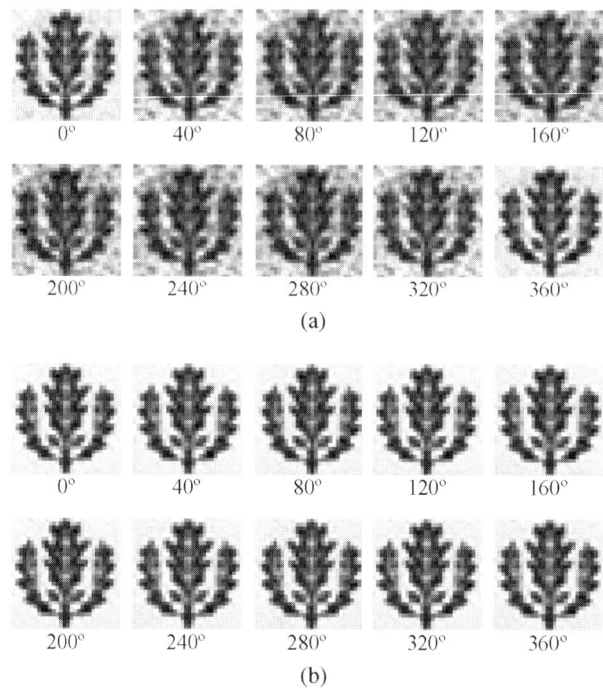

Figure 12.6. Decrypted signatures obtained from rotated versions of the ID tag shown in Fig. 12.5(d) when either (a) the mean value or (b) the median value are computed during the readout process.

From results depicted in Fig. 12.6, one can appreciate that the signature is better deciphered if the median is considered instead of simply the mean value. If the mean value is applied, retrieved signatures are noisy images. When the median is used, however, the pixel values are better preserved for the encrypted signature and the double-phase decryption procedure obtains a signature closer to the original image. To make these differences clearer, the *mse* is graphed versus the rotation angle of the ID tag in Fig. 12.7. This graph points out an improvement in the process of reading out the information from the distortion-invariant ID tag if the median value is computed from a set of pixels. If the mean value is used instead, the difference between the retrieved signature and the original increases, except for rotation angles that do not need interpolation algorithms ($0°, 180°$, and $360°$).

12.6.2 Scale Invariance

Invariance to scale variations is tested by using the lower semicircle of both optical codes of the ID tag (Fig. 12.5(d)). This case occurs, for instance, when the ID tag is captured at different distances from the receiver. We have digitally scaled the ID tag by a factor ranging from 0.2 to 2 in steps of 0.1.

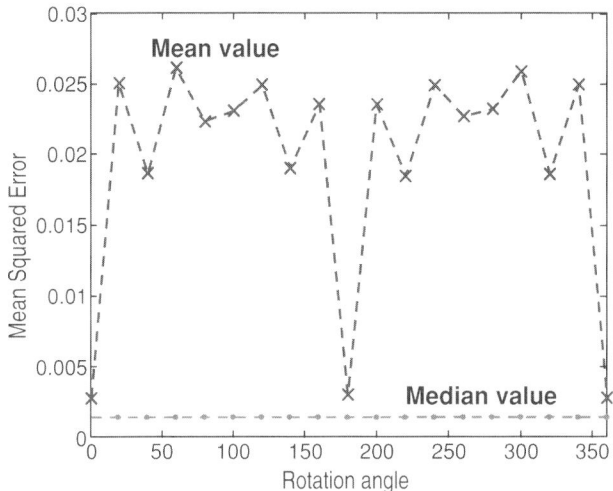

Figure 12.7. Parameter *mse* versus rotation angle of the ID tag. Comparison between the use of the mean value ($\times$) or the median ($\cdot$) in the readout algorithm.

A series of the decrypted signatures obtained from this test are shown in Fig. 12.8. The quality of the recovered signature is visually acceptable in nearly all cases for both procedures, using either the mean value in the readout process or the median. When the ID tag is captured from a long distance (that is, if small scale factors are used), the noise level of the decoded images increases rapidly and the signature is not properly deciphered. In addition, we recall that the system tolerance to scale variations is limited due to the high number of pixels N, which leads to using concentric semicircles in the ID tag.

Figure 12.9 shows the mean-squared error computed between each retrieved signature of Fig. 12.8 and the original function (Fig. 12.5(a)). This graph evidences the differences between the mean and median calculation in the algorithm. If the median is considered, the decoded signature approaches better the original, even though small scale factors are applied. The quality of the decoded signature decreases quickly for scale factors smaller than 0.5.

12.6.3 Simultaneous Rotation and Scale Invariance

Finally, the identification system is tested against rotation and scale distortions when both of them affect simultaneously the two circles of the ID tag. Figure 12.10 displays the output planes of the recognition system along with the decoded signatures obtained for a rotated and scaled version of the ID tag shown in Fig. 12.5(d). The median value is computed in the distortion-invariant readout. In all the cases, the signature has been correctly decoded

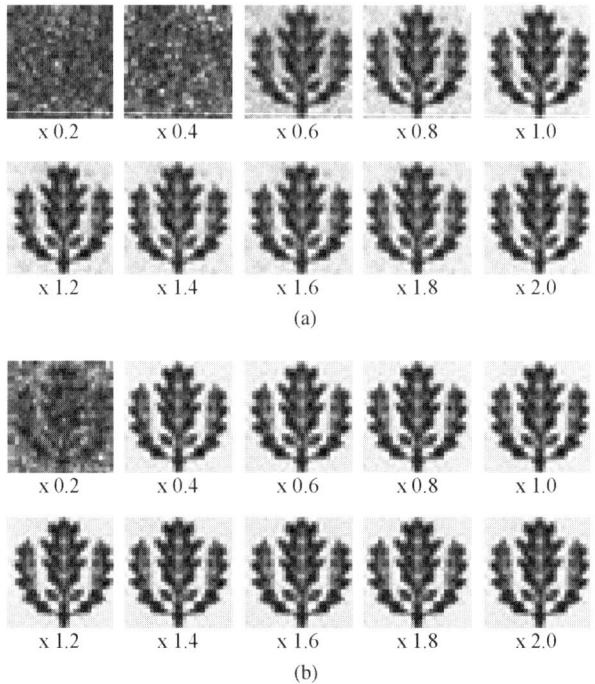

Figure 12.8. Decrypted signatures from scaled versions of the ID tag shown in Fig. 12.5(d) where (a) the mean value or (b) the median value are computed during the readout process.

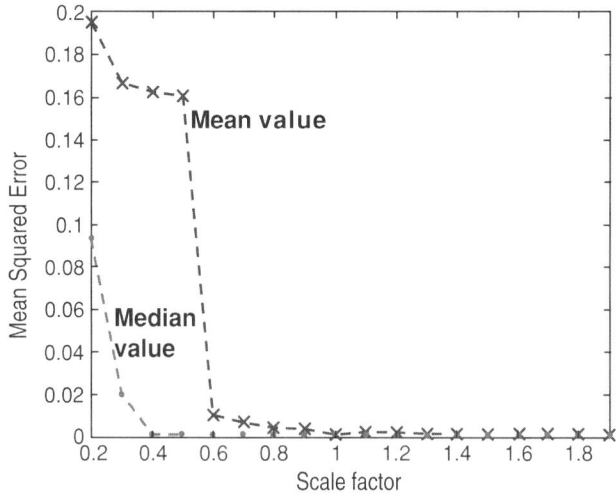

Figure 12.9. Parameter *mse* versus rotation angle of the ID tag. Comparison between the use of the mean value (×) or the median (·) in the readout algorithm.

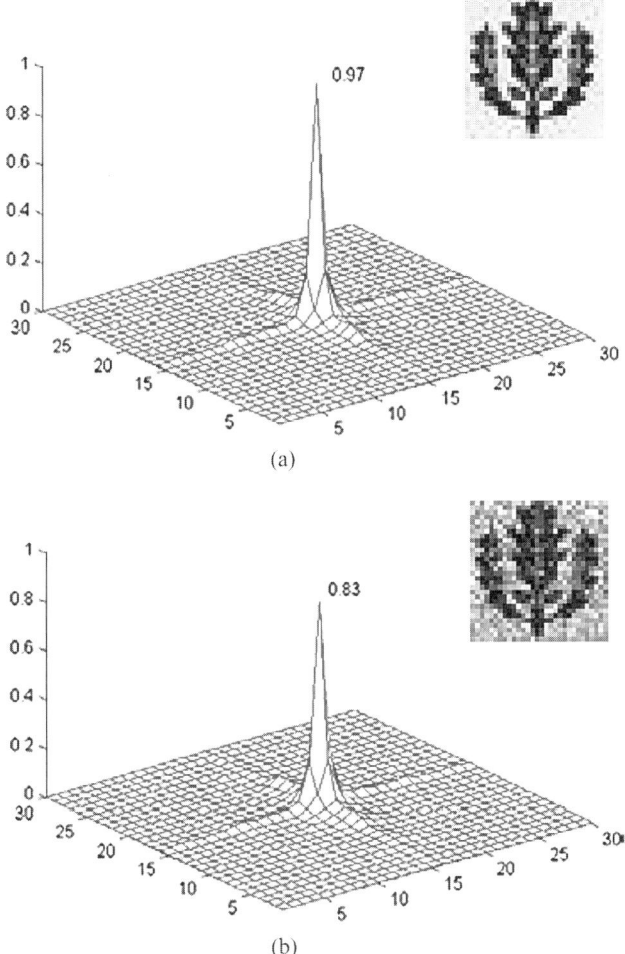

Figure 12.10. Decoded signatures and correlation outputs (with $k = 0.5$) for rotated and scaled versions of the ID tag shown in Fig. 12.5(d). (a) Scale factor $\times 0.6$ and rotation angle $40°$; (b) scale factor $\times 0.5$ and rotation angle $60°$.

and identified using a nonlinearity given by $k = 0.5$ for correlation. However, it can be appreciated that the level of noise increases with the distortion.

To further demonstrate the robustness of the ID tags for verification and identification, let us recover the decrypted information from a rotated ($40°$) and scaled (0.6 scale factor) ID tag, and let us decrypt the encoded information by using a false phase key. As a result, we obtain a noisy image where no signature can be recognized (Fig. 12.11(a)).

It is also important to demonstrate that a different signature, even if it is correctly decrypted with the appropriated phase key, will not be recognized

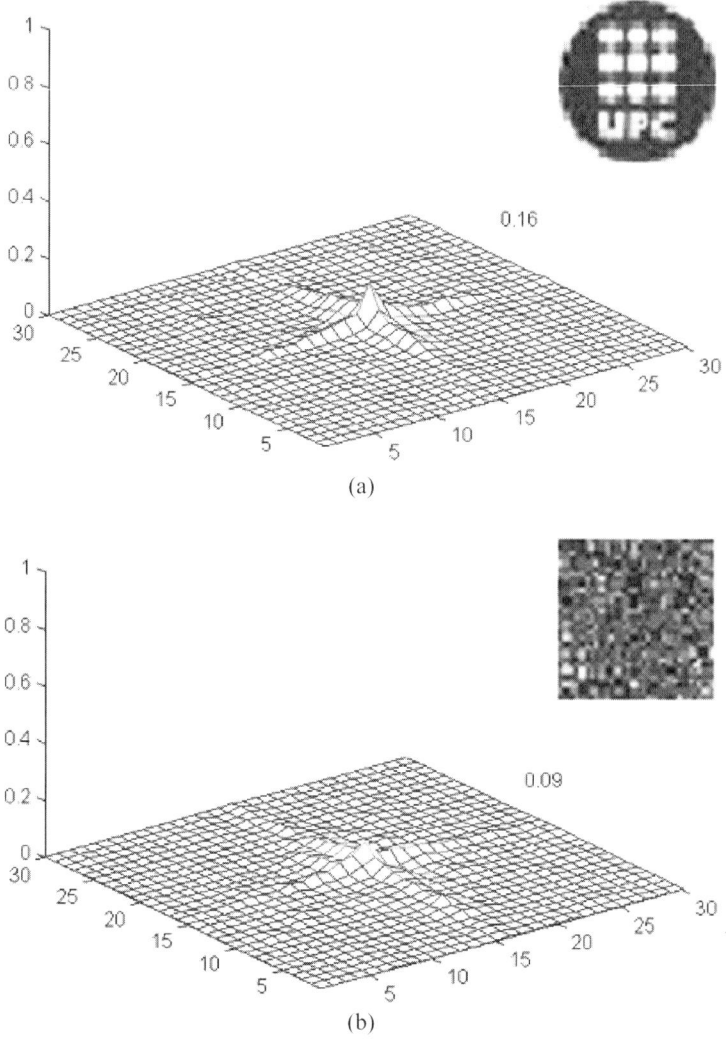

Figure 12.11. (a) Decoded image by using a false key and correlation output for $k = 0.5$. In addition, the ID tag was rotated $40°$ and scaled by a factor of 0.6. (b) Decoding a different signature with the correct phase key. The ID tag was rotated $40°$ and scaled by a factor of 0.6. Correlation output for $k = 0.5$ when the decoded image is compared with the stored reference image (Fig. 12.5(a)).

as the reference image. Figure 12.11(b) presents the decoded signature corresponding to a different logo, and the corresponding correlation output with a low peak which is clearly below the threshold. Thus, the decoded signature is discriminated from the authentic one (Fig. 12.5(a)) used in the previous experiments.

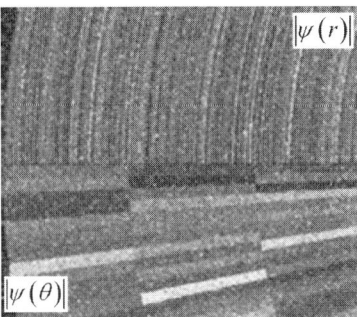

Figure 12.12. Magnified sector of the ID tag corrupted by *Salt & Pepper* noise that modifies 15% of the ID tag pixels.

12.6.4 Noise Robustness

Finally, the proposed verification system is tested in the presence of noise. The captured ID tag is corrupted by *Salt & Pepper* noise, due to the fact that is a common noise for electronic devices which conform the receiver. The corrupted pixels take either the maximum or the minimum value possible in the capturing device.

Figure 12.12 shows a part of the ID tag degraded by *Salt & Pepper* noise in 15% of its pixels. Both optical codes, the one containing the magnitude and the one containing the phase of the encrypted signature $\psi(x,y)$, are equally modified.

Figure 12.13 shows the output results when the noisy ID tag of Fig. 12.12 is captured by the receiver with a rotation of 40° and scaled by × 0.6.

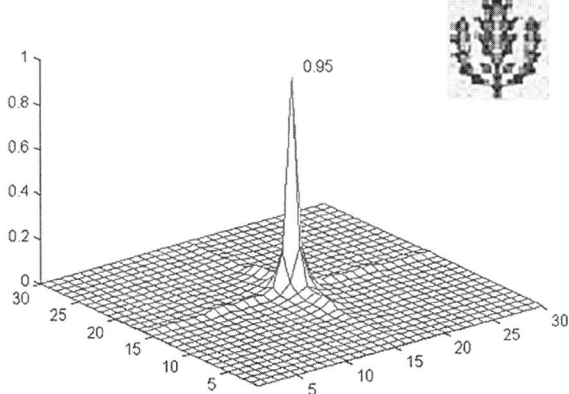

Figure 12.13. Decoded signature and correlation output ($k = 0.5$) for a corrupted ID tag by *Salt & Pepper* noise (15% of pixels) and captured under rotation (40°) and a variation in scale (0.6).

Figure 12.13 depicts both the decoded signature and the corresponding output plane from the nonlinear correlator with $k = 0.5$.

Even in the presence of *Salt & Pepper* noise degrading the signal, the signature is correctly verified.

12.7 Conclusions

We have proposed a method to encode an encrypted signature into an ID tag to provide invariance to rotation and scale distortions. The signature is a characteristic image that allows identification of an object. Thus, identification tags can be used for real-time remote verification and authentication of objects which have diverse applications in transportation and homeland security. The ID tags consist of an optical code containing double-phase encrypted information to increase security. This encrypted information is replicated radially and angularly so as to build an ID tag that can be read even if it appears scaled or rotated. Both the magnitude and the phase of the encrypted signature are codified in grayscale to improve robustness against phase distortions produced by outdoors environmental conditions (rain, air turbulences, etc.)

The designed ID tag can be located on a given object, and is captured by a receiver, which will decode and verify the information. To decipher the information of the tag, two algorithms are used and compared. One uses the mean value of a set of pixels. The other uses the median value instead. Authentication results provided in this chapter show that clear improvement, in terms of recognition performance and noise robustness, is achieved when the median value is applied.

Numerical results demonstrate that the proposed system is able to recover a given signature even if the ID tag is rotated, scaled, or both rotated and scaled simultaneously. The method used for encrypting the signature has been shown to be robust against using a different key for the decryption technique. Also, the receiver is able to discriminate between a reference signature and a different image by using correlation. Moreover, satisfactory authentication results are obtained in the presence of Salt and Pepper noise.

References

[1] B. Javidi. Real-time remote identification and verification of objects using optical ID tags. *Opt. Eng.*, 42(8):1–3, 2003.
[2] E. Pérez-Cabré and B. Javidi. Scale and rotation invariant optical ID tags for automatic vehicle identification and authentication. *IEEE Trans. Veh. Technol.*, 54(4):1295–1303, 2005.
[3] E. Pérez-Cabré, and B. Javidi. Distortion-invariant ID tags for object identification. *Proc. SPIE* 5611:33–41, 2004.
[4] E. Pérez-Cabré, B. Javidi, and M. S. Millán. Detection and authentication of objects by using distortion-invariant optical ID tags, *Proc. SPIE.* 5827:69–80, 2005.

[5] E. Pérez-Cabré, M. S. Millán, and B. Javidi. Remote object authentication using distortion-invariant ID tags. *Proc. SPIE*, 5908:59080N-1–59080N-13, 2005.

[6] Ph. Réfrégier and B. Javidi. Optical image encryption base don input plane and Fourier plane random encoding. *Opt. Lett.* 20(7):767–769, 1995.

[7] E. Pérez-Cabré, M. S. Millán, B. Javidi. Remote optical ID tag recognition and verification using fully spatial phase multiplexing. *Proc. SPIE*. 5986:598602-1–598602-13, 2005.

[8] N. Towghi, B. Javidi, Z. Luo. Fully phase encrypted image processor. *JOSA A*. 16(8):1915-1927, 1999.

[9] J. W. Goodman. *Introduction to Fourier Optics*, 2nd ed. McGraw-Hill, New York, 1996.

[10] B. Javidi, L. Bernard, N. Towghi. Noise performance of double-phase encryption compared with XOR encryption. *Opt. Eng.* 38:9–19, 1999.

[11] F. Goudail, F. Bollaro, Ph. Réfrégier, B. Javidi. Influence of perturbation in a double phase encoding system. *JOSA A*. 15:2629-2638, 1998.

[12] B. Javidi, A. Sergent. Fully phase encoded key and biometrics for security verification. *Opt. Eng.* 36(3):935-942, 1997.

[13] A. Mahalanobis, A review of correlation filters and their application for scene matching. Optoelectronic Devices and Systems for Processing. Critical Review *of Optical Science Technology*, SPIE, Bellingham, WA, pp. 240–260, 1996.

[14] *IEEE Trans. Image Process.* Special issue on Automatic Target Detection and Recognition, 6(1), 1997.

[15] B. Javidi, ed. *Smart Imaging Systems*. SPIE Press, SPIE, Bellingham, WA, 2001.

[16] B. Javidi, ed. *Image Recognition and Classification: Algorithms, Systems and Applications*. Marcel Dekker, New York, 2002.

[17] C. F. Hester, D. Casasent. Multivariant technique for multiclass pattern recognition, *Appl. Opt.* 19(11):1758-1761, 1980.

[18] H. J. Caulfield. Linear combinations of filters for character recognition: a unified treatment, *Appl. Opt.* 19, 3877–3879, 1980.

[19] H. Y. S. Li, Y. Qiao, D. Psaltis. Optical network for real-time face recognition, *Appl. Opt.* 32(26):5026-5035, 1993.

[20] T. D. Wilkinson, Y. Perillot, R. J. Mears, J. L. Bougrenet de la Tocnaye, Scale-invariant optical correlators using ferroelectric liquid-crystal spatial light modulators, *Appl. Opt.* 34(11):1885-1890, 1995.

[21] B. Javidi, D. Painchaud. Distortion-invariant pattern recognition with Fourier-plane nonlinear filters, *Appl. Opt.* 35(2):318–331, 1996.

[22] L. C. Wang, S. Z. Der, N. M. Nasrabadi. Automatic target recognition using feature-decomposition and data-decomposition modular neural networks, *IEEE Trans. Image Process.* 7(8):1113–1121, 1998.

[23] E. Pérez, B. Javidi. Nonlinear distortion-tolerant filters for detection of road signs in background noise, *IEEE Trans. Veh. Technol.* 51(3):567–576, 2002.

[24] J. L. Turin. An introduction to matched filters, *IRE Trans. Inf. Theory*. 6, 311–329, 1960.

[25] B. Javidi. Nonlinear joint power spectrum based optical correlation, *Appl. Opt.* 28(12):2358-2367, 1989.

[26] M. S. Millán, E. Pérez, K. Chalasinska-Macukow. Pattern recognition with variable discrimination capability by dual non-linear optical correlation, *Opt. Commun.* 161, 115–122, 1999.

[27] E. Pérez, M. S. Millán, K. Chalasinska-Macukow. Optical pattern recognition with adjustable sensitivity to shape and texture, *Opt. Commun.* 202, 239–255, 2002.

[28] S. H. Hong, B. Javidi. Optimum nonlinear composite filter for distortion-tolerant pattern recognition, *Appl. Opt.* 41(11):2172–2178, 2003.

[29] B. Javidi, J. L. Horner. *Real-Time Optical Information Processing.* Academic Press, Boston, 1994.

[30] J. L. Horner. Metrics for assessing pattern recognition performance, *Appl. Opt.* 31(2):165–166, 1992.

[31] B. V. K. Vijaya Kumar, L. Hassebrook. Performance measures for correlation filters, *Appl. Opt.* 29(20):2997–3006, 1990.

Speckle Elimination With a Maximum Likelihood Estimation and an Isoline Regularization

Nicolas Bertaux,[1] Yann Frauel,[2] and Bahram Javidi[3]

[1]Institut Fresnel—Equipe Φ-TI—Unité Mixte de Recherche 6133
Domaine Universitaire de Saint Jérôme—13397 Marseille—France
[2]Instituto de Investigaciones en Matemáticas Aplicadas y en Sistemas
Universidad Nacional Autónoma de México, México, DF, Mexico.
[3]Electrical and Computer Engineering Department
University of Connecticut, Storrs, CT 06269-1157, USA

13.1 Presentation

Although most common imaging systems are based on incoherent optics, several important imaging techniques rather use coherent light to illuminate the objects under study. Such coherent imaging systems are useful in a variety of applications. For instance, Synthetic Aperture Radar (SAR) images [1] are used for environmental monitoring, detection of changes between two dates, navigation and guidance, etc. Another class of coherent images is obtained by systems involving a laser illumination. For example, holographic images [2] can be used for three-dimensional object visualization and recognition [3–16].

The main drawback of coherent images is that they are corrupted by a strong random speckle noise [17]. This noise is caused by random interferences that are due to the roughness of the studied objects and that result in high spatial-frequency wavefront deformations. The speckle pattern superimposed on the images makes it very difficult to use these images in many applications. For instance, a correlation between two identical images with different speckle patterns will be corrupted because of the uncorrelated speckle patterns. Similarly, speckle deteriorates most commonly used methods of detection or estimation. For many applications, it is therefore essential to filter out the speckle noise before any further image processing.

Most existing approaches to speckle removal are based on an image model with constant reflectivity in a patchwork of regions also known as mosaic model. This image model is adapted for SAR images. For coherent optical images in general, the reflectivity of the objects in the scene can be continuously variable. In this chapter, we present a speckle removal algorithm designed for image with continuously varying gray-levels [18]. Although we will focus on digital holographic images, the proposed technique can be used for other types of coherent images. The proposed method is based on a maximum likelihood estimation with a general model for the image reflectivity. The image model is based on a lattice of nodes with linear interpolation of the gray-levels between nodes. Moreover, to improve the quality of the result, a constraint on gray-level isolines is introduced. This constraint allows us to obtain smooth results without blurring the edges of the objects in the image.

In this contribution, we will first describe the Maximum Likelihood (ML) estimator derived from the speckle modelization and we will present the image model defined as a continuous lattice based on elementary triangles. Next, we will describe the algorithm designed for ML criterion optimization. Then we will introduce a regularization term based on gray-level isolines of the restored image. Lastly, we will show results of the algorithm and we will analyze its robustness and dependence on the method parameters.

13.2 Description of the Algorithm

13.2.1 Maximum Likelihood Estimator

Active coherent imaging is very helpful for many applications. In this contribution, we use images obtained through digital holography as examples. Digital holographic techniques allow the capture of 3D information about objects. In this case, a hologram is digitally recorded by a camera and then numerically processed to reconstruct arbitrary 2D views of the 3D object [4,6]. The proposed method can be similarly applied to other types of coherent images.

The main difficulty with coherent imaging is the random high spatial-frequency interference pattern (speckle) which cannot be perfectly predicted in general.

The complexity of radar or hologram scenes generally implies that the intensity level of the image must be considered as a random field. The reflectivity of the scene, which is the parameter of interest, has then to be recovered from local parameters such as the local mean, the standard deviation, or the whole probability density function (pdf) of the random field.

Usually, intensity images with speckle correspond to the square modulus of Gaussian circular complex fields. Complex field modelization could be obtained for instance from a wavefront near the object calculated with digital holographic data [4,6]. For SAR images, one can get intensity images by averaging several independent intensity acquisitions. In that case a simple but generally accurate speckle model for intensity images consists in considering that the gray-levels are realizations of independent randoms fields with Gamma pdf and mean value that is proportional to the local reflectivity in the image.

The gamma pdf of the observed gray-level x_n of pixel n is given by

$$p\left(x_n \,|i_n\right) = \left(\frac{L}{i_n^L}\right)\frac{x_n^{L-1}}{\Gamma(L)}\exp\left(-L\frac{x_n}{i_n}\right), \qquad (13.1)$$

where i_n is the mathematical expectation value of the gray-level at pixel n and L the order of the Gamma law. In this case, i_n corresponds to the gray-level of the object one would measure without speckle and is the wanted information. In the following we will assume that L is known. Indeed, the probability density function obtained with a complex Gaussian circular assumption on the amplitudes (which is the case of holography, for example) provides a gamma pdf for the intensity gray-levels with order 1 ($L = 1$). Furthermore, averaging L images leads to gamma pdf of order L.

In the following, we will assume that the gray-level pixels of the image are distributed with gamma pdf (Eq. (13.1)) and that they are statistically independent, i.e., without spatial correlation. With this hypothesis, the log-likelihood of the N pixel image $\mathbf{X} = \{x_n\}_{n\in[1,N]}$ can be written:

$$\ell\left(\mathbf{X}\,|\mathbf{I}\right) = \log\prod_{n=1}^{N}p\left(x_n\,|i_n\right), \qquad (13.2)$$

where $\mathbf{I} = \{i_n\}_{n\in[1,N]}$ and which, with Eq. (13.1), leads to

$$\ell\left(\mathbf{X}\,|\mathbf{I}\right) = C(L,\{x_n\}) - L\sum_{n=1}^{N}\left(\log(i_n) + \frac{x_n}{i_n}\right), \qquad (13.3)$$

where $C(L,\{x_n\})$ is a term independent of $\mathbf{I}$.

The Maximum Likelihood estimation of the intensity image $\mathbf{I} = \{i_n\}$ is then obtained by maximizing the log-likelihood [19] expression of Eq. (13.3):

$$\widehat{\mathbf{I}} = \arg\max_{\mathbf{I}}\ell\left(\mathbf{X}\,|\mathbf{I}\right). \qquad (13.4)$$

One can remark that the maximum likelihood estimation is independent of the L order of the gamma pdf. Indeed, the parameter L in Eq. (13.3) is present only in the additive constant $C(L,\{x_n\})$ and appears as the multiplicative factor of the second term including the parameter of interest $\mathbf{I}$.

13.2.2 Image Model

Equation (13.3) shows that without further hypothesis, the maximum likelihood estimator of the image $\mathbf{I}$ is the trivial result $\widehat{\mathbf{I}} = \mathbf{X}$. Indeed, the first

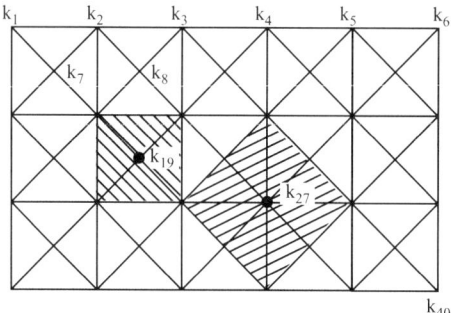

Figure 13.1. Lattice example and its two base polygons.

derivative of the log-likelihood with respect to i_n is

$$\frac{\partial \ell}{\partial i_n} = \frac{1}{i_n} - \frac{x_n}{i_n^2} = \frac{i_n - x_n}{i_n^2} \tag{13.5}$$

and this latter equation provides the solution:

$$\frac{\partial \ell}{\partial i_n} = 0 \Rightarrow i_n = x_n. \tag{13.6}$$

In order to remove the speckle, one can limit the set of interesting solutions to images with slow spatial variations. A first possibility consists in introducing an image model for **I**.

We consider a piecewise linear image model based on elementary triangles. Figure 13.1 shows an example of a lattice with two base polygons drawn arround the nodes k_{19} and k_{27}. The gray-level of each pixel of the modeled image is defined as being a linear interpolation between the gray-levels of the three vertices of the triangle to which it belongs. This linear variation between nodes is a constraint that imposes a slow spatial variation of the intensity. Let $\mathbf{K} = \{k_p\}_{p \in [1,P]}$ be a set of parameters which completely defines the image $\mathbf{I}(\mathbf{K})$. P is the number of nodes in the lattice and is thus equal to the number of parameters of this model. Let each node k_p of the lattice be described by three coordinates $(k_{x_p}, k_{y_p}, k_{z_p})$, where k_{x_p}, k_{y_p} correspond to the coordinates of its location in the image and k_{z_p} is the unknown corresponding gray-level. The parametric intensity image model can be written:

$$\mathbf{I} = \{i_n\} = \mathbf{I}\left(k_{z_1}, \dots, k_{z_p}, \dots, k_{z_P}\right). \tag{13.7}$$

The first advantage of using a simple lattice based on triangles is to lead to a low computational time. Indeed, determining the gray-level for each pixel of the image from the triangle lattice information $(k_{z_1}, \dots, k_{z_P})$ requires only $3N$ additions and $3N$ multiplications. This property is mainly due to the fact that local adjustments of the lattice result in local modifications of the gray-level of the nearest pixels. Indeed, a modification of the gray-level of one node only implies computations on the polygon area which surrounds this node.

For example, the nodes k_{19} and k_{27} of the lattice shown in figure Fig. 13.1 are associated to polygons respectively filled with hatchings.

The second advantage is the continuity of the gray-level of the image model $\mathbf{I} = \{i_n\}$ which is naturally introduced by the lattice design. This continuity property is important to smooth the image.

13.2.3 Optimization of the Log-Likelihood Criterion

In order to optimize the criterion of Eq. (13.3) with a low CPU load, we choose to implement an iterative second order algorithm. This iterative algorithm usually needs a small number of iterations and, moreover, one can remark that the triangular-lattice with linear interpolation is very helpful for quick determination of the criterion derivatives. Indeed, the first derivative of the log-likelihood criterion with respect to the gray-level k_{z_p} of the node k_p, is given by

$$\forall\, k_p \in \text{lattice} \qquad \frac{\partial \ell}{\partial k_{z_p}} = \sum_{n=1}^{N} \frac{\partial \ell}{\partial i_n} \frac{\partial i_n}{\partial k_{z_p}} = -L \sum_{n=1}^{N} \left(\frac{1}{i_n} - \frac{x_n}{i_n^2} \right) \frac{\partial i_n}{\partial k_{z_p}}.$$

Let $\alpha_{p,n} = \frac{\partial i_n}{\partial k_{z_p}}$ be the derivative of the image model with respect to k_{z_p}, we obtain

$$\forall\, k_p \in \text{lattice} \qquad \frac{\partial \ell}{\partial k_{z_p}} = -L \sum_{n=1}^{N} \frac{\alpha_{p,n}}{i_n} \left(1 - \frac{x_n}{i_n} \right). \qquad (13.8)$$

Actually, changes of k_{z_p} affect only the pixels contained in the polygons surrounding the node k_p, which will be denoted $\{p_1 \sim p_q\}$ (see Fig. 13.2). The value of $\alpha_{p,n}$ is zero for pixels outside this polygon. Since the image model is defined by a linear interpolation between each node of the lattice, then the set $\boldsymbol{\alpha}_p = \{\alpha_{p,n}/\text{pixel } n \in \{p_1 \sim p_q\}\}$ forms a pyramid with a polygonal base $\{p_1 \sim p_q\}$—where the value is zero—and a top at node k_p—where the value is 1 (Fig. 13.2). This set of coefficients $\boldsymbol{\alpha}_p$ depends on the location of node k_p and on the location of nodes of the polygonal base associate to k_p. Indeed, it

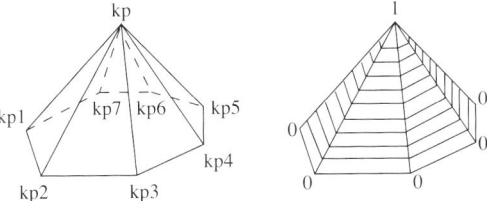

Figure 13.2. Polygonal-based elementary pyramid example used to compute the derivatives.

is easy to show that image model $\mathbf{I}(\mathbf{K})$ can be described as a sum of pyramids $\boldsymbol{\alpha}_p$ weighted by nodal values k_{z_p}:

$$\mathbf{I} = \sum_{p=1}^{P} k_{z_p} \cdot \boldsymbol{\alpha}_p. \tag{13.9}$$

Finally, Eq. (13.8) can thus be written as

$$\forall\, k_p \in \text{lattice} \qquad \frac{\partial \ell}{\partial k_{z_p}} = -L \sum_{n \in \{p_1 \sim p_q\}} \frac{\alpha_{p,n}}{i_n} \left(1 - \frac{x_n}{i_n} \right). \tag{13.10}$$

With similar notations, the second derivative of the ML criterion can be expressed as

$$\forall\, k_p \in \text{lattice} \qquad \frac{\partial^2 \ell}{\partial k_{z_p}^2} = -L \sum_{n \in \{p_1 \sim p_q\}} \frac{\alpha_{p,n}^2}{i_n^2} \left(2\frac{x_n}{i_n} - 1 \right). \tag{13.11}$$

The second-order iterative algorithm that we propose to use in order to determine the k_{z_p} variation at each iteration is based on Newton algorithm and it can be written as

$$\forall\, k_p \in \text{lattice} \qquad \Delta k_{z_p} = - \left(\frac{\partial^2 \ell}{\partial k_{z_p}^2} \right)^{-1} \frac{\partial \ell}{\partial k_{z_p}}. \tag{13.12}$$

Equation (13.12) provides z axis correction for each node k_p. For a pixel n around a node k_p, the correction is $\Delta i_n = \Delta k_{z_p} \alpha_{p,n}$. The iterative process of the correction of the image model can thus be written:

$$\mathbf{I}_t = \mathbf{I}_{t-1} + \sum_{p=1}^{P} \Delta k_{z_p} \cdot \boldsymbol{\alpha}_p. \tag{13.13}$$

Thanks to Eq. (13.13), $\mathbf{I}_t$ at iteration t is determined without extra computations such as for example to reconstruct $\mathbf{I}_t$ from the set of values $\{k_{z_p}\}_{p \in [1,P]}$ known at iteration t. It is necessary only to determine at each iteration Δk_{z_p} with Eq. (13.12) and to modify the image model with Eq. (13.13).

13.2.4 Regularization With Isoline Level

A first regularization of the image $\{i_n\}$ is naturally introduced by the image model since it is based on a parametric model with less degrees of freedom than the pixel number N of the image. However, when small triangles are used, poor speckle noise removal is obtained as shown in Fig. 13.3, while if the triangles are too big, a blurred image is obtained. This figure, a part of a toy car, is a two-dimensional (2D) view near the toy position

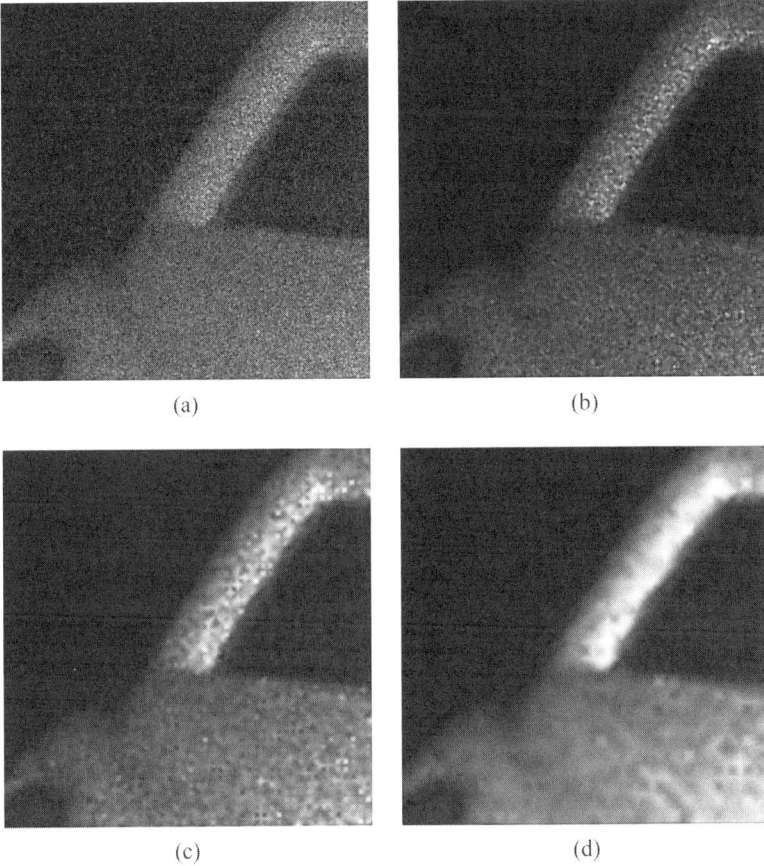

(a)   (b)

(c)   (d)

Figure 13.3. Part of image of a toy car calculated from an experimental digital holo-gram. Results with different triangle sizes of lattice. (a) reference image 401×401 (CPU time, number of nodes), drawn with gamma correction = 2 only for visualiza-tion. (b) 5 pixels base triangle (1.1s, 20201 nodes). (c) 9 pixels base triangle (0.5s, 5101 nodes). (d) 17 pixels base triangle (0.4s, 1301 nodes).

reconstructed from an experimental digital hologram with 12 bits dynamic range. Figure 13.3(a) shows the initial image without speckle removal. Fig-ures 13.3(b)–13.3(d) show the results obtained for different sizes of the ele-mentary triangle. One can see that noise removal is very low with triangles with a base of 5 pixels. In Fig. 13.3(c), the base of the triangle was 9 pix-els and the noise removal effect is still quite low. On the other hand, with triangles with a 17 pixel-base, the speckle noise has been more efficiently re-moved but a blurring effect becomes noticeable and prohibits to use larger triangles.

In order to improve speckle removal, we introduce, with small triangles, a second constraint. This constraint brings a minimal *a priori* information on the smoothness of the result.

Since most common images are continuous images with a small number of edges, images can be decomposed in regular lines with constant gray-levels (Fig. 13.4(a)). Furthermore, these iso-gray-level lines (which will be simply denoted isolines in the following) are in general regular in the sense that their shape does not oscillate a lot. The principle of the proposed isoline regularization thus consists in penalizing irregular sets of constant gray-levels.

More precisely, for each iteration t and for the image model determined at this iteration, starting from a node k_p ($k_p = l_0$) of coordinates (k_{x_p}, k_{y_p}) of the triangle lattice, we consider the circle centered on node k_p with radius equal to the edge size of the triangles. On this circle, we look for the pixel l_1 that has the nearest gray-level to k_{z_p}. Now, let us assume that an isoline has been determined with $q - 1$ pixels. Let $i_{l_0}, \ldots, i_{l_{q-1}}$ denote the gray-level values of the pixels $l_0, \ldots, l_{q-1}$ of this isoline. We can define the mean gray-level value of this isoline $m_{q-1} = \frac{1}{q} \sum_{j=0}^{q-1} i_{l_j}$, where l_0 corresponds to k_p. In order to merge a new pixel l_q to this isoline, a half circle of radius equal to the edge size of the triangles is first defined in the direction $\overrightarrow{l_{q-2}l_{q-1}}$ as shown in Fig. 13.4(b). We thus have to determine in the half circle defined above the pixel l_q which has the gray-level value i_{l_q} nearest to m_{q-1}. If $\left| i_{l_q} - m_{q-1} \right| / m_{q-1}$ is smaller than a threshold β (denoted "constrast threshold") the pixel l_q is merged into the isoline and the process is iterated. The merging process is stopped when this condition cannot be fulfilled anymore or when the maximum length Q of the isoline level, fixed at the beginning, has been reached. A new merging process is then initiated with the next node of the lattice.

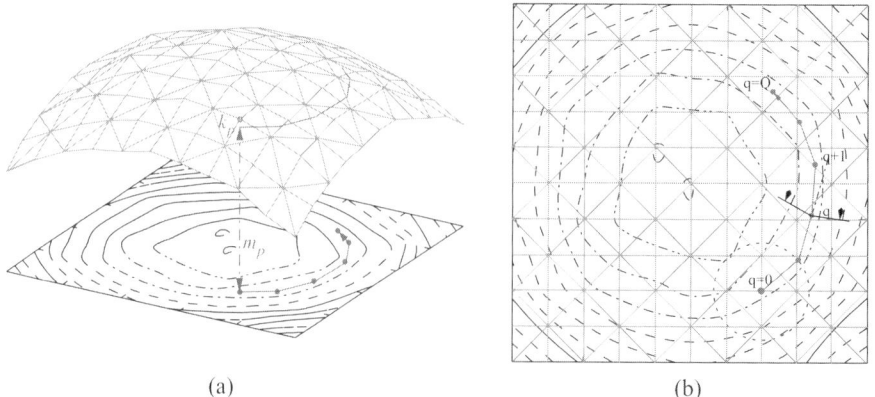

(a)   (b)

Figure 13.4. Isoline level estimation description (a) and determination (b).

Since this process is applied to the restored image obtained at iteration t, the proposed technique thus consists in minimizing

$$J\left(\mathbf{I}(\mathbf{K})\right) = C + L \cdot \sum_{n=1}^{N} \left(\log\left(i_n(\mathbf{K})\right) + \frac{x_n}{i_n(\mathbf{K})} \right)$$

$$+ \mu \cdot L \cdot \sum_{p=1}^{P} N_p \frac{\left(k_{z_p} - m(p, \beta)\right)^2}{m^2(p, \beta)}, \tag{13.14}$$

where μ is a parameter which balances between the likelihood term (first part of the criterion) and the regularization one, N_p is the number of pixels which belong to the polygon $\{p_1 \sim p_q\}$ that include the node k_p, β is the parameter of the regularization method and P is the number of nodes in the lattice. As shown in Eq. (13.14), the proposed technique consists in determining the isoline m_p to which each node belongs in order to penalize the square difference between the gray-level of the nodes and their corresponding isoline. In addition, the square difference in node p is weighted by the isoline gray-level value L/m_p^2 in order to take into account the characteristics of speckle noise which is a multiplicative noise.

The optimization algorithm is similar to the one presented in Section 13.2.3. In order to apply a second-order iterative algorithm, we need to determine the first and second derivatives of the criterion. The first term of criterion (13.14) is identical (multiplied by -1) to the ML criterion (13.3) and its derivatives are provided by Eq. (13.10) and (13.11). The first and second derivatives of the constraint expression are easy to obtain when we remark that the derivative expression of the isoline is

$$\frac{\partial m(p, \beta)}{\partial k_{z_p}} = \frac{1}{Q + 1}, \tag{13.15}$$

since each isoline is calculated on $Q + 1$ pixels by $m = \frac{1}{Q+1} \sum_{j=0}^{Q} i_{l_j}$ as described in Section 13.2.4.

13.3 Results and Influence of the Parameters

13.3.1 Results on Synthetic and Real Images

The synthetic image used for the tests is shown in Fig. 13.5. This image provides borders with strong slopes of 1 pixel and continuous variations of grey-levels. Some grey-level values are shown in Fig. 13.5(a) (horizontal text) and the contrasts for some borders are also reported (vertical text). Figure 13.5(b) shows the noisy image generated with a 10-order speckle noise with a gamma pdf defined by Eq. (13.1).

Figure 13.6 shows the result for the gamma noisy synthetic images. The result of the proposed technique is shown in Fig. 13.6(a) and 13.6(b),

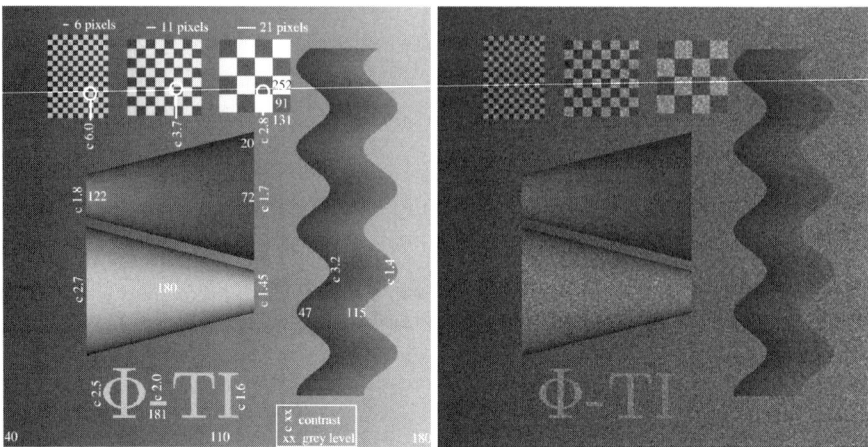

Figure 13.5. Reference synthetic image and 10-order gamma noisy image (512×512).

and has been obtained with a triangle lattice with a 3-pixel base and with regularization by isoline level. As a comparison, Figures 13.6(c) and 13.6(d) show the results of a median filter (with a sliding window of 5×5 pixels).

Results shown in Fig. 13.6(a) demonstrate that the proposed method is able to efficiently remove speckle while keeping sharp edges between the different zones. Of course, estimation of the edges is not perfect for low contrasts, however we can conclude with Figure 13.6(a) that the contrast[1] limit to obtain good results is between 1.1 and 1.2 for a 10-order speckle. For high-contrast edges, one observes with Fig. 13.6(b) that the quality of the restoration is limited by the size of the lattice.

One can also see that with the proposed technique the visual aspect is improved in comparison to the results obtained with a median filter.

The last validation is based on the 2D reconstructed view of an experimental digital hologram where the speckle corresponds to a 1-order speckle. Figure 13.7. provides the result for a lattice with 5 pixels for the base of the triangles and a constraint on the isoline level.

13.3.2 Study of the Influence of the Contrast Threshold β

The contrast threshold β is an important parameter of the algorithm. The value of this parameter allows us to find a trade-off between the smoothness and the loss of low-contrast details in the filtered images. The choice of the

---

[1] The local contrast is defined as the ratio of average gray-levels on both sides of an edge, for example.

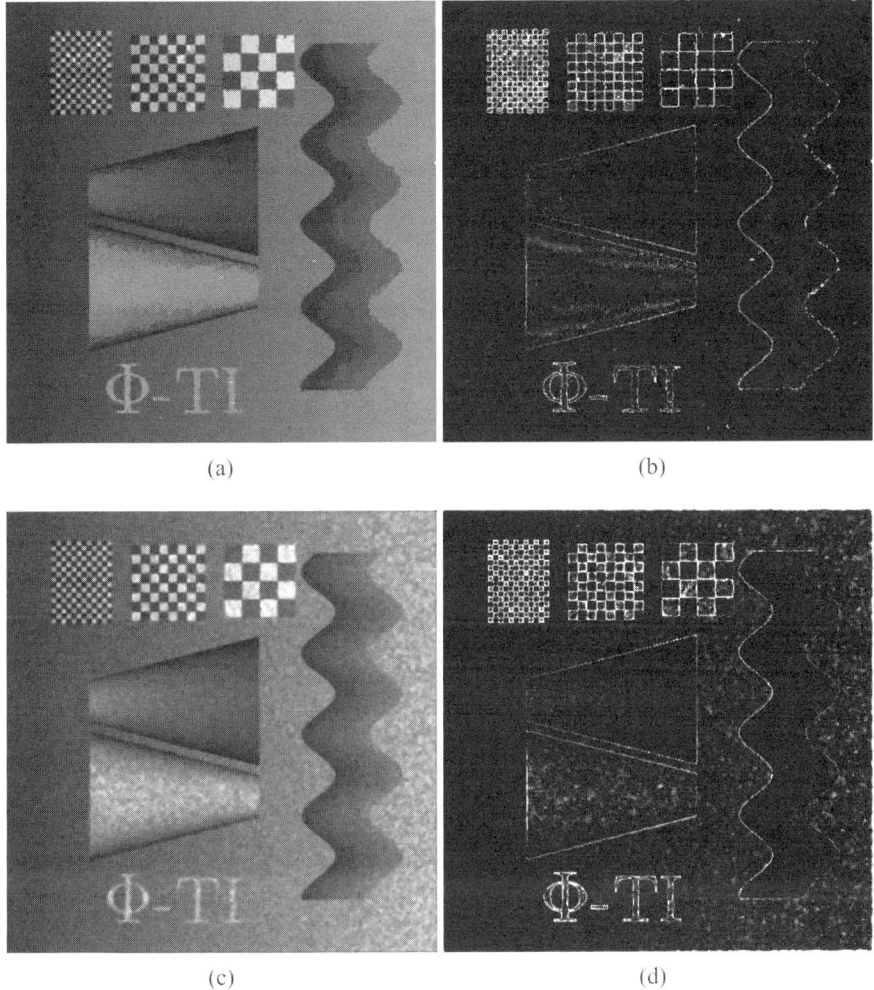

Figure 13.6. Results on synthetic image with 10-order gamma noise (512×512). (a) $\mu = 23$ and $\beta = 0.14$ with 3 pixels base. (46s, PIII-1.1GHz). (b) square error image. mse $= 219$. (c) median filter (5×5). (d) square error image, mse $= 362$.

value of this parameter depends on the contrast levels present in the image and on the speckle order.

In the simulation presented in Fig. 13.8, all the parameters other than β are fixed. We show the filtered images obtained by the proposed algorithm for 9 different values of the parameter β. It can be seen that, for low values of β, the isolines are very short because they are rapidly stopped when finding small contrast steps. The variance of the isoline level is inversely proportional to its number of points. Short isolines are then poorly estimated and cannot be trusted. This explains why the filtered images for low β present many defects

Figure 13.7. (top) Reference image of a toy car reconstructed from an experimental digital hologram 1123 × 1585 with 1-order speckle (or 1 look). Drawn with modification of gray-levels for visualization. White box correspond to the part used in Fig. 13.3 (bottom) Result with 5 pixels base triangle, $\mu = 20.0$ and $\beta = 0.3$. (27s, 222437 nodes, PIII-1.1GHz).

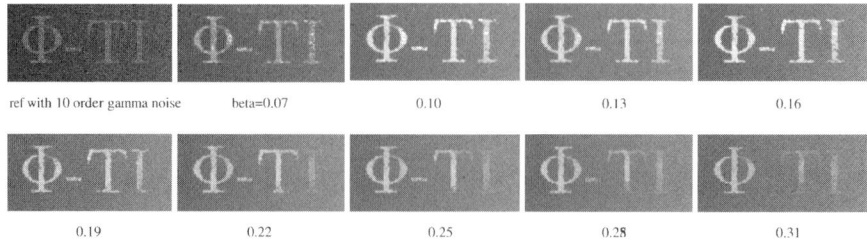

Figure 13.8. Influence of the contrast threshold β for a synthetic image (reference) with 10-order gamma noise. Results for 3 pixels base triangles and $\mu = 23$.

in the supposedly smooth areas. On the other hand, edges are correctly found, even when they have a low contrast, for instance in the case of letter I in Fig. 13.8 for the values of β larger than 0.19. On the contrary, for high values of the parameter β, the filtered images are perfectly smooth but the low-contrast details tend to vanish.

Obviously, the correct value for β depends on the image, the expected result, and the speckle order. For a low order, one needs to increase β in order to force the isolines to extend over larger distances and include more points in the isoline-level estimation. However, in this case, low-contrast details will be lost.

13.3.3 Study of the Influence of the Weight of the Isoline Constraint μ

The weight of the isoline constraint μ is the second important parameter of the presented algorithm. This parameter imposes the regularity of the solution. Since the penalization term uses only information from the estimated image at step $t(\mathbf{I}_t)$ (see Eq. (13.14)), the final solution depends both on the data (noisy image) and on the initial estimation ($\mathbf{I}_{t=0}$). The value of μ determines the respective weights of these two dependences.

In the simulation presented in Fig. 13.9, all parameters other than μ are fixed. The reference image is a synthetic image, two flipped identical plans (gradients) with value between 64 and 192, with 10-order gamma noise. In this particular simulation, the initial estimation ($\mathbf{I}_{t=0}$) is chosen as the result of a 7×7 mean filter of the noisy image, but rotated by 90 degrees (Fig. 13.9). This initialization is an intentionnally poor estimation of the final result. Figure 13.9 shows the results for values between $\mu = 1$ and 200. For small values of μ the regularization has no effect. For weights larger than $\mu = 10$ the isoline constraint is efficient. This relatively low sensitivity can be explained by the fact that isolines exist in any reasonable image, so imposing their existence is not really penalizing. Moreover, for weights larger than $\mu = 20$, we note that the final result depends on the initialization too. We can conclude that if the user has no correct initialization, it is better to choose a weight between $\mu = 10$ and 20. With a correct initialization, larger weights can be used to provide smoother results. Choosing a weigth larger than $\mu = 200$ is not interesting because for these values, the final result depends only on the initialization and not on the data.

Contrary to the contrast threshold β, the weight of the isoline constraint does not depend on the order of the gamma noise.

13.3.4 Robustness of the Algorithm With Respect to Spatial Correlation

Depending on the size of the pixels relatively to the speckle, the speckle pattern can either be spatially uncorrelated (as in Fig. 13.7) or spatially correlated

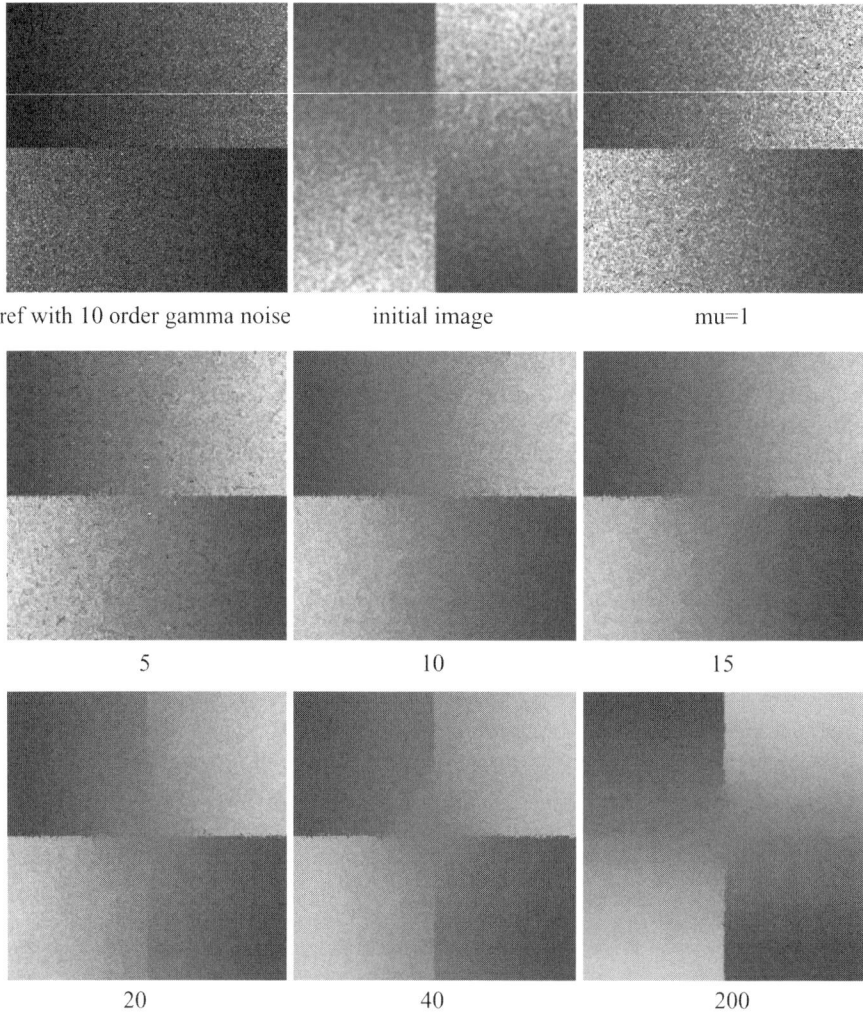

ref with 10 order gamma noise   initial image   mu=1

5   10   15

20   40   200

Figure 13.9. Influence of the weight of the isoline constraint μ. Results for 3 pixels base triangles and $\beta = 0.15$.

(as in Fig. 13.10(b)). For instance, in Fig. 13.10(a) the size of the speckle grain is larger than 1 pixel, which introduces spatial correlation. This image is based on the reconstructed 2D view (1024 × 1024) of an experimental digital holo-gram where speckle has an order equal to 1. Figure 13.10 shows results for different sizes of the base triangles (3 pixels in Fig. 13.10(c), 5 pixels in Fig. 13.10(d) and 9 pixels in Fig. 13.10(e)). We can note that with smaller size triangles the result is not smooth. For larger base triangles, the results are better and similar. The interest of using a lattice with small size base trian-gles is the sensitivity to small details. Since the likelihood term does not hold

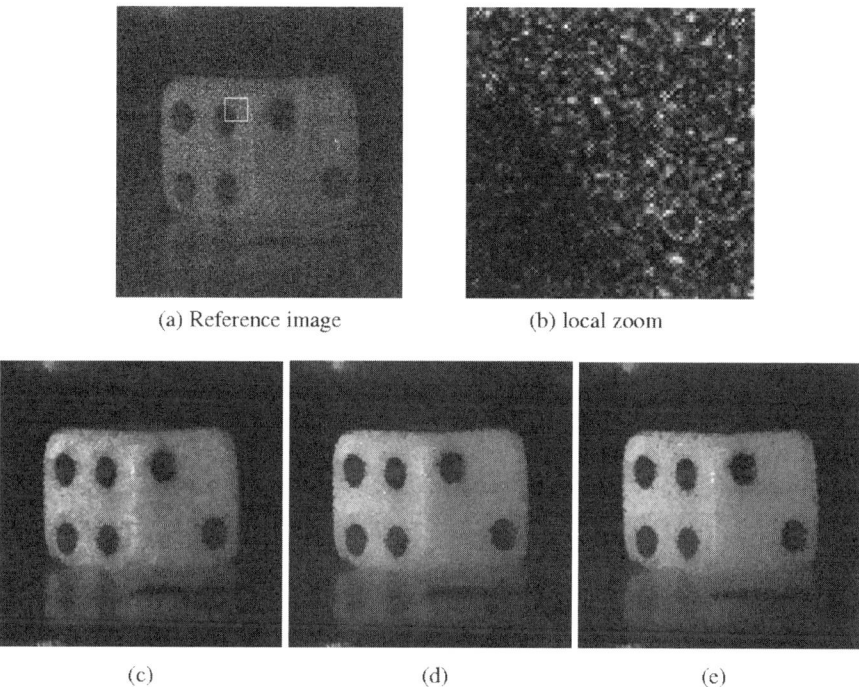

(a) Reference image   (b) local zoom

(c)   (d)   (e)

Figure 13.10. Influence of the spatial correlation. Results for 3(c). 5(d) and 9(e) pixels base triangles and $\beta = 0.3, \mu = 23$.

account of the spatial correlation, each grain of speckle is thus regarded as details.

As can be seen in Figures 13.7 and 13.10, the quality of the filtered images is similar for both uncorrelated and correlated speckle when using similar parameters. The proposed technique is thus robust to spatial correlation of the speckle pattern if the size of the base triangles is bigger than the speckle size.

13.4 Conclusion

In this chapter, we have presented a new technique to remove the speckle pattern from coherent images. Although we provided example images obtained from digital holography, the proposed approach can be applied to other types of coherent images.

Our technique is based on a maximum likelihood estimation of the reflectivity of the object and on a regularization whose aim is to obtain smooth filtered images. We have first presented the maximum likelihood estimator of the image, and then we introduced a regularization in the form of a lattice modelization of the image. In order to further improve the results, we have added a second regularization by gray-level isolines. The resulting technique allows one to obtain smooth surfaces while keeping sharp edges of the objects. Lastly, we have presented results of speckle removal and we have investigated the robustness of the technique with respect to the parameter values.

Acknowledgment

We thank Pr. Philippe Réfrégier for his advices and his availability.

References

[1] C. Oliver and S. Quegan. *Understanding Synthetic Aperture Radar Images*, Artech House Publishers, Boston, 1998.

[2] H. J. Caulfield, *Handbook of Optical Holography*, Academic Press, London, 1979.

[3] O. Matoba, T. Naughton, Y. Frauel, N. Bertaux, and B. Javidi. Real-time three-dimensional object reconstruction by use of a phase-encoded digital hologram, *Appl. Opt.* 41:6187–6192, 2002.

[4] Y. Frauel, E. Tajahuerce, M.-A. Castro, and B. Javidi. Distortion-tolerant three-dimensional object recognition with digital holography. *Appl. Opt.* 40:3887–3893, 2001.

[5] Y. Frauel, E. Tajahuerce, O. Matoba, A. Castro, and B. Javidi. Comparison of passive ranging integral imaging and active imaging digital holography for three-dimensional object recognition. *Appl. Opt.* 43:452–462, 2004.

[6] U. Schnars and W. P. O. Juptner. Direct recording of holograms by a CCD target and numerical reconstruction, *Appl. Opt.* 33:179–181, 1994.

[7] B. Javidi and E. Tajahuerce. Three dimensional image recognition using digital holography. *Opt. Lett.* 25(9):610–612, May 2000.

[8] B. Javidi. Real-time Remote Identification and Verification of Objects Using Optical ID Tags, *Opt. Eng.* 42(8), August 2003.

[9] S. Shin and B. Javidi, Three Dimensional Object Recognition by use of a Photorefractive Volume Holographic Processor, *Opt. Lett.* 26(15):1161–1164, August 1, 2001.

[10] S. Shin and B. Javidi, Viewing angle enhancement of speckle reduced volume holographic three-dimensional display by use of integral imaging, *App. Opt.* 41(26):5562–5567, 10 September, 2002.

[11] J. W. Goodman and R. W. Lawrence, Digital image formation from electronically detected holograms, *Appl. Phys. Lett.* 11:77–79, 1967.

[12] L. Onural and P. D. Scott, Digital decoding of in-line holograms, *Opt. Eng.* 26:1124–1132, 1987.

[13] R. W. Kronrod, N.S. Merzlyakov, and L.P., Yaroslavskii, Reconstruction of a hologram with a computer, *Sov. Phys. Tech. Phys.* 17:333–334, 1972.

[14] M. Jacquot, P. Sandoz, and G. Tribillon, High resolution digital holography, *Opt. Comm.* 190:87–94, 2001.

[15] J. H. Massig, Digital off-axis holography with a synthetic aperture, *Opt. Lett.* 27:2179–2181, 2002.

[16] E. Tajahuerce, O. Matoba, and B. Javidi, Shift-invariant three-dimensional object recognition by means of digital holography. *Appl. Opt.* 40:3877–3886, 2001.

[17] M. Françon, Laser Speckle and Application in Optics, Academic Press, New York, 1979.

[18] N. Bertaux, Y. Frauel, Ph. Réfrégier, B. Javidi, Speckle removal using a maximum-likelihood technique with isoline gray-level regularization, *JOSA A*, 21(12):2283–2291, Dec. 2004.

[19] H.L. Van Trees, Detection Estimation and Modulation theory, New York, Wiley, 1968.

Index

Printed in the United States of America